Simulation Modeling - Recent Advances, New Perspectives, and Applications

Edited by Abdo Abou Jaoudé

Published in London, United Kingdom

Simulation Modeling - Recent Advances, New Perspectives, and Applications
http://dx.doi.org/10.5772/intechopen.1000371
Edited by Abdo Abou Jaoudé

Contributors
Abdo Abou Jaoudé, Adriano O. Solis, Alexander Talavirya, Alissa Dubgorn, Anas Nassar, Christian Schultz, Firas Makahleh, Gregory Zacharewicz, Jalal Possik, Lars Lewerentz, Michael Laskin, Rakhab Mehta, Ralf Schneider, Seddigheh Rezapour, Seçkin Karagöz, Stefan Kemnitz, Valery Borisovich Morozov

First published in London, United Kingdom, 2024 by IntechOpen
IntechOpen is the global imprint of INTECHOPEN LIMITED, registered in England and Wales, registration number: 11086078, 167-169 Great Portland Street, London, W1W 5PF, United Kingdom

British Library Cataloguing-in-Publication Data
A catalogue record for this book is available from the British Library

Additional hard and PDF copies can be obtained from orders@intechopen.com

Simulation Modeling - Recent Advances, New Perspectives, and Applications
Edited by Abdo Abou Jaoudé
p. cm.
Print ISBN 978-1-83769-706-9
Online ISBN 978-1-83769-705-2
eBook (PDF) ISBN 978-1-83769-707-6

For EU product safety concerns:
IN TECH d.o.o., Prolaz Marije Krucifikse Kozulić 3, 51000 Rijeka, Croatia,
info@intechopen.com or visit our website at intechopen.com.

Meet the editor

Abdo Abou-Jaoudé has been teaching for many years and has a passion for researching mathematics. He is currently an Associate Professor of Mathematics and Statistics at Notre Dame University-Louaizé (NDU), Lebanon. He holds a BSc and an MSc in Computer Science from NDU, and three Ph.D.'s in Applied Mathematics, Computer Science, and Applied Statistics and Probability from Bircham International University, Spain. He also holds two Ph.D.'s in Mathematics and Prognostics from the Lebanese University, Lebanon, and Aix-Marseille University, France. Dr. Abou-Jaoudé's broad research interests are in the fields of pure and applied mathematics. He has published twenty-three international journal articles and six contributions to conference proceedings, in addition to seventeen books on prognostics, mathematics, physics, and computer science.

Contents

Preface

It gives me great pleasure to introduce this work on simulation modeling. In this book, *Simulation Modeling – Recent Advances, New Perspectives, and Applications*, we discuss some fundamental aspects of the theory of numerical methods of simulation and explore their use to solve a large array of problems. Therefore, we cover many topics on deterministic and random techniques.

A simulation is the imitation of the operation of a real-world process or system over time. Simulations require the use of models; the model represents the key characteristics or behaviors of the selected system or process, whereas the simulation represents the evolution of the model over time. Often, computers are used to execute the simulation.

Additionally, simulation is used in many contexts, such as simulation of technology for performance tuning or optimizing, safety engineering, testing, training, education, and video games. Simulation is also used with scientific modeling of natural systems or human systems to gain insight into their functioning, as in economics. Simulation can be used to show the eventual real effects of alternative conditions and courses of action. Simulation is also used when the real system cannot be engaged because it may not be accessible, or it may be dangerous or unacceptable to engage, or it is being designed but not yet built, or it may simply not exist.

Moreover, key issues in modeling and simulation include the acquisition of valid sources of information about the relevant selection of key characteristics and behaviors used to build the model, the use of simplifying approximations and assumptions within the model, and fidelity and validity of the simulation outcomes. Procedures and protocols for model verification and validation are an ongoing field of academic study, refinement, research, and development in simulations technology or practice, particularly in the work of computer simulation.

Finally, this book discusses some important aspects of simulation modeling and explores its use to solve a large set of problems. As such, the book will be of interest to all scholars, researchers, and undergraduate and graduate students in pure and applied mathematics, physical sciences, engineering and technology, computer science, numerical analysis, scientific computing, and science in general.

Abdo Abou Jaoudé, Ph.D.
Notre Dame University-Louaizé,
Zouk Mosbeh, Lebanon

Section 1

Simulation and the Complex Probability Paradigm

Chapter 1

The Paradigm of Complex Probability and Quantum Mechanics: The Quantum Harmonic Oscillator with Gaussian Initial Condition – The Position Wavefunction

Abdo Abou Jaoudé

"Necessity... the mother of invention."
Plato.

"Whatsoever is, is in God, and without God nothing can be, or be conceived."
Baruch Spinoza.

"Wonder is the feeling of the philosopher, and philosophy begins in wonder."
Plato

"Not ignorance, but ignorance of ignorance, is the death of knowledge."
Alfred North Whitehead.

Abstract

In the current work, we extend and incorporate the five-axioms probability system of *Andrey Nikolaevich Kolmogorov,* set up in 1933 the imaginary set of numbers, and this by adding three supplementary axioms. Consequently, any stochastic experiment can thus be achieved in the extended complex probabilities set $\mathcal{C}$ which is the sum of the real probabilities set $\mathcal{R}$ and the imaginary probabilities set $\mathcal{M}$. The purpose here is to evaluate the complex probabilities by considering additional novel imaginary dimensions to the experiment occurring in the "real" laboratory. Therefore, the random phenomenon outcome and result in $\mathcal{C} = \mathcal{R} + \mathcal{M}$ can be predicted absolutely and perfectly no matter what the random distribution of the input variable in $\mathcal{R}$ is since the associated probability in the entire set $\mathcal{C}$ is constantly and permanently equal to one. Thus, the following consequence indicates that chance and randomness in $\mathcal{R}$ are replaced now by absolute and total determinism in $\mathcal{C}$ as a result of subtracting from

IntechOpen

the degree of our knowledge of the chaotic factor in the probabilistic experiment. Moreover, I will apply to the established theory of quantum mechanics my original complex probability paradigm (*CPP*) in order to express the quantum mechanics problem considered here completely deterministically in the universe of probabilities $\mathcal{C} = \mathcal{R} + \mathcal{M}$.

Keywords: degree of our knowledge, chaotic factor, complex random vector, probability norm, complex probability set $\mathcal{C}$, distribution function, density function, expectations, variances

1. Introduction

The theory of quantum mechanics provides a description of nature physical properties at the scale of atoms and subatomic particles and is a fundamental theory in physics. Quantum mechanics is the foundation of all quantum physics, including quantum field theory, quantum chemistry, quantum information science, and quantum technology [1–13].

Classical physics differs from quantum mechanics since in the latter, we have that angular momentum, momentum, energy, and other quantities of a bound system are limited to discrete values (quantization), and there are restrictions to how accurately the physical quantity value can be determined and predicted prior to its measurement given a complete set of initial conditions (the uncertainty principle), and objects have characteristics of waves and particles (wave–particle duality).

The theory of quantum mechanics was developed progressively from theories to explain observations that could not be explained by classical physics, such as the black-body radiation problem solution proposed by Max Planck's in 1900 and the explanation of the photoelectric by Albert Einstein's 1905 paper as a correspondence between energy and frequency. The full development of quantum mechanics in the mid-1920s by Niels Bohr, Max Born, Werner Heisenberg, Erwin Schrödinger, and others were the early attempts to explain and understand microscopic phenomena and which is now known as the "old quantum theory." Various specially developed mathematical formalisms formulate the modern theory of quantum mechanics. In one of these invented formalisms, a mathematical entity named the wave function provides in the form of probability amplitudes the information about what measurements of a particle's momentum, energy, and other physical properties may give.

Moreover, the quantum-mechanical analog of the classical harmonic oscillator is the quantum harmonic oscillator. It is one of the most important model systems in quantum mechanics because an arbitrary smooth potential can generally be estimated as a harmonic potential at the neighborhood of a stable equilibrium point. Furthermore, since an exact, analytical solution is known, it is one of the few quantum-mechanical systems for which this kind of solution is provided.

Consequently, I will relate my complex probability paradigm (*CPP*) to this well-known and important problem in quantum mechanics in order to express it completely deterministically.

In the end, and to conclude, this research work is organized as follows: in Section 1, we will present the introduction, then in Section 2, we will explain the advantages and the purpose of the present work. Afterward, in Section 3, we will explain and summarize the extended Kolmogorov's axioms and hence present the original parameters and interpretation of the complex probability paradigm. Additionally, in Section 4, the quantum harmonic oscillators with the Gaussian initial condition problem will be related to the new paradigm after applying *CPP* in the first chapter to the position wavefunction and then in the following second chapter to the momentum wavefunction of the problem. Hence, some corresponding simulations will be achieved, and subsequently, the characteristics of these random distributions will be evaluated in the probabilities sets $\mathcal{R}$, $\mathcal{M}$, and $\mathcal{C}$. Finally, in Section 5, a comprehensive summary concludes the work. Then, we will present the list of references mentioned and cited in the current research work.

2. The purpose and the advantages of the current publication

Computing probabilities is all our work in the classical theory of probability. Adding new dimensions to our stochastic experiment is an innovative idea in the current paradigm, making the study absolutely deterministic. As a matter of fact, the theory of probability is a nondeterministic theory by essence which means that all the random events outcome is due to luck and chance. Hence, we make the study deterministic by adding new imaginary dimensions to the phenomenon occurring in the "real" laboratory, which is $\mathcal{R}$, and therefore a stochastic experiment will have a certain outcome in the complex probabilities set $\mathcal{C}$. It is of great significance that random systems become completely predictable since we will be perfectly knowledgeable to predict the outcome of all stochastic and chaotic phenomena that occur in nature, for example, in all stochastic processes, in statistical mechanics, or in the well-established field of quantum mechanics. Consequently, the work that should be done is to add the contributions of $\mathcal{M}$, which is the set of imaginary probabilities, to the set of real probabilities $\mathcal{R}$ that will make the random phenomenon in $\mathcal{C} = \mathcal{R} + \mathcal{M}$ completely deterministic. Since this paradigm is found to be fruitful, then a new theory in prognostic and stochastic sciences is established, and this is to understand deterministically those events that used to be stochastic events in $\mathcal{R}$. This is what I coined by the term "The Complex Probability Paradigm" that was elaborated and initiated in my 23 previous papers [14–36].

To summarize, the advantages and the purposes of this current work and chapter are to:

1. Relate probability theory to the field of complex variables and analysis in mathematics and, therefore, to extend the theory of classical probability to the set of complex numbers. This task was elaborated and initiated in my 23 previous papers.

2. Apply the novel probability axioms and *CPP* paradigm to quantum mechanics, specifically to the quantum harmonic oscillators with Gaussian initial condition problem.

3. Demonstrate that any stochastic and random event and experiment can be expressed deterministically in the complex probabilities set $\mathcal{C}$.

4. Quantify both the chaos magnitude and the degree of our knowledge of the wavefunction position distribution and *CPP* in the sets $\mathcal{R}$, $\mathcal{M}$, and $\mathcal{C}$.

5. Represent graphically and illustrate the parameters and functions of the original paradigm related to this quantum mechanics problem.

6. Evaluate all the characteristics of the wavefunction position distribution.

7. Demonstrate that the classical concepts of the stochastic system have a probability of occurring permanently equal to one in the complex set; consequently, no ignorance, no unpredictability, no stochasticity, no disorder, no randomness, no nondeterminism, and no chaos exist in:

$$\mathcal{C}\ (\text{complex set}) = \mathcal{R}\ (\text{real set}) + \mathcal{M}\ (\text{imaginary set}) \tag{1}$$

8. Prepare to apply the novel paradigm to other topics in stochastic processes, in statistical mechanics, and to the field of prognostics in science, engineering, and quantum mechanics. This will be the task in my following research work and publications.

Compared with existing literature, the major contribution of the current research work is to apply the novel paradigm of *CPP* to quantum mechanics and to express it completely deterministically. And concerning some applications of the novel developed paradigm, and as a future work, it can be applied to any nondeterministic phenomenon in quantum mechanics. The next figure displays the major purposes of the complex probability paradigm (*CPP*) (**Figure 1**).

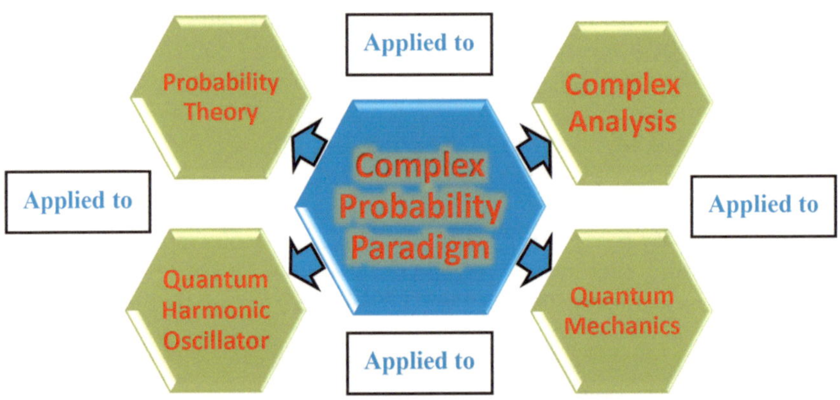

Figure 1.
The diagram of the Complex Probability Paradigm applied to Quantum Mechanics major purposes and goals.

3. The complex probability paradigm

3.1 The original Andrey Nikolaevich Kolmogorov system of axioms

The simplicity of Kolmogorov's system of axioms may be surprising [14–36]. Let E be a collection of elements $\{E_1, E_2, \ldots\}$ called elementary events, and let F be a set of subsets of E called random events [37–41]. The five axioms for a finite set E are:

Axiom 1: F is a field of sets.

Axiom 2: F contains the set E.

Axiom 3: A nonnegative real number $P_{rob}(A)$, called the probability of A, is assigned to each set A in F. We always have $0 \leq P_{rob}(A) \leq 1$.

Axiom 4: $P_{rob}(E)$ equals 1.

Axiom 5: If A and B have no elements in common, the number assigned to their union is:

$$P_{rob}(A \cup B) = P_{rob}(A) + P_{rob}(B) \tag{2}$$

hence, we say that A and B are disjoint; otherwise, we have:

$$P_{rob}(A \cup B) = P_{rob}(A) + P_{rob}(B) - P_{rob}(A \cap B) \tag{3}$$

And we also say that: $P_{rob}(A \cap B) = P_{rob}(A) \times P_{rob}(B/A) = P_{rob}(B) \times P_{rob}(A/B)$ which is the conditional probability. If both A and B are independent, then: $P_{rob}(A \cap B) = P_{rob}(A) \times P_{rob}(B)$.

Moreover, we can generalize and say that for N disjoint (mutually exclusive) events $A_1, A_2, \ldots, A_j, \ldots, A_N$ (for $1 \leq j \leq N$), we have the following additivity rule:

$$P_{rob}\left(\bigcup_{j=1}^{N} A_j\right) = \sum_{j=1}^{N} P_{rob}(A_j) \tag{4}$$

And we also say that for N independent events $A_1, A_2, \ldots, A_j, \ldots, A_N$ (for $1 \leq j \leq N$), we have the following product rule:

$$P_{rob}\left(\bigcap_{j=1}^{N} A_j\right) = \prod_{j=1}^{N} P_{rob}(A_j) \tag{5}$$

3.2 Adding the imaginary part $\mathcal{M}$

Now, we can add to this system of axioms an imaginary part such that:

Axiom 6: Let $P_m = i \times (1 - P_r)$ be the probability of an associated complementary event in $\mathcal{M}$ (the imaginary part or universe) to the event A in $\mathcal{R}$ (the real part or universe). It follows that $P_r + P_m/i = 1$ where i is the imaginary number with $i = \sqrt{-1}$ or $i^2 = -1$.

Axiom 7: We construct the complex number or vector $Z = P_r + P_m = P_r + i(1 - P_r)$ having a norm $|Z|$ such that:

$$|Z|^2 = P_r^2 + (P_m/i)^2. \tag{6}$$

Axiom 8: Let Pc denote the probability of an event in the complex probability set and universe $\mathcal{C}$ where $\mathcal{C} = \mathcal{R} + \mathcal{M}$. We say that Pc is the probability of an event A in $\mathcal{R}$ with its associated and complementary event in $\mathcal{M}$ such that:

$$Pc^2 = (P_r + P_m/i)^2 = |Z|^2 - 2iP_rP_m \text{ and is always equal to 1.} \tag{7}$$

We can see that by taking into consideration the set of imaginary probabilities we added three new and original axioms and consequently the system of axioms defined by Kolmogorov was hence expanded to encompass the set of imaginary numbers and realm [42–65].

3.3 A concise interpretation of the original *CPP* paradigm

To conclude and to summarize, we state that the degree of our certain knowledge is undesirably incomplete and imperfect and thus unsatisfactory in the real probability universe $\mathcal{R}$. Hence, we extend our study to the set of complex numbers $\mathcal{C}$, which includes the contributions of both the set of real probabilities which is $\mathcal{R}$ and the set of complementary imaginary probabilities which is $\mathcal{M}$. Consequently, this will result to a perfect and absolute degree of our knowledge in the probability universe $\mathcal{C} = \mathcal{R} + \mathcal{M}$ because $Pc = 1$ continuously. In fact, the study in the complex universe $\mathcal{C}$ leads to a certain prediction of any stochastic and random event and experiment since in $\mathcal{C}$, we subtract and eliminate the measured chaotic factor from the computed degree of our knowledge. This will result in a probability permanently equal to 1 in the universe $\mathcal{C}$ as it is shown in the following equation deduced from *CPP*:
$Pc^2 = DOK - Chf = DOK + MChf = 1 = Pc$. Many numerous discrete and continuous probability distributions were illustrated in my 23 previous research works, and that confirm this hypothesis and original paradigm [14–36]. The Extended Kolmogorov Axioms (*EKA* for short) or the Complex Probability Paradigm (*CPP* for short) can be summarized and shown in the next illustration (**Figure 2**):

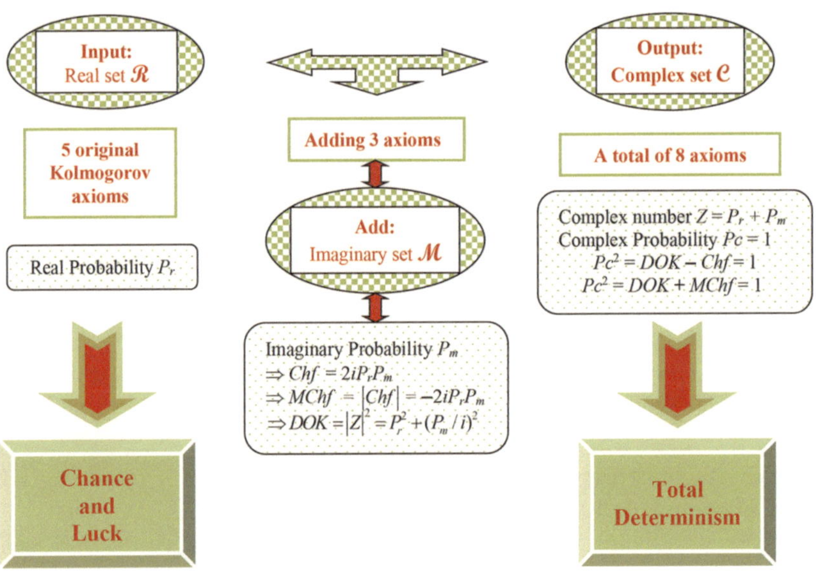

Figure 2.
The EKA *or the* CPP *diagram.*

4. The Quantum Harmonic Oscillator with Gaussian Initial Condition and the Complex Probability Paradigm (*CPP*) Parameters – The Position Wavefunction and *CPP*

In this section, we will relate and link quantum mechanics to the complex probability paradigm with all its parameters by applying it to the quantum harmonic oscillators with Gaussian initial condition and this by using the four *CPP* concepts, which are: the real probability P_r in the real probability set $\mathcal{R}$, the imaginary probability P_m in the imaginary probability set $\mathcal{M}$, the complex random vector or number Z in the complex probability set $\mathcal{C} = \mathcal{R} + \mathcal{M}$, and the deterministic real probability Pc also in the probability set $\mathcal{C}$ [1–36, 66–92].

4.1 The position wavefunction solution

One significant quantum–theoretic periodic wave packet which has a ubiquitous presence in the arena of quantum optics and quantum electronics is the *coherent state* of the simple harmonic oscillator. This state evolves from the initial condition [1–3],

$$\Psi(x, t = 0) = \left(\frac{m\Omega}{\pi\hbar}\right)^{1/4} \exp\left(-\frac{m\Omega(x - x_0)^2}{2\hbar}\right) \tag{8}$$

where w is the angular frequency and Ω is the width of the initial state, and where Ω is not necessarily equal to w. This equation has the form of simple Gaussian modeling of the ground (stationary) state of the harmonic oscillator but with an added feature: its center has x_0 the amount of displacement. Note that a coherent state is a Gaussian wave packet that does not flatten out over time since all the terms are in phase. Coherent states also sport another interesting feature: *they satisfy the* ***minimum*** *uncertainty relation!*

Integrating over the propagator eventually delivers [1–3]:

$$|\psi(x,t)|^2 \sim \mathbb{N}\left(x_0\cos(wt), \sqrt{\frac{\hbar}{2m\Omega}\left(\cos^2(wt) + \frac{\Omega^2}{w^2}\sin^2(wt)\right)}\right) \tag{9}$$

where the notation $\mathbb{N}(\mu_x, \sigma_x)$ is deployed, designating a normal distribution of mean $\mu_x = x_0\cos(wt)$ with standard deviation $\sigma_x = \sqrt{\frac{\hbar}{2m\Omega}\left(\cos^2(wt) + \frac{\Omega^2}{w^2}\sin^2(wt)\right)}$. Knowing that we have taken in this study $\Omega = n \times w$, where n is a simple multiplier and can be equal to ¼, or ½, or 1, or 2, or 50, or 100, or 300, etc., as it will be shown afterward in the simulations section.

4.2 The position wavefunction probability distribution and *CPP*

For the quantum harmonic oscillators with Gaussian initial condition, the wavefunction position probability density function (*PDF*) is given by:

$$f(x) = |\psi(x,t)|^2 = \mathbb{N}\left(x_0\cos(wt), \sqrt{\frac{\hbar}{2m\Omega}\left(\cos^2(wt) + \frac{\Omega^2}{w^2}\sin^2(wt)\right)}\right) \tag{10}$$

Therefore, the wavefunction position cumulative probability distribution function (CDF), which is equal to $P_r(T)$ in $\mathcal{R}$ is:

$$\begin{aligned} P_r(T) &= F(t_j) = P_{rob}(T \le t_j) = \int_0^{t_j} |\psi(x,t)|^2 dt \\ &= \int_0^{t_j} \mathbb{N}\left(x_0 \cos(wt), \sqrt{\frac{\hbar}{2m\Omega}\left(\cos^2(wt) + \frac{\Omega^2}{w^2}\sin^2(wt)\right)}\right) dt \end{aligned} \tag{11}$$

And the real complementary probability to $P_r(T)$ in $\mathcal{R}$ which is $P_m(T)/i$ is:

$$\begin{aligned} P_m(T)/i &= 1 - P_r(T) = 1 - F(t_j) = 1 - P_{rob}(T \le t_j) = P_{rob}(T > t_j) = 1 - \int_0^{t_j} |\psi(x,t)|^2 dt \\ &= \int_{t_j}^{+\infty} |\psi(x,t)|^2 dt \\ &= 1 - \int_0^{t_j} \mathbb{N}\left(x_0 \cos(wt), \sqrt{\frac{\hbar}{2m\Omega}\left(\cos^2(wt) + \frac{\Omega^2}{w^2}\sin^2(wt)\right)}\right) dt \\ &= \int_{t_j}^{+\infty} \mathbb{N}\left(x_0 \cos(wt), \sqrt{\frac{\hbar}{2m\Omega}\left(\cos^2(wt) + \frac{\Omega^2}{w^2}\sin^2(wt)\right)}\right) dt \end{aligned} \tag{12}$$

Consequently, the imaginary complementary probability to $P_r(T)$ in $\mathcal{M}$ which is $P_m(T)$ is:

$$\begin{aligned} P_m(T) &= i[1 - P_r(T)] = i[1 - F(t_j)] = i[1 - P_{rob}(T \le t_j)] = iP_{rob}(T > t_j) \\ &= i\left[1 - \int_0^{t_j} |\psi(x,t)|^2 dt\right] = i\int_{t_j}^{+\infty} |\psi(x,t)|^2 dt \\ &= i\left[1 - \int_0^{t_j} \mathbb{N}\left(x_0 \cos(wt), \sqrt{\frac{\hbar}{2m\Omega}\left(\cos^2(wt) + \frac{\Omega^2}{w^2}\sin^2(wt)\right)}\right) dt\right] \\ &= i\int_{t_j}^{+\infty} \mathbb{N}\left(x_0 \cos(wt), \sqrt{\frac{\hbar}{2m\Omega}\left(\cos^2(wt) + \frac{\Omega^2}{w^2}\sin^2(wt)\right)}\right) dt \end{aligned} \tag{13}$$

Furthermore, the complex random number or vector in $\mathcal{C} = \mathcal{R} + \mathcal{M}$ which is $Z(T)$ is:

$$
\begin{aligned}
Z(T) &= P_r(T) + P_m(T) = P_r(T) + i[1 - P_r(T)] = F(t_j) + i[1 - F(t_j)] \\
&= P_{rob}(T \le t_j) + i[1 - P_{rob}(T \le t_j)] = P_{rob}(T \le t_j) + iP_{rob}(T > t_j) \\
&= \left[\int_0^{t_j} |\psi(x,t)|^2 dt\right] + i\left[1 - \int_0^{t_j} |\psi(x,t)|^2 dt\right] = \int_0^{t_j} |\psi(x,t)|^2 dt + i\int_{t_j}^{+\infty} |\psi(x,t)|^2 dt \\
&= \left[\int_0^{t_j} \mathbb{N}\left(x_0\cos(wt), \sqrt{\frac{\hbar}{2m\Omega}\left(\cos^2(wt) + \frac{\Omega^2}{w^2}\sin^2(wt)\right)}\right)dt\right] \\
&\quad + i\left[1 - \int_0^{t_j} \mathbb{N}\left(x_0\cos(wt), \sqrt{\frac{\hbar}{2m\Omega}\left(\cos^2(wt) + \frac{\Omega^2}{w^2}\sin^2(wt)\right)}\right)dt\right] \\
&= \left[\int_0^{t_j} \mathbb{N}\left(x_0\cos(wt), \sqrt{\frac{\hbar}{2m\Omega}\left(\cos^2(wt) + \frac{\Omega^2}{w^2}\sin^2(wt)\right)}\right)dt\right] \\
&\quad + i\left[\int_{t_j}^{+\infty} \mathbb{N}\left(x_0\cos(wt), \sqrt{\frac{\hbar}{2m\Omega}\left(\cos^2(wt) + \frac{\Omega^2}{w^2}\sin^2(wt)\right)}\right)dt\right]
\end{aligned}
\tag{14}
$$

Additionally, the degree of our knowledge which is $DOK(T)$ is:

$$
\begin{aligned}
DOK(T) &= [P_r(T)]^2 + [P_m(T)/i]^2 = [P_r(T)]^2 + [1 - P_r(T)]^2 = [F(t_j)]^2 + [1 - F(t_j)]^2 \\
&= [P_{rob}(T \le t_j)]^2 + [1 - P_{rob}(T \le t_j)]^2 = [P_{rob}(T \le t_j)]^2 + [P_{rob}(T > t_j)]^2 \\
&= \left[\int_0^{t_j} |\psi(x,t)|^2 dt\right]^2 + \left[1 - \int_0^{t_j} |\psi(x,t)|^2 dt\right]^2 \\
&= \left[\int_0^{t_j} |\psi(x,t)|^2 dt\right]^2 + \left[\int_{t_j}^{+\infty} |\psi(x,t)|^2 dt\right]^2 \\
&= \left[\int_0^{t_j} \mathbb{N}\left(x_0\cos(wt), \sqrt{\frac{\hbar}{2m\Omega}\left(\cos^2(wt) + \frac{\Omega^2}{w^2}\sin^2(wt)\right)}\right)dt\right]^2 \\
&\quad + \left[1 - \int_0^{t_j} \mathbb{N}\left(x_0\cos(wt), \sqrt{\frac{\hbar}{2m\Omega}\left(\cos^2(wt) + \frac{\Omega^2}{w^2}\sin^2(wt)\right)}\right)dt\right]^2 \\
&= \left[\int_0^{t_j} \mathbb{N}\left(x_0\cos(wt), \sqrt{\frac{\hbar}{2m\Omega}\left(\cos^2(wt) + \frac{\Omega^2}{w^2}\sin^2(wt)\right)}\right)dt\right]^2 \\
&\quad + \left[\int_{t_j}^{+\infty} \mathbb{N}\left(x_0\cos(wt), \sqrt{\frac{\hbar}{2m\Omega}\left(\cos^2(wt) + \frac{\Omega^2}{w^2}\sin^2(wt)\right)}\right)dt\right]^2
\end{aligned}
\tag{15}
$$

Moreover, the chaotic factor which is $Chf(T)$ is:

$$\begin{aligned}
Chf(T) &= 2iP_r(T)P_m(T) = 2iP_r(T)\times i[1-P_r(T)] = -2P_r(T)[1-P_r(T)] \\
&= -2F(t_j)\left[1-F(t_j)\right] \\
&= -2P_{rob}(T\le t_j)\left[1-P_{rob}(T\le t_j)\right] = -2P_{rob}(T\le t_j)P_{rob}(T>t_j) \\
&= -2\left[\int_0^{t_j}|\psi(x,t)|^2dt\right]\times\left[1-\int_0^{t_j}|\psi(x,t)|^2dt\right] \\
&= -2\left[\int_0^{t_j}|\psi(x,t)|^2dt\right]\times\left[\int_{t_j}^{+\infty}|\psi(x,t)|^2dt\right] \\
&= -2\left[\int_0^{t_j}\mathbb{N}\left(x_0\cos(wt),\sqrt{\frac{\hbar}{2m\Omega}\left(\cos^2(wt)+\frac{\Omega^2}{w^2}\sin^2(wt)\right)}\right)dt\right] \\
&\quad\times\left[1-\int_0^{t_j}\mathbb{N}\left(x_0\cos(wt),\sqrt{\frac{\hbar}{2m\Omega}\left(\cos^2(wt)+\frac{\Omega^2}{w^2}\sin^2(wt)\right)}\right)dt\right] \\
&= -2\left[\int_0^{t_j}\mathbb{N}\left(x_0\cos(wt),\sqrt{\frac{\hbar}{2m\Omega}\left(\cos^2(wt)+\frac{\Omega^2}{w^2}\sin^2(wt)\right)}\right)dt\right] \\
&\quad\times\left[\int_{t_j}^{+\infty}\mathbb{N}\left(x_0\cos(wt),\sqrt{\frac{\hbar}{2m\Omega}\left(\cos^2(wt)+\frac{\Omega^2}{w^2}\sin^2(wt)\right)}\right)dt\right]
\end{aligned} \tag{16}$$

In addition, the magnitude of the chaotic factor which is $MChf(T)$ is:

$$\begin{aligned}
MChf(T) &= |Chf(T)| = -2iP_r(T)P_m(T) = -2iP_r(T)\times i[1-P_r(T)] = 2P_r(T)[1-P_r(T)] \\
&= 2F(t_j)\left[1-F(t_j)\right] \\
&= 2P_{rob}(T\le t_j)\left[1-P_{rob}(T\le t_j)\right] = 2P_{rob}(T\le t_j)P_{rob}(T>t_j) \\
&= 2\left[\int_0^{t_j}|\psi(x,t)|^2dt\right]\times\left[1-\int_0^{t_j}|\psi(x,t)|^2dt\right] \\
&= 2\left[\int_0^{t_j}|\psi(x,t)|^2dt\right]\times\left[\int_{t_j}^{+\infty}|\psi(x,t)|^2dt\right] \\
&= 2\left[\int_0^{t_j}\mathbb{N}\left(x_0\cos(wt),\sqrt{\frac{\hbar}{2m\Omega}\left(\cos^2(wt)+\frac{\Omega^2}{w^2}\sin^2(wt)\right)}\right)dt\right] \\
&\quad\times\left[1-\int_0^{t_j}\mathbb{N}\left(x_0\cos(wt),\sqrt{\frac{\hbar}{2m\Omega}\left(\cos^2(wt)+\frac{\Omega^2}{w^2}\sin^2(wt)\right)}\right)dt\right] \\
&= 2\left[\int_0^{t_j}\mathbb{N}\left(x_0\cos(wt),\sqrt{\frac{\hbar}{2m\Omega}\left(\cos^2(wt)+\frac{\Omega^2}{w^2}\sin^2(wt)\right)}\right)dt\right] \\
&\quad\times\left[\int_{t_j}^{+\infty}\mathbb{N}\left(x_0\cos(wt),\sqrt{\frac{\hbar}{2m\Omega}\left(\cos^2(wt)+\frac{\Omega^2}{w^2}\sin^2(wt)\right)}\right)dt\right]
\end{aligned} \tag{17}$$

Finally, the real probability in the complex probability universe $\mathcal{C} = \mathcal{R} + \mathcal{M}$ which is $Pc(T)$ is:

$$
\begin{aligned}
Pc^2(T) &= \{[P_r(T)] + [P_m(T)/i]\}^2 = \{[P_r(T)] + [1 - P_r(T)]\}^2 \\
&= \{[F(t_j)] + [1 - F(t_j)]\}^2 = \{P_{rob}(T \le t_j) + [1 - P_{rob}(T \le t_j)]\}^2 \\
&= \{P_{rob}(T \le t_j) + P_{rob}(T > t_j)\}^2 \\
&= \left\{\int_0^{t_j} |\psi(x,t)|^2 dt + \left[1 - \int_0^{t_j} |\psi(x,t)|^2 dt\right]\right\}^2 \\
&= \left\{\int_0^{t_j} |\psi(x,t)|^2 dt + \int_{t_j}^{+\infty} |\psi(x,t)|^2 dt\right\}^2 = \left\{\int_0^{+\infty} |\psi(x,t)|^2 dt\right\}^2 \\
&= \left\{\left[\int_0^{t_j} \mathbb{N}\left(x_0\cos(wt), \sqrt{\frac{\hbar}{2m\Omega}\left(\cos^2(wt) + \frac{\Omega^2}{w^2}\sin^2(wt)\right)}\right) dt\right] + \left[1 - \int_0^{t_j} \mathbb{N}\left(x_0\cos(wt), \sqrt{\frac{\hbar}{2m\Omega}\left(\cos^2(wt) + \frac{\Omega^2}{w^2}\sin^2(wt)\right)}\right) dt\right]\right\}^2 \\
&= \left\{\left[\int_0^{t_j} \mathbb{N}\left(x_0\cos(wt), \sqrt{\frac{\hbar}{2m\Omega}\left(\cos^2(wt) + \frac{\Omega^2}{w^2}\sin^2(wt)\right)}\right) dt\right] + \left[\int_{t_j}^{+\infty} \mathbb{N}\left(x_0\cos(wt), \sqrt{\frac{\hbar}{2m\Omega}\left(\cos^2(wt) + \frac{\Omega^2}{w^2}\sin^2(wt)\right)}\right) dt\right]\right\}^2 \\
&= \left\{\int_0^{+\infty} \mathbb{N}\left(x_0\cos(wt), \sqrt{\frac{\hbar}{2m\Omega}\left(\cos^2(wt) + \frac{\Omega^2}{w^2}\sin^2(wt)\right)}\right) dt\right\}^2 \\
&= 1^2 = 1 \\
&= Pc(T)
\end{aligned} \tag{18}
$$

And, $Pc(T)$ can be computed using *CPP* as follows:

$$
\begin{aligned}
Pc^2(T) &= DOK(T) - Chf(T) = [P_r(T)]^2 + [P_m(T)/i]^2 - 2iP_r(T)P_m(T) \\
&= [P_r(T)]^2 + [1 - P_r(T)]^2 + 2P_r(T)[1 - P_r(T)] = \{P_r(T) + [1 - P_r(T)]\}^2 \\
&= \left\{\int_0^{t_j} |\psi(x,t)|^2 dt + \left[1 - \int_0^{t_j} |\psi(x,t)|^2 dt\right]\right\}^2 = \left\{\int_0^{t_j} |\psi(x,t)|^2 dt + \int_{t_j}^{+\infty} |\psi(x,t)|^2 dt\right\}^2 = \left\{\int_0^{+\infty} |\psi(x,t)|^2 dt\right\}^2 \\
&= 1^2 = 1 \\
&= Pc(T)
\end{aligned} \tag{19}
$$

and $Pc(T)$ can always be computed using *CPP* as follows also:

$$
\begin{aligned}
Pc^2(T) &= DOK(T) + MChf(T) = [P_r(T)]^2 + [P_m(T)/i]^2 + [-2iP_r(T)P_m(T)] \\
&= [P_r(T)]^2 + [1 - P_r(T)]^2 + 2P_r(T)[1 - P_r(T)] = \{P_r(T) + [1 - P_r(T)]\}^2 \\
&= \left\{ \int_0^{t_j} |\psi(x,t)|^2 dt + \left[1 - \int_0^{t_j} |\psi(x,t)|^2 dt \right] \right\}^2 \\
&= \left\{ \int_0^{t_j} |\psi(x,t)|^2 dt + \int_{t_j}^{+\infty} |\psi(x,t)|^2 dt \right\}^2 = \left\{ \int_0^{+\infty} |\psi(x,t)|^2 dt \right\}^2 = 1^2 = 1 = Pc(T)
\end{aligned}
\tag{20}
$$

Hence, the prediction of all the wavefunction position probabilities of the quantum harmonic oscillators with the Gaussian initial condition problem in the universe $\mathcal{C} = \mathcal{R} + \mathcal{M}$ is permanently certain and perfectly deterministic.

4.3 The new model simulations

Figures 3–37 illustrate all the calculations done above.

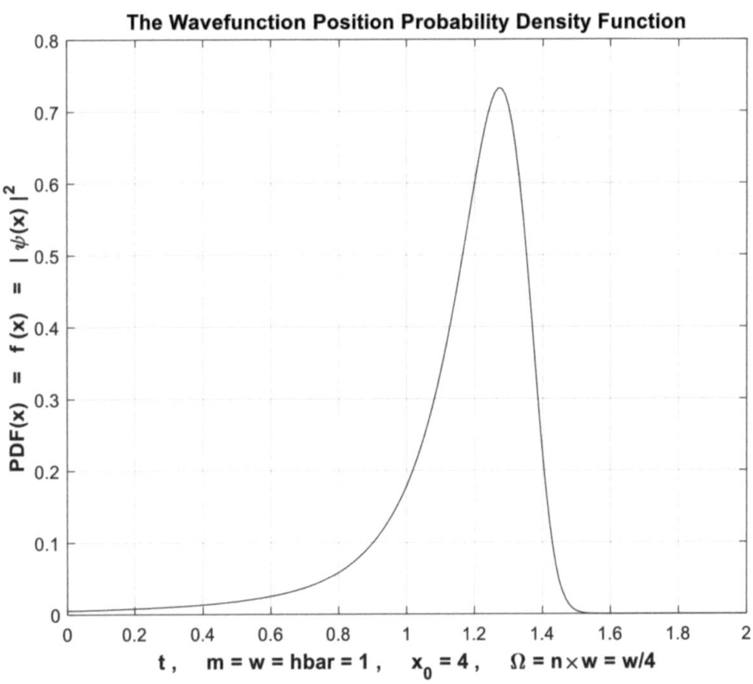

Figure 3.
The graph of the PDF *as a function of the random variable* T *of the wavefunction position probability density for* n = 1/4.

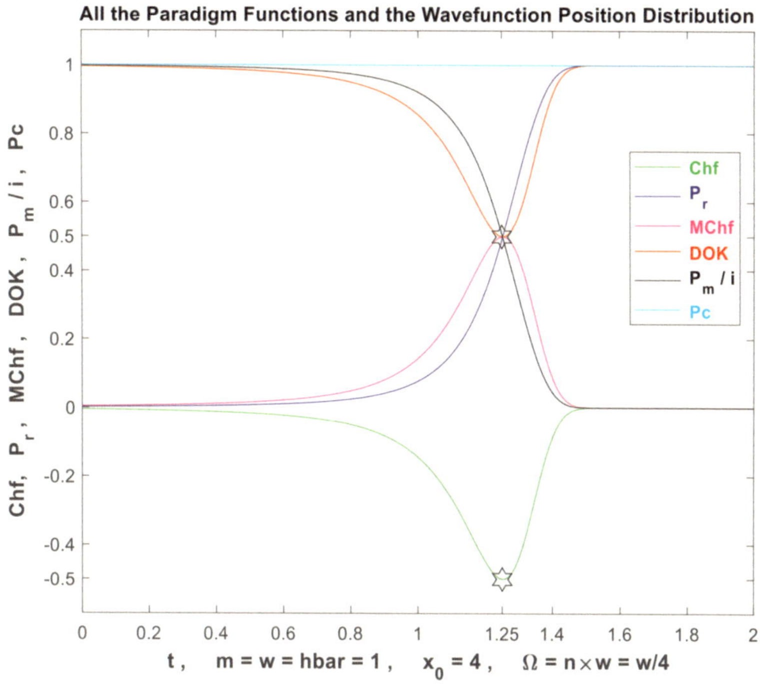

Figure 4.
The graphs of all the CPP *parameters for the wavefunction position probability distribution as functions of the random variable* T *for* n = 1/4.

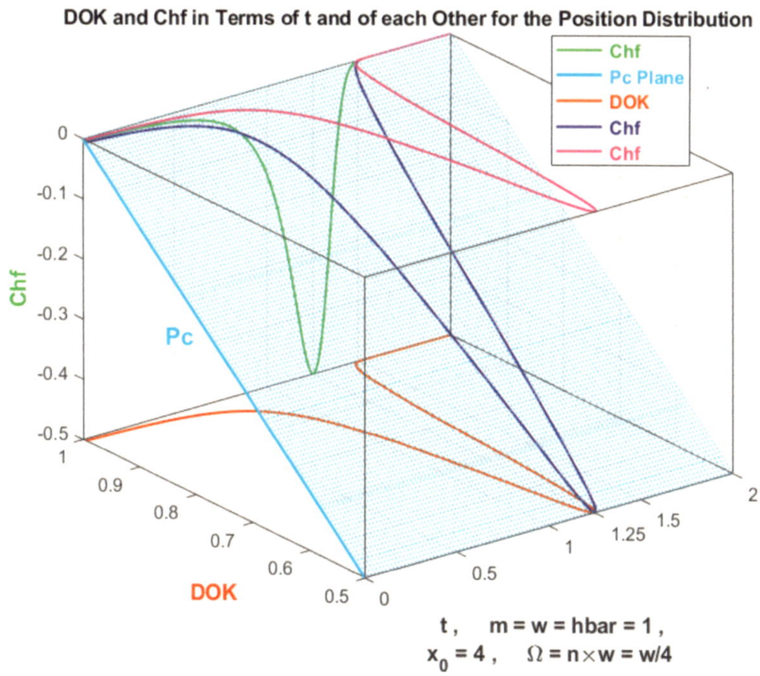

Figure 5.
The graphs of DOK *and* Chf *and the deterministic probability* Pc *for the wavefunction position probability distribution in terms of* T *and of each other for* n = 1/4.

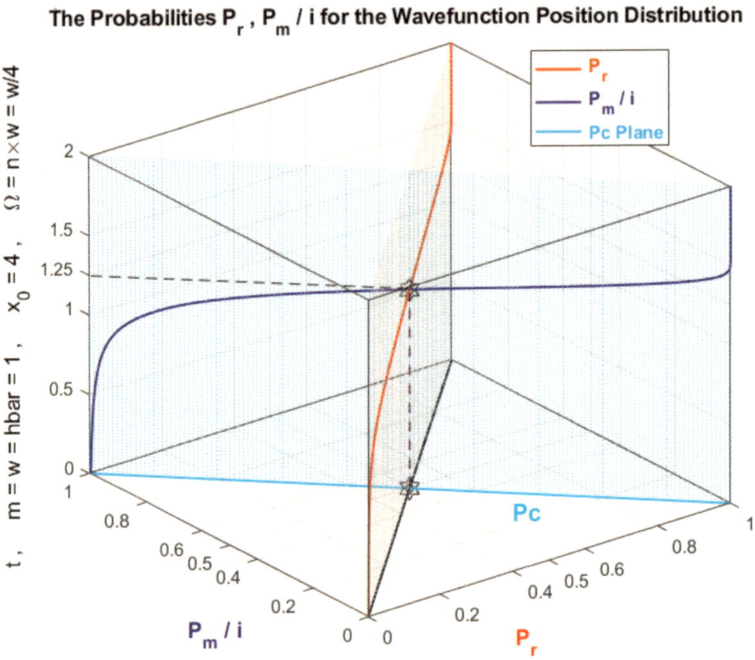

Figure 6.
The graphs of P_r *and* P_m/i *and* Pc *for the wavefunction position probability distribution in terms of* T *and of each other for* n = 1/4.

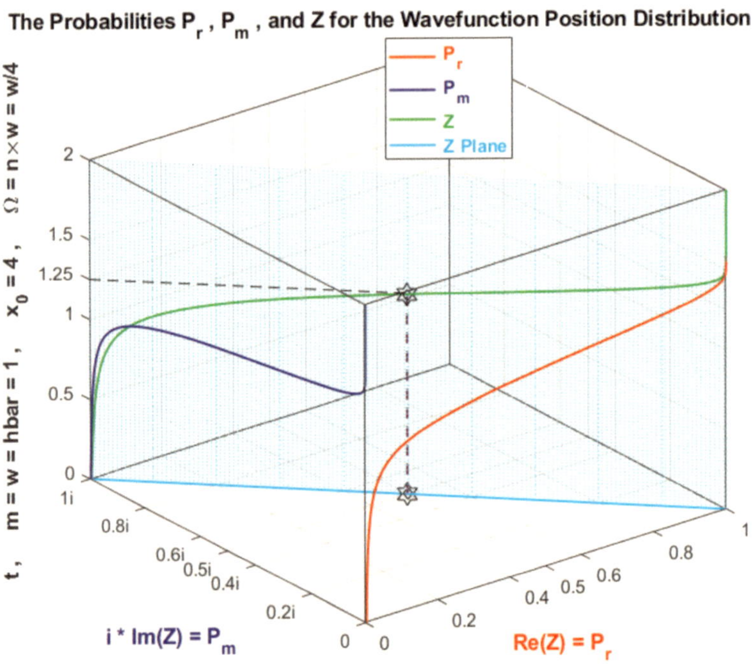

Figure 7.
The graphs of the probabilities P_r *and* P_m *and* Z *for the wavefunction position probability distribution in terms of* T *for* n = 1/4.

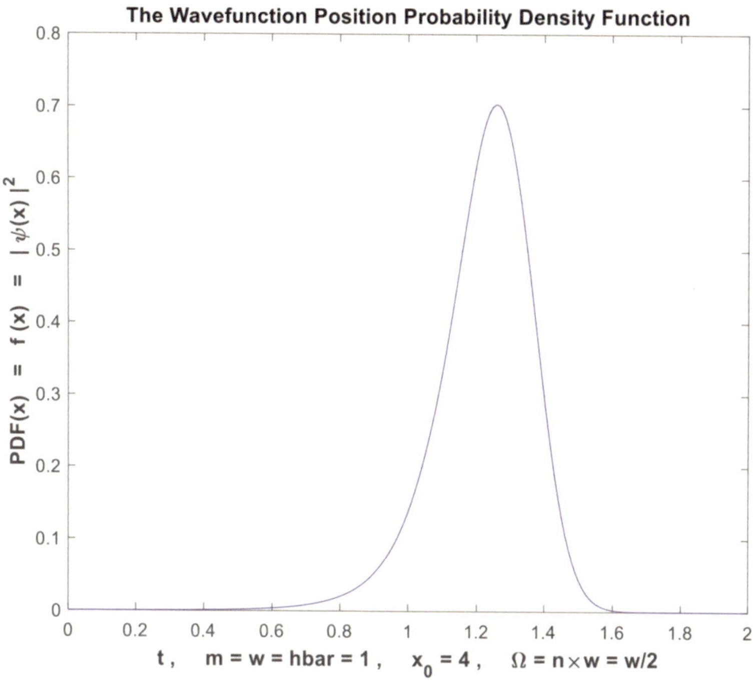

Figure 8.
The graph of the PDF *as a function of the random variable* T *of the wavefunction position probability density for* n = 1/2.

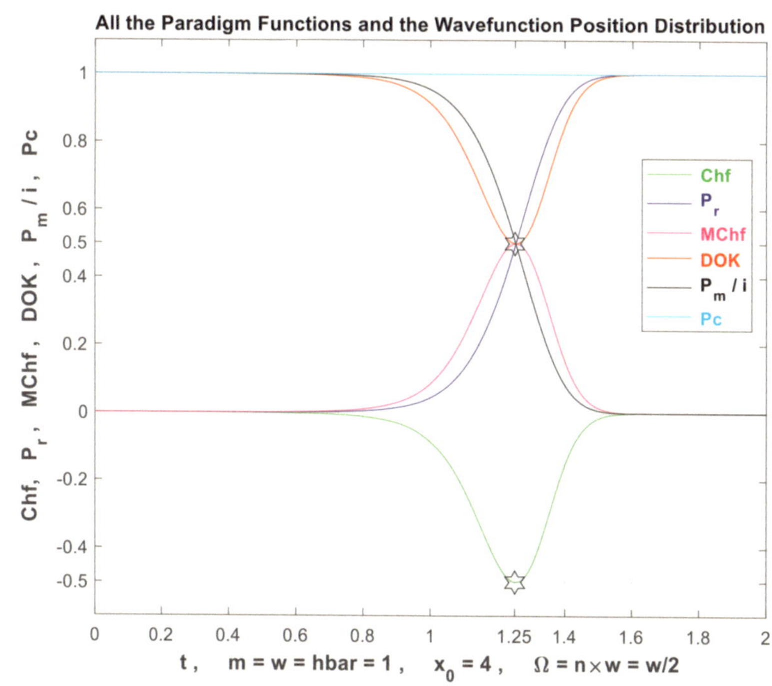

Figure 9.
The graphs of all the CPP *parameters for the wavefunction position probability distribution as functions of the random variable* T *for* n = 1/2.

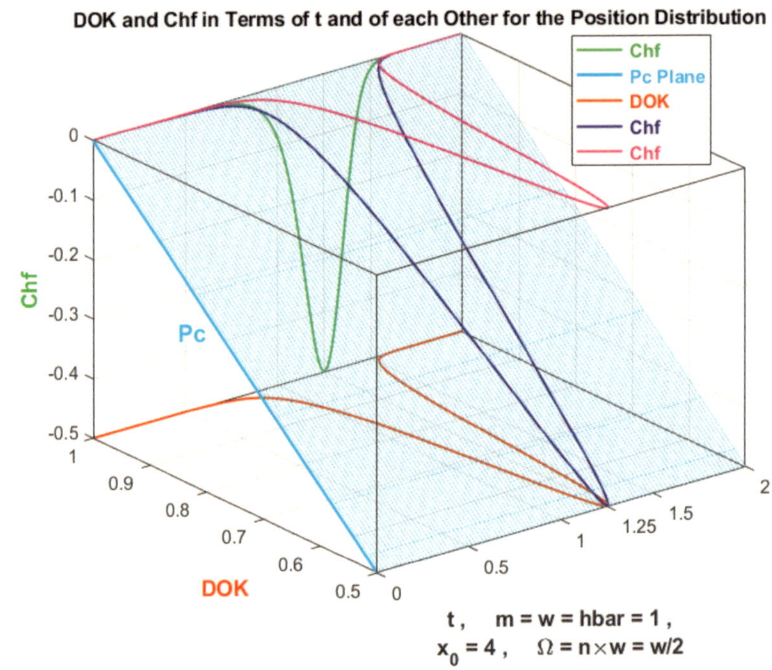

Figure 10.
The graphs of DOK *and* Chf *and the deterministic probability* Pc *for the wavefunction position probability distribution in terms of* T *and of each other for* n = 1/2.

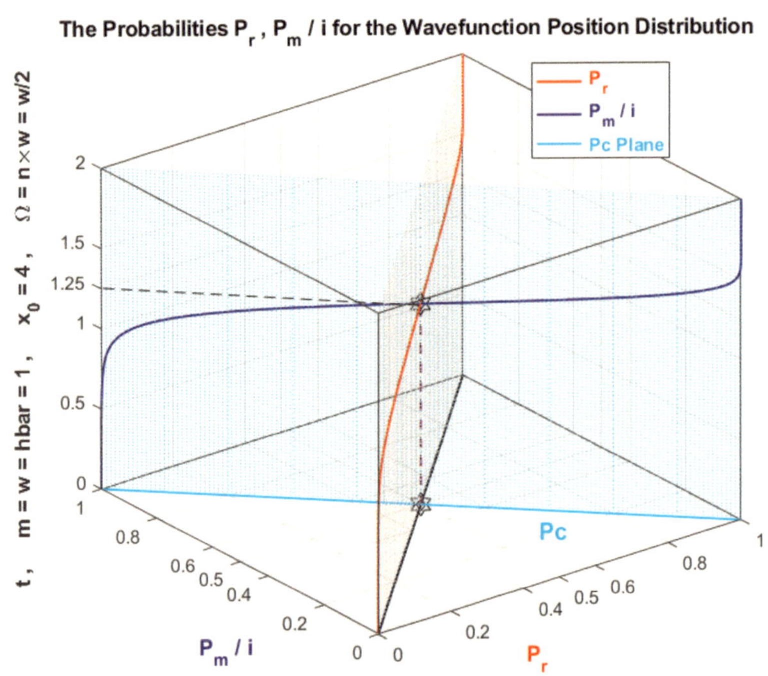

Figure 11.
The graphs of P_r *and* P_m/i *and* Pc *for the wavefunction position probability distribution in terms of* T *and of each other for* n = 1/2.

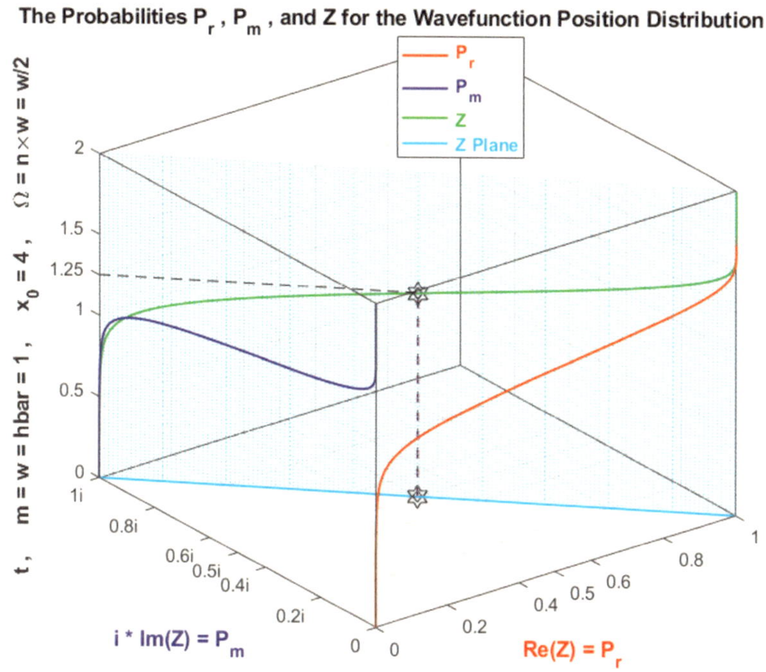

Figure 12.
The graphs of the probabilities P_r *and* P_m *and* Z *for the wavefunction position probability distribution in terms of* T *for* n = 1/2.

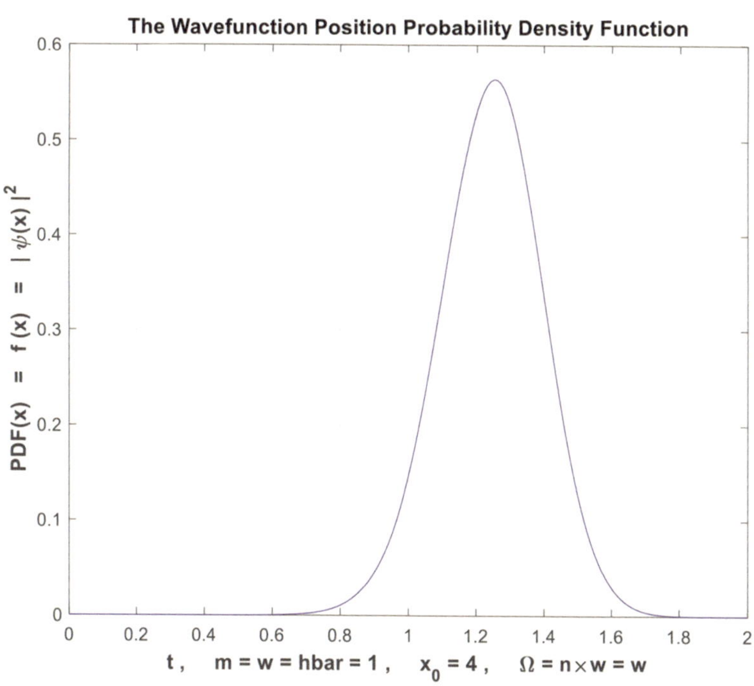

Figure 13.
The graph of the PDF *as a function of the random variable* T *of the wavefunction position probability density for* n = 1.

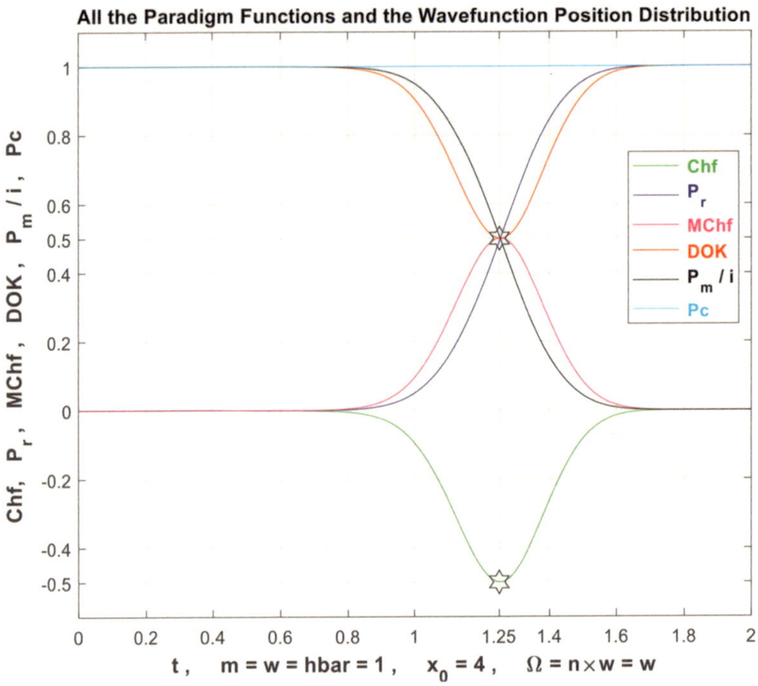

Figure 14.
The graphs of all the CPP *parameters for the wavefunction position probability distribution as functions of the random variable* T *for* n = 1.

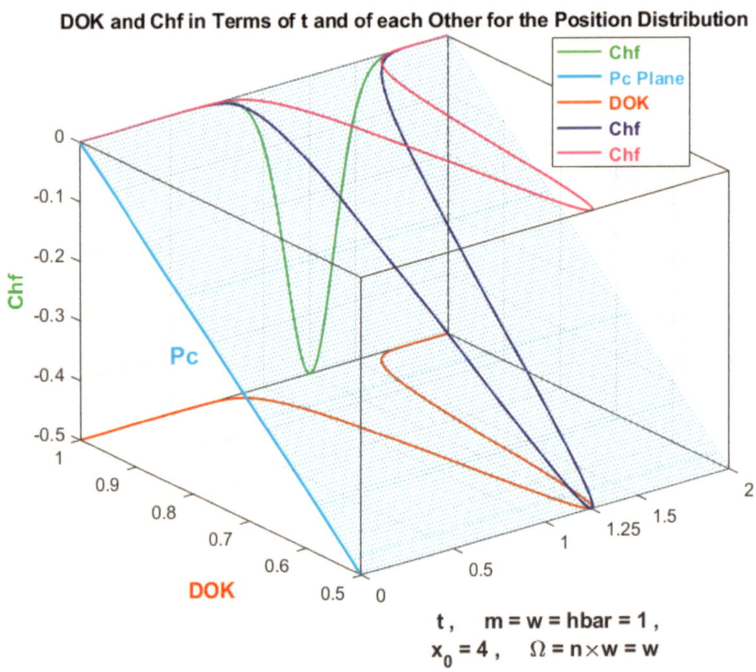

Figure 15.
The graphs of DOK *and* Chf *and the deterministic probability* Pc *for the wavefunction position probability distribution in terms of* T *and of each other for* n = 1.

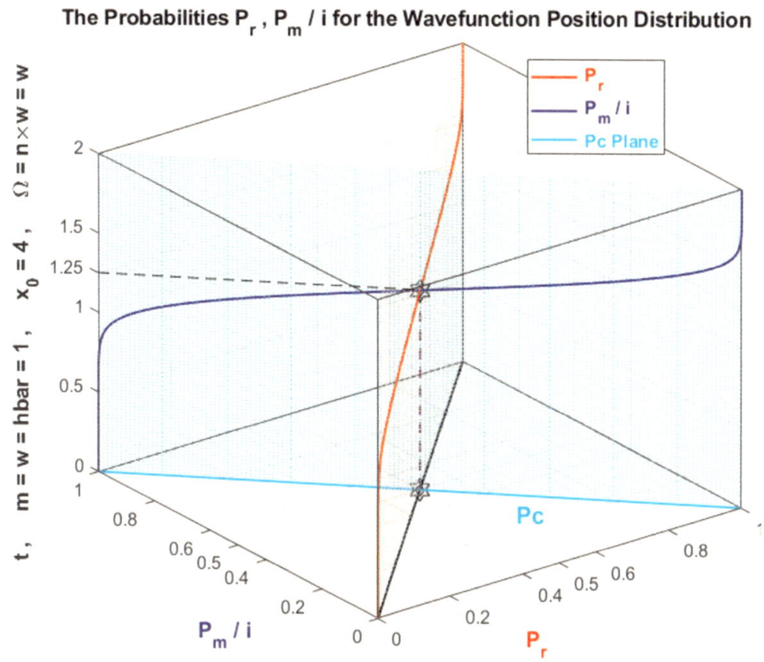

Figure 16.
The graphs of P_r *and* P_m/i *and* Pc *for the wavefunction position probability distribution in terms of* T *and of each other for* n = 1.

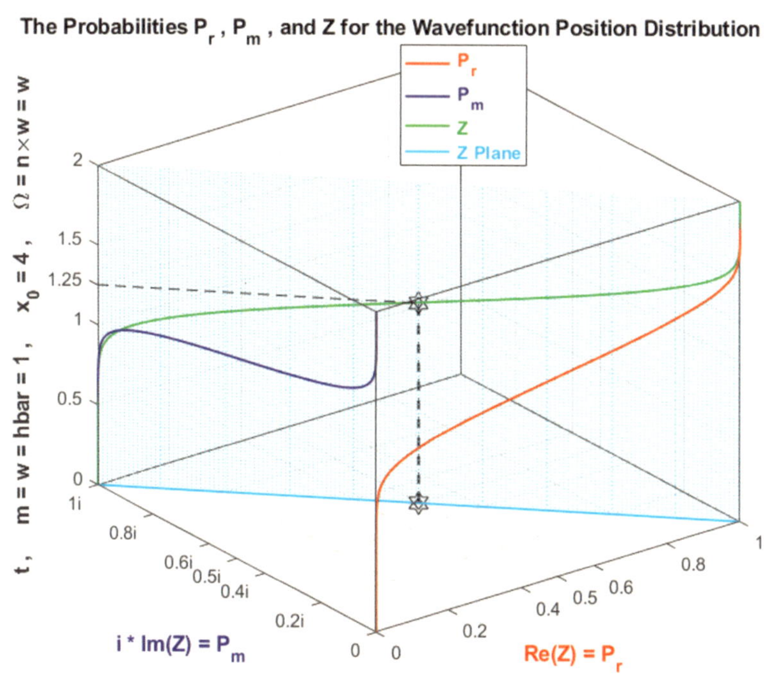

Figure 17.
The graphs of the probabilities P_r *and* P_m *and* Z *for the wavefunction position probability distribution in terms of* T *for* n = 1.

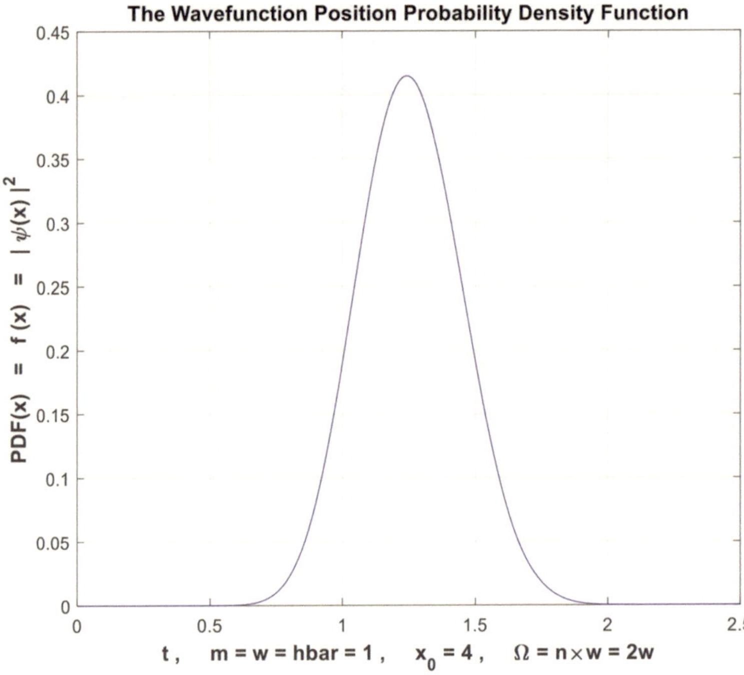

Figure 18.
The graph of the PDF *as a function of the random variable* T *of the wavefunction position probability density for* n = 2.

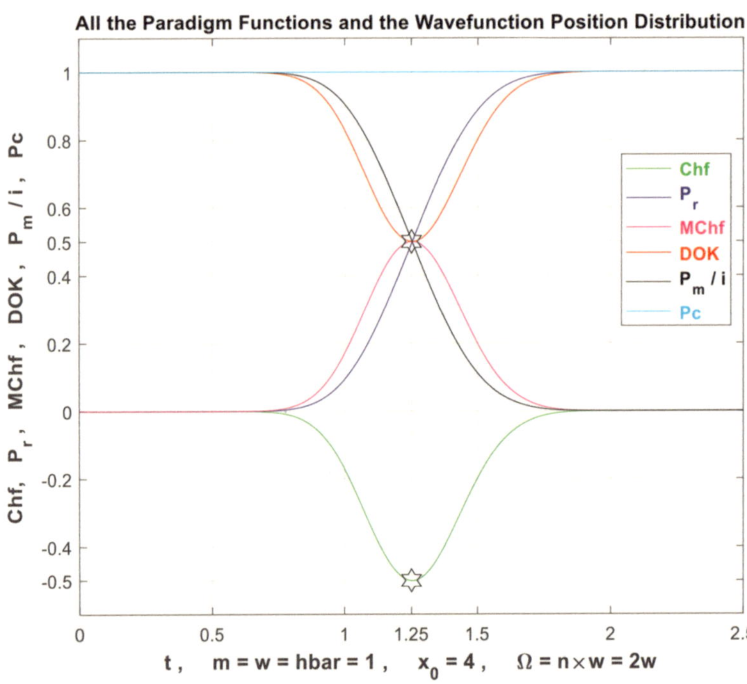

Figure 19.
The graphs of all the CPP *parameters for the wavefunction position probability distribution as functions of the random variable* T *for* n = 2.

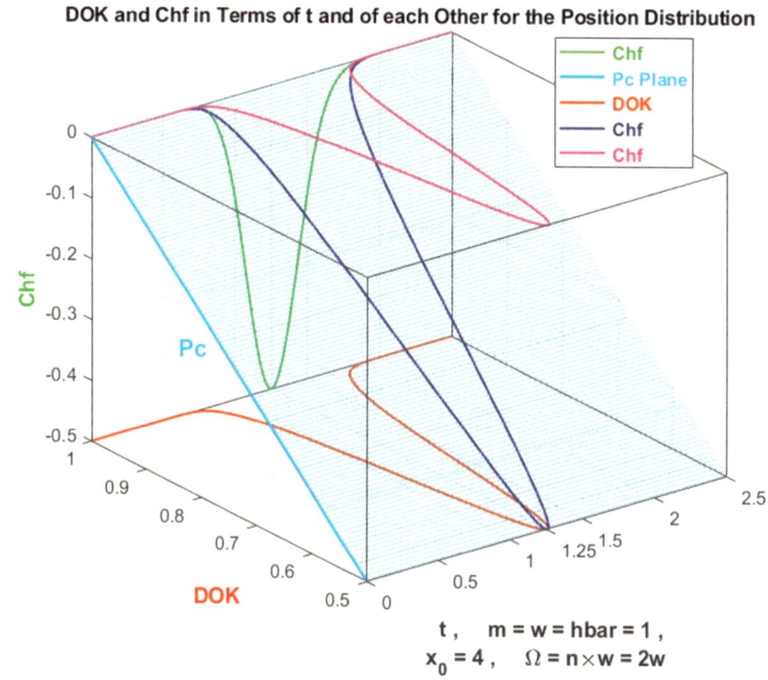

Figure 20.
The graphs of DOK *and* Chf *and the deterministic probability* Pc *for the wavefunction position probability distribution in terms of* T *and of each other for* n = 2.

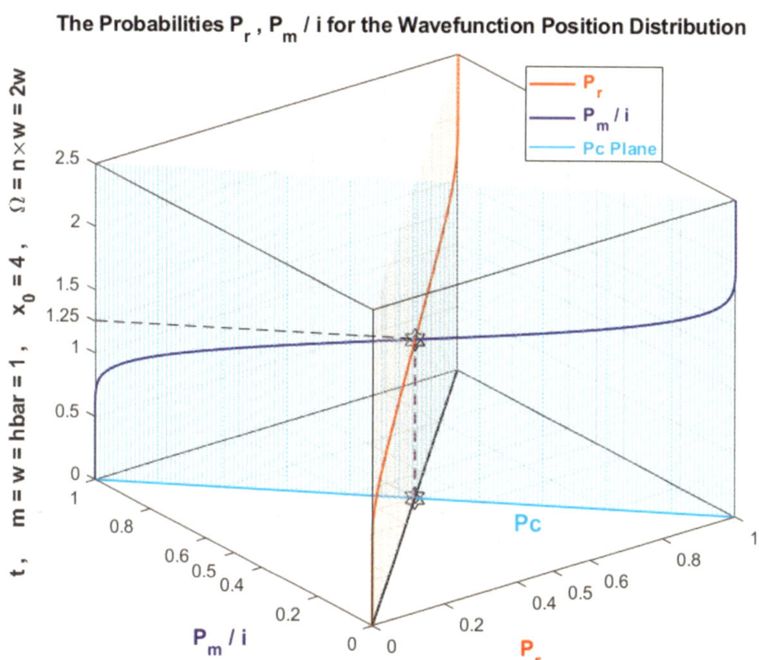

Figure 21.
The graphs of P_r *and* P_m/i *and* Pc *for the wavefunction position probability distribution in terms of* T *and of each other for* n = 2.

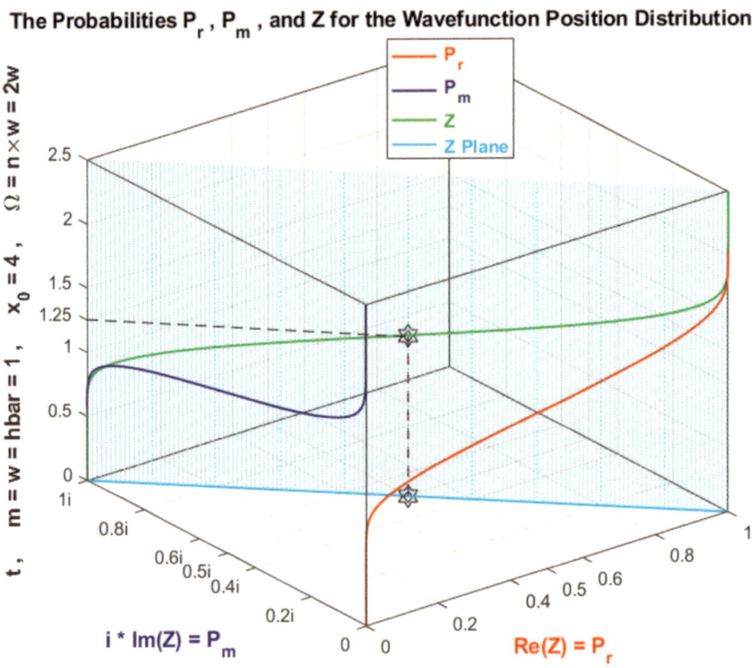

Figure 22.
The graphs of the probabilities P_r *and* P_m *and* Z *for the wavefunction position probability distribution in terms of* T *for* n = 2.

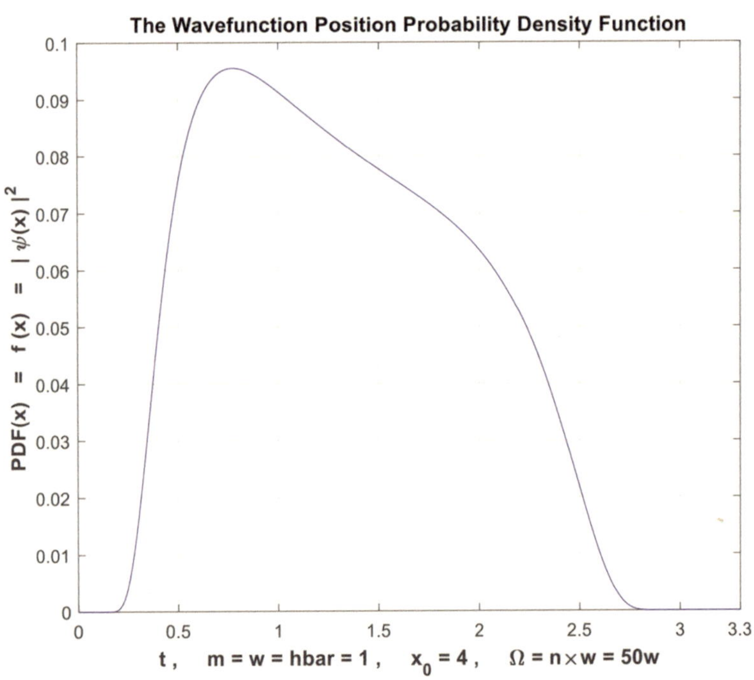

Figure 23.
The graph of the PDF *as a function of the random variable* T *of the wavefunction position probability density for* n = 50.

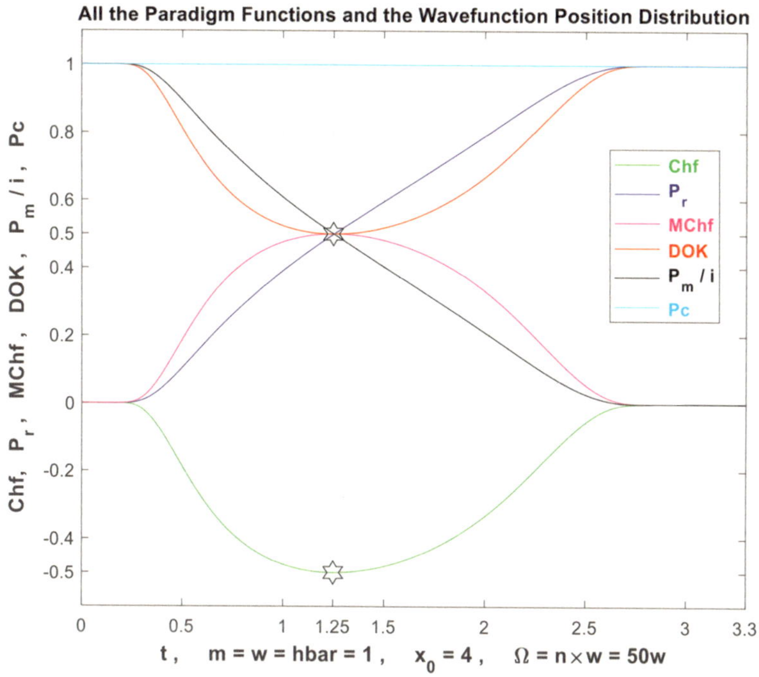

Figure 24.
The graphs of all the CPP *parameters for the wavefunction position probability distribution as functions of the random variable* T *for* n = 50.

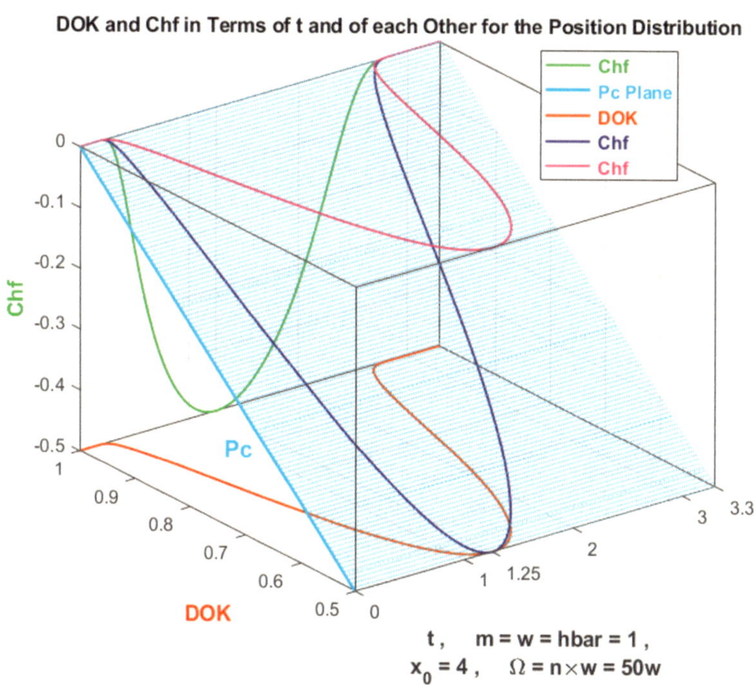

Figure 25.
The graphs of DOK *and* Chf *and the deterministic probability* Pc *for the wavefunction position probability distribution in terms of* T *and of each other for* n = 50.

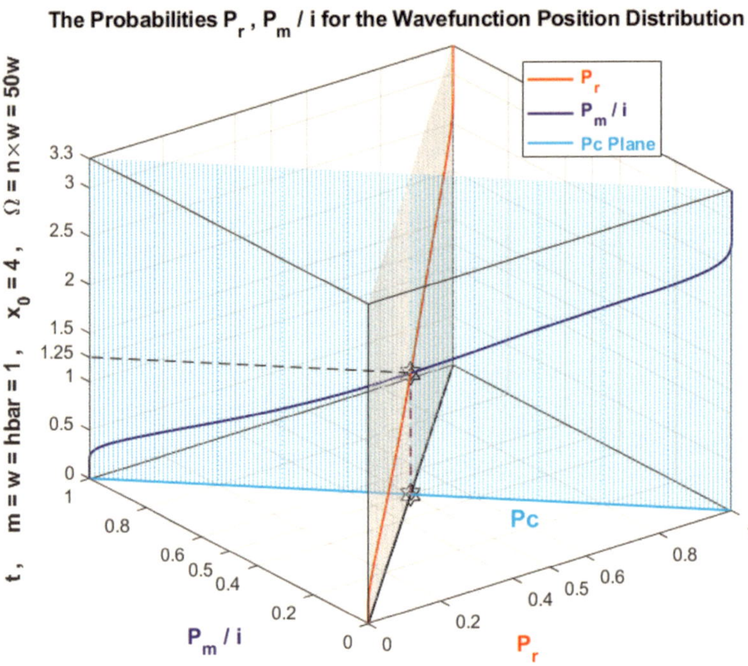

Figure 26.
The graphs of P_r *and* P_m/i *and* Pc *for the wavefunction position probability distribution in terms of* T *and of each other for* n = 50.

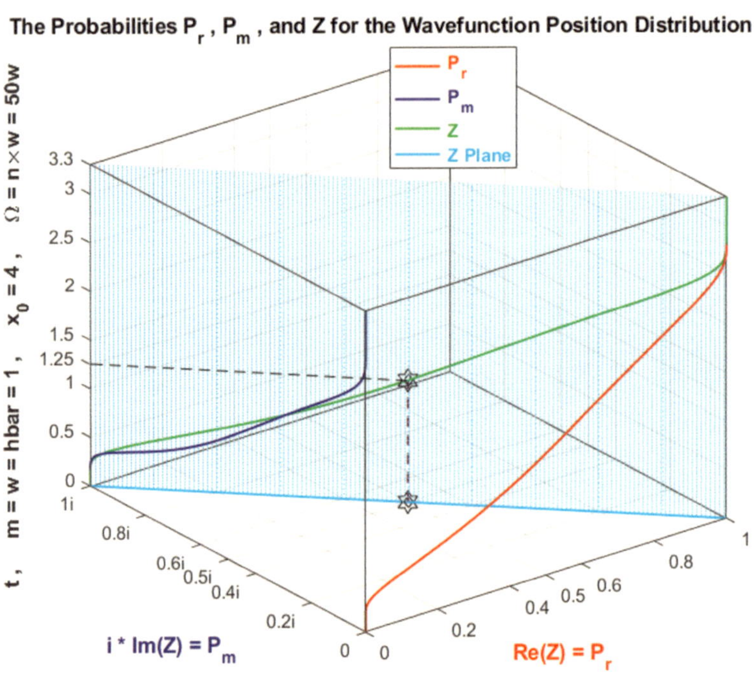

Figure 27.
The graphs of the probabilities P_r *and* P_m *and* Z *for the wavefunction position probability distribution in terms of* T *for* n = 50.

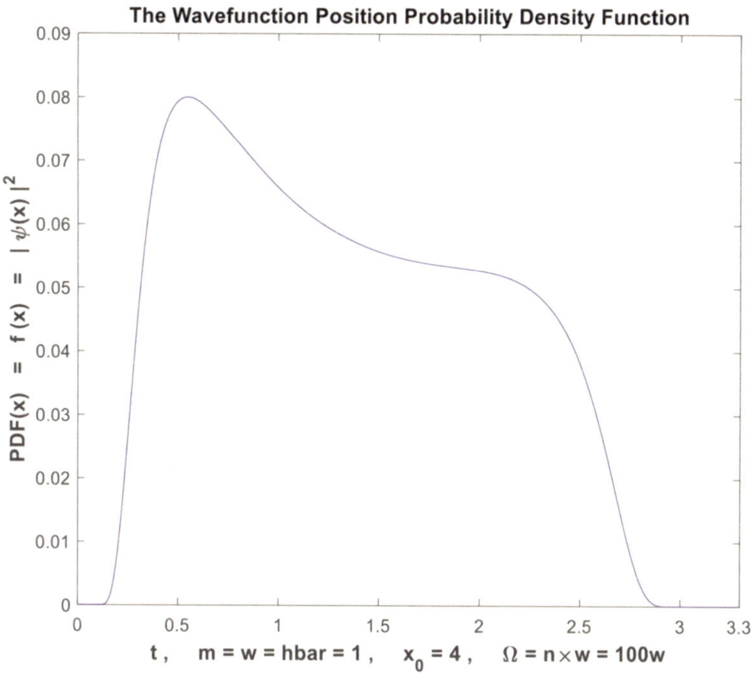

Figure 28.
The graph of the PDF *as a function of the random variable* T *of the wavefunction position probability density for* n = 100.

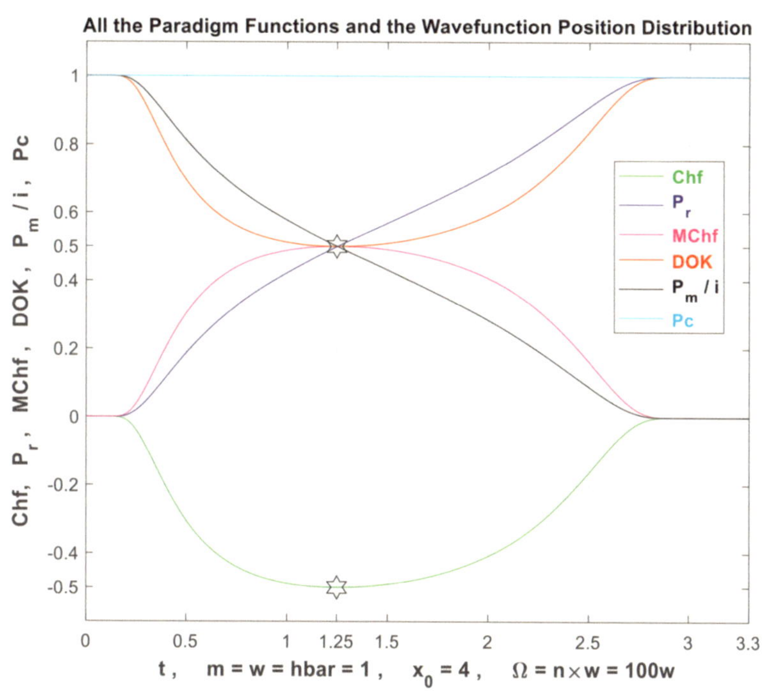

Figure 29.
The graphs of all the CPP *parameters for the wavefunction position probability distribution as functions of the random variable* T *for* n = 100.

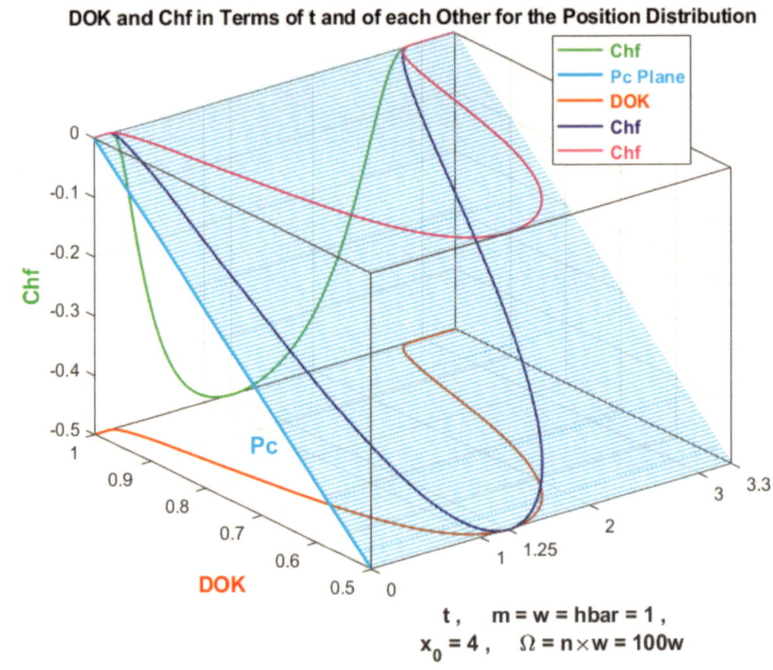

Figure 30.
The graphs of DOK *and* Chf *and the deterministic probability* Pc *for the wavefunction position probability distribution in terms of* T *and of each other for* n = 100.

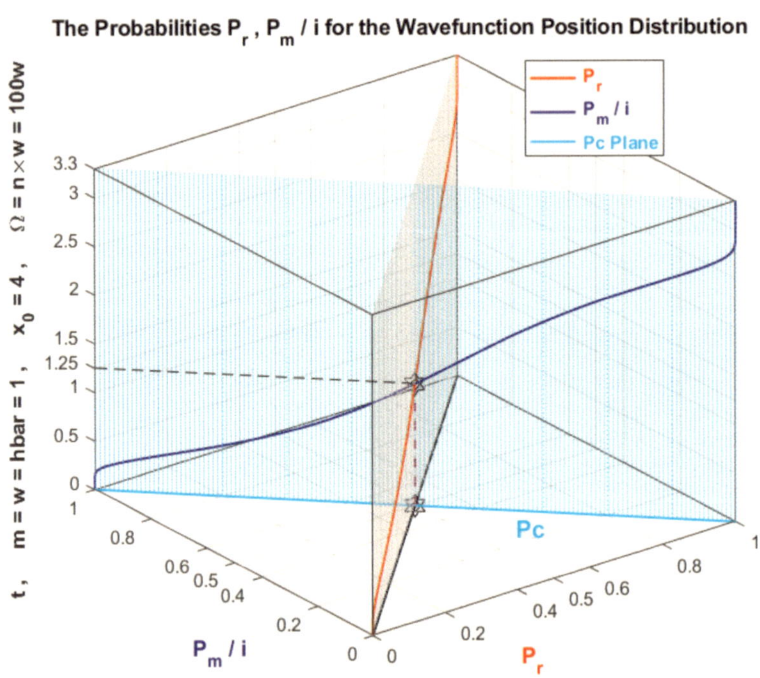

Figure 31.
The graphs of P_r *and* P_m/i *and* Pc *for the wavefunction position probability distribution in terms of* T *and of each other for* n = 100.

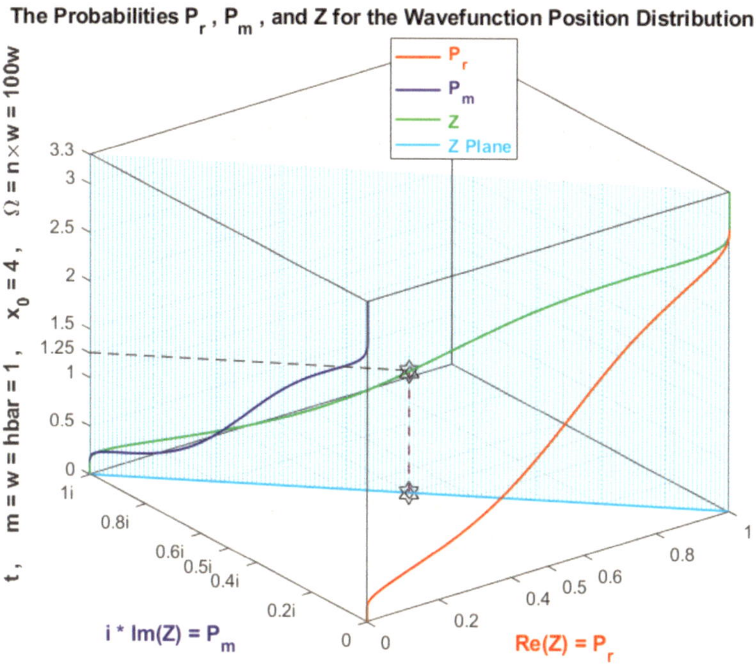

Figure 32.
The graphs of the probabilities P_r *and* P_m *and* Z *for the wavefunction position probability distribution in terms of* T *for* n = *100.*

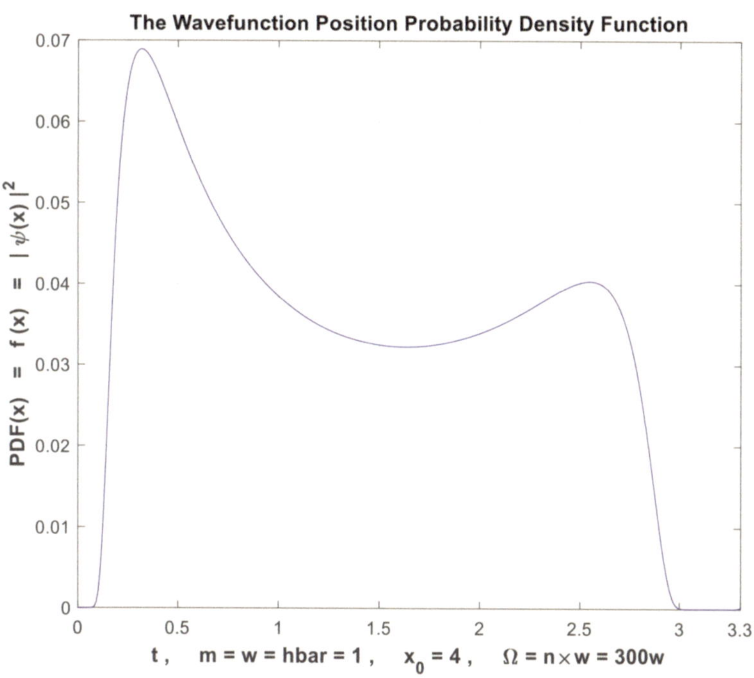

Figure 33.
The graph of the PDF *as a function of the random variable* T *of the wavefunction position probability density for* n = *300.*

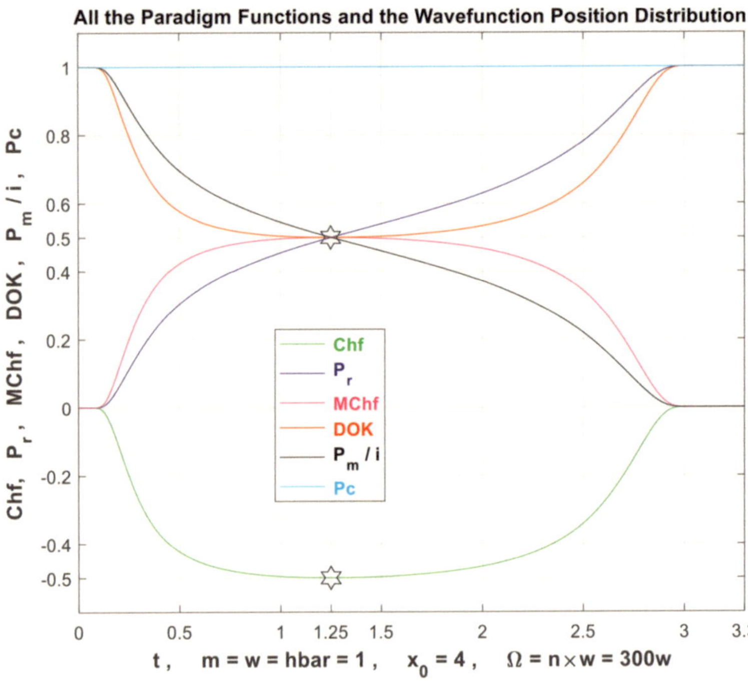

Figure 34.
The graphs of all the CPP *parameters for the wavefunction position probability distribution as functions of the random variable* T *for* n = 300.

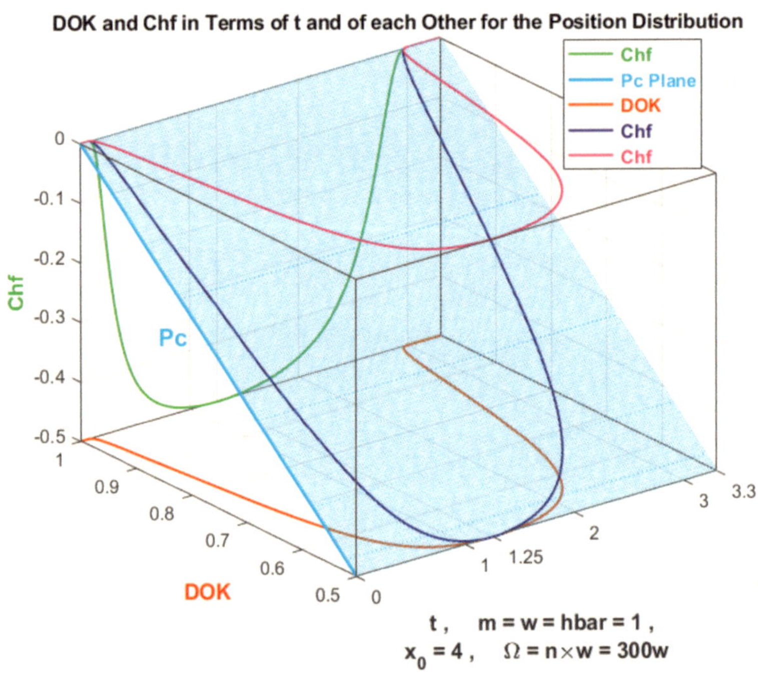

Figure 35.
The graphs of DOK *and* Chf *and the deterministic probability* Pc *for the wavefunction position probability distribution in terms of* T *and of each other for* n = 300.

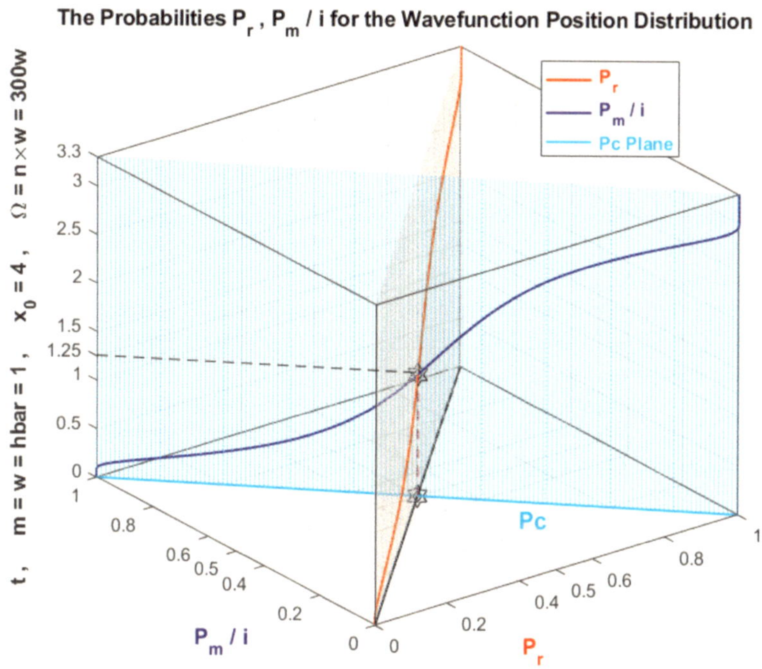

Figure 36.
The graphs of P_r *and* P_m/i *and* Pc *for the wavefunction position probability distribution in terms of* T *and of each other for* n = 300.

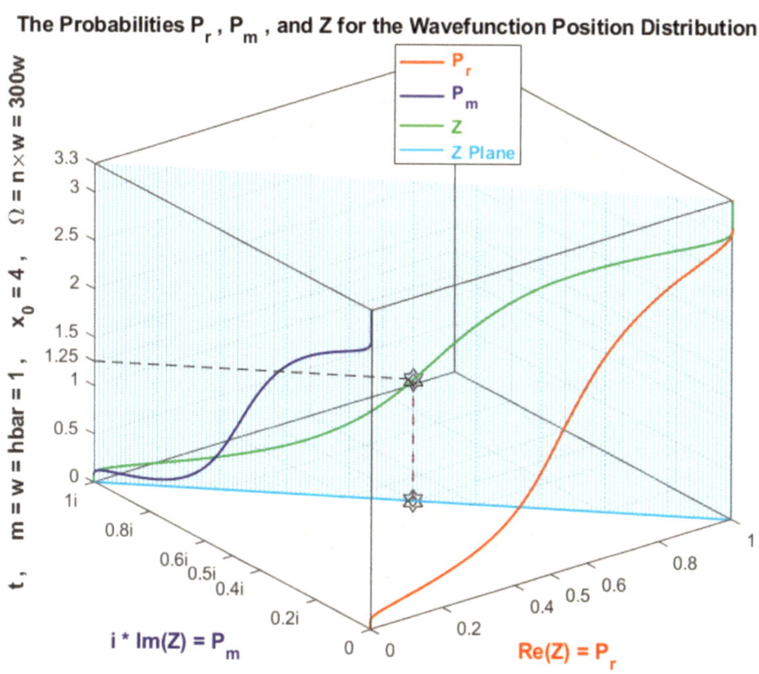

Figure 37.
The graphs of the probabilities P_r *and* P_m *and* Z *for the wavefunction position probability distribution in terms of* T *for* n = 300.

4.3.1 Simulations interpretation

In **Figures 3, 8, 13, 18, 23, 28,** and **33,** we can see the graphs of the probability density functions (*PDF*) of the wavefunction position probability distribution for this problem as functions of the time random variable $T: 0 \leq T \leq 3.3$ for $n = 1/4, 1/2, 1, 2, 50, 100$, and 300.

In **Figures 4, 9, 14, 19, 24, 29,** and **34,** we can also see the graphs and the simulations of all the *CPP* parameters (*Chf*, *MChf*, *DOK*, P_r, P_m/i, and *Pc*) as functions of the time random variable *T* for the wavefunction position probability distribution of the quantum harmonic oscillators with Gaussian initial condition problem for $n = 1/4, 1/2, 1, 2, 50, 100$, and 300. Hence, we can visualize all the new paradigm functions for this problem.

In the cubes (**Figures 5, 10, 15, 20, 25, 30,** and **35**), the simulation of *DOK* and *Chf* as functions of each other and of the time random variable *T* for the quantum harmonic oscillators with Gaussian initial condition problem wavefunction position probability distribution can be seen. The thick line in cyan is the projection of the plane $Pc^2(T) = DOK(T) - Chf(T) = 1 = Pc(T)$ on the plane $T = L_b$ = lower bound of $T = 0$. This thick line starts at the point ($DOK = 1$, $Chf = 0$) when $T = L_b = 0$, reaches the point ($DOK = 0.5$, $Chf = -0.5$) when $T = 1.25$, and returns at the end to ($DOK = 1$, $Chf = 0$) when $T = U_b$ = upper bound of *T*. The other curves are the graphs of $DOK(T)$ (red) and $Chf(T)$ (green, blue, pink) in different simulation planes. Notice that they all have a minimum at the point ($DOK = 0.5$, $Chf = -0.5, T = 0$). The last simulation point corresponds to ($DOK = 1$, $Chf = 0, T = U_b$).

In the cubes (**Figures 6, 11, 16, 21, 26, 31,** and **36**), we can notice the simulation of the real probability $P_r(T)$ in $\mathcal{R}$ and its complementary real probability $P_m(T)/i$ in $\mathcal{R}$ also in terms of the time random variable *T* for the quantum harmonic oscillators with Gaussian initial condition problem wavefunction position probability distribution. The thick line in cyan is the projection of the plane $Pc^2(T) = P_r(T) + P_m(T)/i = 1 = Pc(T)$ on the plane $T = L_b$ = lower bound of $T = 0$. This thick line starts at the point ($P_r = 0$, $P_m/i = 1$) and ends at the point ($P_r = 1$, $P_m/i = 0$). The red curve represents $P_r(T)$ in the plane $P_r(T) = P_m(T)/i$ in light gray. This curve starts at the point ($P_r = 0$, $P_m/i = 1, T = L_b$ = lower bound of $T = 0$), reaches the point ($P_r = 0.5$, $P_m/i = 0.5, T = 1.25$), and gets at the end to ($P_r = 1$, $P_m/i = 0, T = U_b$ = upper bound of *T*). The blue curve represents $P_m(T)/i$ in the plane in cyan $P_r(T) + P_m(T)/i = 1 = Pc(T)$. Notice the importance of the point, which is the intersection of the red and blue curves at $T = 1.25$ and when $P_r(T) = P_m(T)/i = 0.5$.

In the cubes (**Figures 7, 12, 17, 22, 27, 32,** and **37**), we can notice the simulation of the complex probability $Z(T)$ in $\mathcal{C} = \mathcal{R} + \mathcal{M}$ as a function of the real probability $P_r(T) = \text{Re}(Z)$ in $\mathcal{R}$ and of its complementary imaginary probability $P_m(T) = i \times \text{Im}(Z)$ in $\mathcal{M}$, and this in terms of the time random variable *T* for the quantum harmonic oscillators with Gaussian initial condition problem wavefunction position probability distribution. The red curve represents $P_r(T)$ in the plane $P_m(T) = 0$, and the blue curve represents $P_m(T)$ in the plane $P_r(T) = 0$. The green curve represents the complex probability $Z(T) = P_r(T) + P_m(T) = \text{Re}(Z) + i \times \text{Im}(Z)$ in the plane $P_r(T) = iP_m(T) + 1$ or $Z(T)$ plane in cyan. The curve of $Z(T)$ starts at the point ($P_r = 0$, $P_m = i, T = L_b$ = lower bound of $T = 0$) and ends at the point ($P_r = 1$, $P_m = 0, T = U_b$ = upper bound of *T*). The thick line in cyan is $P_r(T = L_b = 0) = iP_m(T = L_b = 0) + 1$, and it is the projection of the $Z(T)$ curve on the complex probability plane whose equation is: $T = L_b = 0$. This projected thick line starts at the point ($P_r = 0$, $P_m = i, T = L_b = 0$) and ends at the point ($P_r = 1$, $P_m = 0, T = L_b = 0$). Notice the importance of the point corresponding to $T = 1.25$ and $Z = 0.5 + 0.5i$ when $P_r = 0.5$ and $P_m = 0.5i$.

4.4 The characteristics of the position probability distribution

In this quantum mechanics problem, the average, or expectation value of the position of a particle is given by Ref. [20]:

$$\langle x\rangle = \int_0^{+\infty} t|\psi(x,t)|^2 dt = \int_0^{+\infty} t\mathbb{N}\left(x_0\cos(wt), \sqrt{\frac{\hbar}{2m\Omega}\left(\cos^2(wt)+\frac{\Omega^2}{w^2}\sin^2(wt)\right)}\right)dt \quad (21)$$

For this problem, it can be shown that the average position is always $\langle x\rangle = x_0\cos(wt)$.

So, in the probability set and universe $\mathcal{R}$, we have:

$$\langle x\rangle_R = \langle x\rangle = x_0\cos(wt) \quad (22)$$

The variance in the position is a measure of the uncertainty in the position of the particle, so in the probability set and universe $\mathcal{R}$, we have:

$$\begin{aligned}
\text{Var}_{x,R} &= \text{Var}(x) = \langle x^2\rangle_R - \langle x\rangle_R^2 \\
&= \left\{\int_0^{+\infty} t^2|\psi(x,t)|^2 dt\right\} - \left\{\int_0^{+\infty} t|\psi(x,t)|^2 dt\right\}^2 \\
&= \left\{\int_0^{+\infty} t^2\mathbb{N}\left(x_0\cos(wt), \sqrt{\frac{\hbar}{2m\Omega}\left(\cos^2(wt)+\frac{\Omega^2}{w^2}\sin^2(wt)\right)}\right)dt\right\} - [x_0\cos(wt)]^2 \\
&= \frac{\hbar}{2m\Omega}\left(\cos^2(wt)+\frac{\Omega^2}{w^2}\sin^2(wt)\right)
\end{aligned} \quad (23)$$

In the probability set and universe $\mathcal{M}$, we have:

$$\begin{aligned}
\langle x\rangle_M &= \int_0^{+\infty} t\left\{i\left[1-|\psi(x,t)|^2\right]\right\}dt \\
&= i\int_0^{+\infty} t\left\{1-\mathbb{N}\left(x_0\cos(wt), \sqrt{\frac{\hbar}{2m\Omega}\left(\cos^2(wt)+\frac{\Omega^2}{w^2}\sin^2(wt)\right)}\right)\right\}dt \\
&= i\left\{\int_0^{U_b} t\,dt - \int_0^{+\infty} t\mathbb{N}\left(x_0\cos(wt), \sqrt{\frac{\hbar}{2m\Omega}\left(\cos^2(wt)+\frac{\Omega^2}{w^2}\sin^2(wt)\right)}\right)dt\right\} \\
&= i\left\{\left[\frac{t^2}{2}\right]_0^{U_b} - \langle x\rangle_R\right\} = i\left\{\frac{U_b^2}{2} - x_0\cos(wt)\right\}
\end{aligned} \quad (24)$$

where U_b is the upper bound of the definite integral above. Practically, the standard normal distribution probability is very nearly equal to 1.0000 (0.99997 exactly) for $U_b = 4$.

Note that, If $U_b \to +\infty \Leftrightarrow \langle x \rangle_M \to +\infty$.
Moreover,

$$\mathrm{Var}_{x,M} = \langle x^2 \rangle_M - \langle x \rangle_M^2$$

$$= \left\{ \int_0^{+\infty} t^2 \left\{ i\left[1 - |\psi(x,t)|^2\right] \right\} dt \right\} - \left\{ \int_0^{+\infty} t \left\{ i\left[1 - |\psi(x,t)|^2\right] \right\} dt \right\}^2$$

$$= i \int_0^{+\infty} t^2 \left\{ 1 - \mathbb{N}\left(x_0 \cos(wt), \sqrt{\frac{\hbar}{2m\Omega}\left(\cos^2(wt) + \frac{\Omega^2}{w^2} \sin^2(wt) \right)} \right) \right\} dt$$

$$- \left\{ i\left[\frac{U_b^2}{2} - x_0 \cos(wt) \right] \right\}^2$$

$$= i\left\{ \int_0^{U_b} t^2 dt - \int_0^{+\infty} t^2 \left\{ \mathbb{N}\left(x_0 \cos(wt), \sqrt{\frac{\hbar}{2m\Omega}\left(\cos^2(wt) + \frac{\Omega^2}{w^2} \sin^2(wt) \right)} \right) \right\} dt \right\} + \left[\frac{U_b^2}{2} - x_0 \cos(wt) \right]^2$$

$$= i\left\{ \int_0^{U_b} t^2 dt - \mathrm{Var}_{x,R} \right\} + \left[\frac{U_b^2}{2} - x_0 \cos(wt) \right]^2 = i\left\{ \left[\frac{t^3}{3} \right]_0^{U_b} - \mathrm{Var}_{x,R} \right\} + \left[\frac{U_b^2}{2} - x_0 \cos(wt) \right]^2 \tag{25}$$

$$= i\left\{ \frac{U_b^3}{3} - \frac{\hbar}{2m\Omega}\left(\cos^2(wt) + \frac{\Omega^2}{w^2} \sin^2(wt) \right) \right\} + \left[\frac{U_b^2}{2} - x_0 \cos(wt) \right]^2$$

Note that, if $U_b \to +\infty \Leftrightarrow \mathrm{Var}_{x,M} \to +\infty$.
In the probability set and the universe $\mathcal{C} = \mathcal{R} + \mathcal{M}$, we have from *CPP*:

$$\langle x \rangle_C = \int_0^{+\infty} t[z(x,t)] dt = \int_0^{+\infty} t\left\{ |\psi(x,t)|^2 + i\left[1 - |\psi(x,t)|^2\right] \right\} dt$$

$$= \int_0^{+\infty} t |\psi(x,t)|^2 dt + \int_0^{+\infty} t\left\{ i\left[1 - |\psi(x,t)|^2\right] \right\} dt$$

$$= \int_0^{+\infty} t \mathbb{N}\left(x_0 \cos(wt), \sqrt{\frac{\hbar}{2m\Omega}\left(\cos^2(wt) + \frac{\Omega^2}{w^2} \sin^2(wt) \right)} \right) dt$$

$$+ i \int_0^{+\infty} t \left\{ 1 - \mathbb{N}\left(x_0 \cos(wt), \sqrt{\frac{\hbar}{2m\Omega}\left(\cos^2(wt) + \frac{\Omega^2}{w^2} \sin^2(wt) \right)} \right) \right\} dt$$

$$= \langle x \rangle_R + \langle x \rangle_M = x_0 \cos(wt) + i\left\{ \frac{U_b^2}{2} - x_0 \cos(wt) \right\} \tag{26}$$

Note that, If $U_b \to +\infty \Leftrightarrow \langle x \rangle_C \to +\infty$.

$$
\begin{aligned}
\mathrm{Var}_{x,C} &= \langle x^2\rangle_C - \langle x\rangle_C^2 = \left[\int_0^{+\infty} t^2[z(x,t)]dt\right] - [\langle x\rangle_R + \langle x\rangle_M]^2 \\
&= \left[\int_0^{+\infty} t^2\left\{|\psi(x,t)|^2 + i\left[1-|\psi(x,t)|^2\right]\right\}dt\right] - [\langle x\rangle_R + \langle x\rangle_M]^2 \\
&= \left[\int_0^{+\infty} t^2|\psi(x,t)|^2 dt + \int_0^{+\infty} t^2\left\{i\left[1-|\psi(x,t)|^2\right]\right\}dt\right] - [\langle x\rangle_R + \langle x\rangle_M]^2 \\
&= \left[\langle x^2\rangle_R + \langle x^2\rangle_M\right] - [\langle x\rangle_R + \langle x\rangle_M]^2 \\
&= \left[\langle x^2\rangle_R + \langle x^2\rangle_M\right] - \left[\langle x\rangle_R^2 + \langle x\rangle_M^2 + 2\langle x\rangle_R\langle x\rangle_M\right] \\
&= \left[\langle x^2\rangle_R - \langle x\rangle_R^2\right] + \left[\langle x^2\rangle_M - \langle x\rangle_M^2\right] - 2\langle x\rangle_R\langle x\rangle_M \\
&= \mathrm{Var}_{x,R} + \mathrm{Var}_{x,M} - 2\langle x\rangle_R\langle x\rangle_M \\
&= \frac{\hbar}{2m\Omega}\left(\cos^2(wt) + \frac{\Omega^2}{w^2}\sin^2(wt)\right) + i\left\{\frac{U_b^3}{3} - \frac{\hbar}{2m\Omega}\left(\cos^2(wt) + \frac{\Omega^2}{w^2}\sin^2(wt)\right)\right\} \\
&\quad + \left[\frac{U_b^2}{2} - x_0\cos(wt)\right]^2 - 2\{x_0\cos(wt)\} \times \left\{i\left\{\frac{U_b^2}{2} - x_0\cos(wt)\right\}\right\}
\end{aligned}
\tag{27}
$$

Note that, If $U_b \to +\infty \Leftrightarrow \mathrm{Var}_{x,C} \to +\infty$.

Tables 1–4 compute the position distribution characteristics for $x_0 = 4$, $m = w = \hbar = 1$, $n = 1/4$ where $\Omega = n \times w$, and $U_b = 4, 20, 50, 1000$.

Position distribution characteristics	$x_0 = 4, m = w = \hbar = 1, \Omega = w/4, U_b = 4$
$\langle x\rangle_R$	−2.6146
$\mathrm{Var}_{x,R}$	0.9261
$\langle x\rangle_M$	$i\times$10.6146
$\mathrm{Var}_{x,M}$	112.6692+$i\times$20.4072
$\langle x\rangle_C = \langle x\rangle_R + \langle x\rangle_M$	−2.6146+$i\times$10.6146
$\mathrm{Var}_{x,C} = \mathrm{Var}_{x,R} + \mathrm{Var}_{x,M} - 2\langle x\rangle_R\langle x\rangle_M$	113.5953+$i\times$75.9124

Table 1.
The position distribution characteristics for $U_b = 4$ and $n = 1/4$.

Position distribution characteristics	$x_0 = 4, m = w = \hbar = 1, \Omega = w/4, U_b = 20$
$\langle x\rangle_R$	1.6323
$\mathrm{Var}_{x,R}$	0.4372
$\langle x\rangle_M$	$i\times$198.3677
$\mathrm{Var}_{x,M}$	3.9350e+04+$i\times$2.6662e+03
$\langle x\rangle_C = \langle x\rangle_R + \langle x\rangle_M$	1.6323+$i\times$198.3677
$\mathrm{Var}_{x,C} = \mathrm{Var}_{x,R} + \mathrm{Var}_{x,M} - 2\langle x\rangle_R\langle x\rangle_M$	3.9350e+04+$i\times$2.0186e+03

Table 2.
The position distribution characteristics for $U_b = 20$ and $n = 1/4$.

Position distribution characteristics	$x_0 = 4, m = w = \hbar = 1, \Omega = w/4, U_b = 50$
$\langle x \rangle_R$	3.8599
$\text{Var}_{x,R}$	1.8709
$\langle x \rangle_M$	$i\times$1.2461e+03
$\text{Var}_{x,M}$	1.5529e+06+$i\times$4.1665e+04
$\langle x \rangle_C = \langle x \rangle_R + \langle x \rangle_M$	3.8599+ $i\times$1.2461e+03
$\text{Var}_{x,C} = \text{Var}_{x,R} + \text{Var}_{x,M} - 2\langle x \rangle_R \langle x \rangle_M$	1.5529e+06+$i\times$3.2045e+04

Table 3.
The position distribution characteristics for $U_b = 50$ *and* $n = 1/4$.

Position distribution characteristics	$x_0 = 4, m = w = \hbar = 1, \Omega = w/4, U_b = 1000$
$\langle x \rangle_R$	2.2495
$\text{Var}_{x,R}$	0.7180
$\langle x \rangle_M$	$i\times$5.0000e+05
$\text{Var}_{x,M}$	2.5000e+11+$i\times$3.3333e+08
$\langle x \rangle_C = \langle x \rangle_R + \langle x \rangle_M$	2.2495+ $i\times$5.0000e+05
$\text{Var}_{x,C} = \text{Var}_{x,R} + \text{Var}_{x,M} - 2\langle x \rangle_R \langle x \rangle_M$	2.5000e+11+$i\times$3.3108e+08

Table 4.
The position distribution characteristics for $U_b = 1000$ *and* $n = 1/4$.

5. Conclusion and perspectives

In the present research chapter, the novel extended system of eight axioms (*EKA*) of A. N. Kolmogorov was applied and linked to the quantum harmonic oscillators with the Gaussian initial condition problem in quantum mechanics theory. Thus, a strong bond between quantum mechanics and the novel paradigm (*CPP*) was accomplished. Hence, the Paradigm of "Complex Probability" was more elaborated beyond the scope of my 23 previous research publications on this topic.

Furthermore, as it was verified and proved in the original paradigm, before the beginning of the simulation of the random event and at its end we have the chaotic factor (*Chf* and *MChf*) is 0 and the degree of our knowledge (*DOK*) is 1 since the stochastic and probabilistic effects and fluctuations have either not started yet or they have finished and terminated their task on the random phenomenon. During the execution of the nondeterministic experiment and process we also have: $-0.5 \le Chf < 0$, $0 < MChf \le 0.5$, and $0.5 \le DOK < 1$. We can see that during the whole process, we have constantly and incessantly $Pc^2 = DOK - Chf = DOK + MChf = 1 = Pc$, that shows that the simulation which behaved probabilistically and randomly in the real universe and set $\mathcal{R}$ is now deterministic and certain in the complex probability universe and set $\mathcal{C} = \mathcal{R} + \mathcal{M}$, and this after adding to the stochastic phenomenon performed in the real universe and set $\mathcal{R}$ the contributions of the imaginary universe and set $\mathcal{M}$ and thus after subtracting and eliminating from the degree of our knowledge the chaotic factor.

This is certainly very wonderful, fascinating, and fruitful and shows and proves once again the rewards of extending the five axioms of the probability of A. N. Kolmogorov and thus the benefits and novelty of my original and novel model in the fields of applied mathematics, prognostics, and quantum mechanics that can be called verily: "The Complex Probability Paradigm." As future and prospective challenges and research works, we intend to elaborate more on the original probability paradigm developed and to apply it to a large array of stochastic phenomena and nondeterministic experiments encountered in the theory of quantum mechanics.

Data availability

The data used to support the findings of this study are available from the author upon request.

Conflicts of interest

The author declares that there are no conflicts of interest regarding the publication of this research work.

Nomenclature

$\mathcal{R}$	probabilities and events real set
$\mathcal{M}$	probabilities and events imaginary set
$\mathcal{C}$	probabilities and events complex set
i	the imaginary number where $i^2 = -1$ or $i = \sqrt{-1}$
EKA	Extended Kolmogorov's Axioms
CPP	Complex Probability Paradigm
P_{rob}	any event probability
P_r	probability of an event in the real set $\mathcal{R}$
P_m	probability of an event in the imaginary set $\mathcal{M}$ associated with the real probability in $\mathcal{R}$
Pc	probability of an event in $\mathcal{R}$ with its corresponding complementary event in $\mathcal{M}$ = probability in the complex probability set $\mathcal{C} = \mathcal{R} + \mathcal{M}$
Z	complex probability number = sum of P_r and P_m = complex random vector
$DOK = \|Z\|^2$	the random experiment or system degree of our knowledge; it is the square of the norm of Z
Chf	the chaotic factor of Z
MChf	the magnitude of the chaotic factor of Z
$\|\psi(x,t)\|^2$	probability density function of the position wavefunction
$\|\phi(p,t)\|^2$	probability density function of the momentum wavefunction
$\langle x \rangle_R, \langle x \rangle_M, \langle x \rangle_C$	averages, or expectations, or means of the position wavefunction probability density function in $\mathcal{R}$, $\mathcal{M}$, and $\mathcal{C}$, respectively
$\mathrm{Var}_{x,R}, \mathrm{Var}_{x,M}, \mathrm{Var}_{x,C}$	variances of the position wavefunction probability density function in $\mathcal{R}$, $\mathcal{M}$, and $\mathcal{C}$, respectively

$\langle p \rangle_R, \langle p \rangle_M, \langle p \rangle_C$	averages, or expectations, or means of the momentum wavefunction probability density function in $\mathcal{R}$, $\mathcal{M}$, and $\mathcal{C}$, respectively
$\text{Var}_{p,R}, \text{Var}_{p,M}, \text{Var}_{p,C}$	variances of the momentum wavefunction probability density function in $\mathcal{R}$, $\mathcal{M}$, and $\mathcal{C}$, respectively
H_x^R	entropy in the real universe $\mathcal{R}$ of the particle position
$NegH_x^R$	negative entropy in the real universe $\mathcal{R}$ of the particle position
$\overline{H}_x^R$	complementary entropy in the real universe $\mathcal{R}$ of the particle position
H_x^M	entropy in the imaginary universe $\mathcal{M}$ of the particle position
H_x^C	entropy in the complex universe $\mathcal{C}$ of the particle position
H_p^R	entropy in the real universe $\mathcal{R}$ of the particle momentum
$NegH_p^R$	negative entropy in the real universe $\mathcal{R}$ of the particle momentum
$\overline{H}_p^R$	complementary entropy in the real universe $\mathcal{R}$ of the particle momentum
H_p^M	entropy in the imaginary universe $\mathcal{M}$ of the particle momentum
H_p^C	entropy in the complex universe $\mathcal{C}$ of the particle momentum

Author details

Abdo Abou Jaoudé
Faculty of Natural and Applied Sciences, Department of Mathematics and Statistics, Notre Dame University-Louaize, Lebanon

*Address all correspondence to: abdoaj@idm.net.lb

References

[1] Wikipedia. The free encyclopedia, Quantum Mechanics. Available from: https://en.wikipedia.org/

[2] Wikipedia. The free encyclopedia, Uncertainty Principle. Available from: https://en.wikipedia.org/

[3] Wikipedia. The free encyclopedia, Quantum Harmonic Oscillator. Available from: https://en.wikipedia.org/

[4] Griffiths DJ. Introduction to Quantum Mechanics. 2nd ed. New Jersey, United States: Prentice Hall; 2004

[5] Liboff RL. Introductory Quantum Mechanics. Boston, United States: Addison–Wesley; 2002

[6] Rashid MA. Transition Amplitude For Time-Dependent Linear Harmonic Oscillator With Linear Time-Dependent Terms Added To The Hamiltonian (PDF-Microsoft PowerPoint). M.A. Rashid–Center for Advanced Mathema tics and Physics. Islamabad, Islamabad Capital Territory, Pakistan: National Center for Physics. 2006

[7] Hall BC. Quantum Theory for Mathematicians. Graduate Texts in Mathematics. Vol. 267. New York City, United States: Springer; 2013

[8] Pauli W. Wave Mechanics: Volume 5 of Pauli Lectures on Physics. New York, United States: Dover Books on Physics; 2000

[9] Condon EU. Immersion of the Fourier Transform in a continuous group of functional transformations. Proceedings of the National Academic Science USA. 1937;**23**:158-164

[10] Albert M. Quantum Mechanics. North-Holland; 1967. p. 456

[11] Fradkin DM. Three-Dimensional Isotropic Harmonic Oscillator and SU3. American Journal of Physics. 1965;**33**(3): 207-211

[12] Mahan GD. Many Particles Physics. New York: Springer; 1981

[13] Quantum Harmonic Oscillator. Hyperphysics [Accessed: September 24, 2009]

[14] Abou Jaoude A, El-Tawil K, Kadry S. Prediction in complex dimension using Kolmogorov's set of axioms. Journal of Mathematics and Statistics, Science Publications. 2010; **6**(2):116-124

[15] Abou Jaoude A. The complex statistics paradigm and the law of large numbers. Journal of Mathematics and Statistics, Science Publications. 2013; **9**(4):289-304

[16] Abou Jaoude A. The theory of complex probability and the first order reliability method. Journal of Mathematics and Statistics, Science Publications. 2013;**9**(4):310-324

[17] Abou Jaoude A. Complex probability theory and prognostic. Journal of Mathematics and Statistics, Science Publications. 2014;**10**(1):1-24

[18] Abou Jaoude A. The complex probability paradigm and analytic linear prognostic for vehicle suspension systems. American Journal of Engineering and Applied Sciences, Science Publications. 2015;**8**(1):147-175

[19] Abou Jaoude A. The paradigm of complex probability and the Brownian motion. Systems Science and Control Engineering. 2015;**3**(1):478-503

[20] Abou Jaoude A. The paradigm of complex probability and Chebyshev's inequality. Systems Science and Control Engineering. 2016;**4**(1):99-137

[21] Abou Jaoude A. The paradigm of complex probability and analytic nonlinear prognostic for vehicle suspension systems. Systems Science and Control Engineering. 2016;**4**(1): 99-137

[22] Abou Jaoude A. The paradigm of complex probability and analytic linear prognostic for unburied petrochemical pipelines. Systems Science and Control Engineering. 2017;**5**(1):178-214

[23] Abou Jaoude A. The paradigm of complex probability and Claude Shannon's information theory. Systems Science and Control Engineering. 2017; **5**(1):380-425

[24] Abou Jaoude A. The paradigm of complex probability and analytic nonlinear prognostic for unburied petrochemical pipelines. Systems Science and Control Engineering. 2017;**5**(1): 495-534

[25] Abou Jaoude A. The paradigm of complex probability and Ludwig Boltzmann's entropy. Systems Science and Control Engineering. 2018;**6**(1): 108-149

[26] Abou Jaoude A. The paradigm of complex probability and Monte Carlo methods. Systems Science and Control Engineering. 2019;**7**(1):407-451

[27] Abou Jaoude A. Analytic prognostic in the linear damage case applied to buried petrochemical pipelines and the complex probability paradigm. Fault Detection, Diagnosis and Prognosis. 2020;**1**(5):65-103. DOI: 10.5772/intechopen.90157

[28] Abou Jaoude A. The Monte Carlo Techniques and The Complex Probability Paradigm. Forecasting in Mathematics – Recent Advances, New Perspectives and Applications. 2020;**1**(1):1-29. DOI: 10.5772/intechopen.93048

[29] Abou Jaoude A. The paradigm of complex probability and prognostic using FORM. London Journal of Research in Science: Natural and Formal (LJRS), London, United Kingdom. 2020; **20**(4):1-65

[30] Abou Jaoude A. The paradigm of complex probability and the central limit theorem. London Journal of Research in Science: Natural and Formal (LJRS), London, United Kingdom. 2020;**20**(5): 1-57

[31] Abou Jaoude A. The paradigm of complex probability and Thomas Bayes' Theorem. The Monte Carlo Methods – Recent Advances, New Perspectives and Applications. 2021;**1**(1):1-44. DOI: 10.5772/intechopen.98340

[32] Abou Jaoude A. The paradigm of complex probability and Isaac Newton's classical mechanics: On the foundation of statistical physics. The Monte Carlo Methods – Recent Advances, New Perspectives and Applications. 2021;**1**(2):45-116. DOI: 10.5772/intechopen.98341

[33] Abou Jaoude A. The paradigm of complex probability and quantum mechanics: The infinite potential well problem – The position wave function. Applied Probability Theory – New Perspectives, Recent Advances and Trends. 2022;**1**(1):1-44. DOI: 10.5772/intechopen.107300

[34] Abou Jaoude A. The paradigm of complex probability and quantum mechanics: The infinite potential well problem – The momentum

wavefunction and the wavefunction entropies. Applied Probability Theory – New Perspectives, Recent Advances and Trends. 2022;**1**(2):45-88. DOI: 10.5772/intechopen.107665

[35] Abou Jaoude A. The paradigm of complex probability and the theory of metarelativity – A simplified model of MCPP. Operator Theory – Recent Advances, New Perspectives and Applications. 2023;**1**(1):1-36. DOI: 10.5772/intechopen.110378

[36] Abou Jaoude A. The paradigm of complex probability and the theory of metarelativity – The general model and some consequences of MCPP. Operator Theory – Recent Advances, New Perspectives and Applications. 2023;**1**(2):37-78. DOI: 10.5772/intechopen.110377

[37] Benton W. Probability, Encyclopedia Britannica. Vol. 18. Chicago: Encyclopedia Britannica Inc; 1966. pp. 570-574

[38] Benton W. Mathematical Probability, Encyclopedia Britannica. Vol. 18. Chicago: Encyclopedia Britannica Inc.; 1966. pp. 574-579

[39] Feller W. An Introduction to Probability Theory and Its Applications. 3rd ed. New York: Wiley; 1968

[40] Walpole R, Myers R, Myers S, Ye K. Probability and Statistics for Engineers and Scientists. 7th ed. New Jersey: Prentice Hall; 2002

[41] Freund JE. Introduction to Probability. New York: Dover Publications; 1973

[42] Srinivasan SK, Mehata KM. Stochastic Processes. 2nd ed. New Delhi: McGraw-Hill; 1988

[43] Barrow JD. The Book of Nothing. New York, New York, United States: Vintage; 2002

[44] Becker K, Becker M, Schwarz JH. String Theory and M-Theory. Cambridge, United Kingdom: Cambridge University Press; 2007

[45] De Broglie L. La Physique Nouvelle et les Quanta. Paris, France: Flammarion; 1937

[46] Einstein A. Traduction Française : Comment Je Vois le Monde. Paris, France: Maurice Solovine et Régis Hanrion: Flammarion; 1979

[47] Einstein A. La Relativité. Paris, France: Petite Bibliothèque Payot; 2001

[48] Feynmann R. Traduction Française : La Nature de la Physique. Paris, Le Seuil: Hélène Isaac, Jean-Marc Lévy-Leblond, Françoise Balibar; 1980

[49] Balibar F. Albert Einstein : Physique, Philosophie. Paris, Le Seuil: Politique; 2002

[50] Gates E. Einstein's Telescope. New York, United States: Norton; 2010

[51] Greene B. The Elegant Universe. New York, New York, United States: Vintage; 2003

[52] Gribbin J. Traduction Française : A la Poursuite du Big Bang. Michel Cassé: Flammarion; 1994

[53] Gubser SS. The Little Book of String Theory. Princeton; 2010

[54] Hawking S. The Essential Einstein, His Greatest Works. Penguin Books; 2007

[55] Hoffmann B, Dukas H. Albert Einstein, Creator and Rebel. New York:

Viking; 1972; traduction Française : Albert Einstein, créateur et rebelle. M. Manly, Paris, Le Seuil (1975)

[56] Luminet J-P. Les Trous Noirs. Paris, France: Le Seuil; 1992

[57] Penrose R. Traduction Française : Les Deux Infinis et L'Esprit Humain. Roland Omnès: Flammarion; 1999

[58] Penrose R. The Road to Reality. New York, New York, United States: Vintage; 2004

[59] Planck M. Traduction Française : Initiations à la Physique. J. du Plessis de Grenédan: Flammarion; 1993

[60] Poincaré H. La Science et L'Hypothèse. Paris: Flammarion; 1968

[61] Proust D, Vanderriest C. Les Galaxies et la Structure de L'Univers. Paris, France: Le Seuil; 1997

[62] Reeves H. Patience dans L'Azur. L'évolution cosmique. Paris, France: Le Seuil; 1988

[63] Ronan C. Traduction Française : Histoire Mondiale des Sciences. Claude Bonnafont. Paris, France: Le Seuil; 1988

[64] Sagan C. Traduction Française : Cosmic Connection ou L'appel des étoiles. Paris, France: Le Seuil: Vincent Bardet; 1975

[65] Weinberg S. Dreams of a Final Theory. New York, New York, United States: Vintage; 1993

[66] Stewart I. Does God Play Dice? 2nd ed. Oxford: Blackwell Publishing; 2002

[67] Barrow J. Pi in the Sky. Oxford: Oxford University Press; 1992

[68] Bogdanov I, Bogdanov G. Au Commencement du Temps. Paris: Flammarion; 2009

[69] Bogdanov I, Bogdanov G. Le Visage de Dieu. Paris: Editions Grasset et Fasquelle; 2010

[70] Bogdanov I, Bogdanov G. La Pensée de Dieu. Paris: Editions Grasset et Fasquelle; 2012

[71] Bogdanov I, Bogdanov G. La Fin du Hasard. Paris: Editions Grasset et Fasquelle; 2013

[72] Bell ET. The Development of Mathematics. New York: Dover Publications, Inc., United States of America; 1992

[73] Boursin J-L. Les Structures du Hasard. Paris: Editions du Seuil; 1986

[74] Dacunha-Castelle D. Chemins de l'Aléatoire. Paris: Flammarion; 1996

[75] Dalmedico-Dahan A, Chabert J-L, Chemla K. Chaos Et Déterminisme. Paris: Edition du Seuil; 1992

[76] Ekeland I. Au Hasard. La Chance, la Science et le Monde. Paris: Editions du Seuil; 1991

[77] Gleick J. Chaos, Making a New Science. New York: Penguin Books; 1997

[78] Davies P. The Mind of God. London: Penguin Books; 1993

[79] Gillies D. Philosophical Theories of Probability. London: Routledge; 2000

[80] Hawking S. On the Shoulders of Giants. London: Running Press; 2002

[81] Pickover C. Archimedes to Hawking. Oxford: Oxford University Press; 2008

[82] Abou Jaoude A. The Computer Simulation of Monté Carlo Methods and Random Phenomena. United

Kingdom: Cambridge Scholars Publishing; 2019

[83] Abou Jaoude A. The Analysis of Selected Algorithms for the Stochastic Paradigm. United Kingdom: Cambridge Scholars Publishing; 2019

[84] Abou Jaoude A. The Analysis of Selected Algorithms for the Statistical Paradigm, Volume 1. The Republic of Moldova: Generis Publishing; 2021

[85] Abou Jaoude A. The Analysis of Selected Algorithms for the Statistical Paradigm, Volume 2. The Republic of Moldova: Generis Publishing; 2021

[86] Abou Jaoude A. Forecasting in Mathematics – Recent Advances, New Perspectives and Applications. London, UK, London: IntechOpen; 2021

[87] Abou Jaoude A. The Monte Carlo Methods – Recent Advances, New Perspectives and Applications. London, UK, London: IntechOpen; 2022

[88] Abou Jaoude A. Applied Probability Theory – New Perspectives, Recent Advances and Trends. London, UK, London: IntechOpen; 2023

[89] Abou Jaoude A. Operator Theory – Recent Advances, New Perspectives and Applications. London, UK, London: IntechOpen; 2023. In Press

[90] Abou Jaoude A. Ph.D. Thesis in Applied Mathematics: Numerical Methods and Algorithms for Applied Mathematicians. Bircham International University. 2004. Available from: http://www.bircham.edu

[91] Abou Jaoude A. Ph.D. Thesis in Computer Science: Computer Simulation of Monté Carlo Methods and Random Phenomena. Bircham International University. 2005. Available from: http://www.bircham.edu

[92] Abou Jaoude A. Ph.D. Thesis in Applied Statistics and Probability: Analysis and Algorithms for the Statistical and Stochastic Paradigm. Bircham International University. 2007. Available from: http://www.bircham.edu

Chapter 2

The Paradigm of Complex Probability and Quantum Mechanics: The Quantum Harmonic Oscillator with Gaussian Initial Condition – The Momentum Wavefunction and the Wavefunction Entropies

Abdo Abou Jaoudé

"Ignorance, the root and stem of all evil."
Plato.

"The energy of the mind is the essence of life."
Aristotle.

"I was taught that the way of progress was neither swift nor easy."
Marie Curie.

"It takes an extraordinary intelligence to contemplate the obvious."
Alfred North Whitehead.

Abstract

The system of probability axioms of *Andrey Nikolaevich Kolmogorov* put forward in 1933 can be developed to encompass the set of imaginary numbers after adding to his established five axioms a supplementary three axioms. Therefore, any probabilistic phenomenon can thus be performed in what is now the set of complex probabilities $\mathcal{C}$ which is the sum of the real set of probabilities $\mathcal{R}$ and the complementary and associated and corresponding imaginary set of probabilities $\mathcal{M}$. The aim here is to compute the complex probabilities by taking into consideration additional novel imaginary dimensions to the phenomenon that occurs in the "real" laboratory. Hence, the corresponding probability in the entire probability set $\mathcal{C} = \mathcal{R} + \mathcal{M}$ is, whatever the random distribution of the input random variable considered in $\mathcal{R}$, permanently and constantly equal to 1. Thus, the result of the stochastic experiment in $\mathcal{C}$ can be foretold perfectly and completely. Subsequently, the consequence shows that luck and chance in $\mathcal{R}$ is substituted now by absolute determinism in $\mathcal{C}$. Accordingly, this is the consequence of the fact that the probability in $\mathcal{C}$ is got by subtracting from the degree of our knowledge of the random system the chaotic factor. Henceforth, I will apply to the established and well-known theory of quantum mechanics my innovative and original Complex Probability Paradigm (*CPP*) which will yield a completely deterministic expression of quantum theory in the universe of probabilities $\mathcal{C} = \mathcal{R} + \mathcal{M}$.

Keywords: degree of our knowledge, chaotic factor, probability norm, complex random vector, complex probability set $\mathcal{C}$, real entropy, imaginary entropy, complex entropy

1. Introduction

Firstly, classical physics explains energy and matter only on a familiar to human experience scale, and that includes the astronomical bodies behavior such as the planets or the moon [1–13]. By contrast, quantum mechanics studies matter and its interactions with energy on the subatomic particles and atomic scales. Knowing that, classical physics is still adopted in much of modern technology and science. However, towards the end of the nineteenth century, it was found by scientists that classical physics could not explain numerous phenomena discovered in both the macro (large) and the micro (small) worlds. Hence, the theory of relativity and the theory of quantum mechanics were developed to resolve inconsistencies between classical theory and observed phenomena. Thus, this has led to these two major revolutions in physics that resulted to a shift in the original scientific paradigm. Therefore, physicists discovered the limitations of classical physics and developed the main concepts of the quantum theory that replaced it in the early decades of the twentieth century. They described these concepts in roughly the order in which they were first discovered.

Moreover, light behaves in some aspects like waves and in other aspects like particles. Matter which is the "stuff" of the universe is made up of particles such as protons, electrons, neutrons, and atoms and which show wavelike behavior also. Additionally, neon lights, like some light sources, exhibit only certain definite frequencies of light, which is a small set of distinct pure colors fixed by the atomic structure of neon. Quantum mechanics proves that light, along with all other forms of electromagnetic radiation, comes in photons that are discrete units, and calculates the spectral energies that correspond to pure colors, and it computes as well its light beams intensities. The smallest observable particle of the electromagnetic field is a single photon also called a quantum. Knowing that, we have never experimentally observed a partial photon. More broadly, many properties of objects, such as position, speed, and angular momentum, that appeared continuous in the zoomed-out view of classical mechanics, turn out to be quantized in the very tiny, zoomed-in scale of quantum mechanics as it was shown by quantum theory. Such elementary particle properties are necessary to take on one of a set of discrete and small allowable values. But since the gap between these discrete values is similarly small, then the discontinuities are only noticed at very tiny atomic scales.

Furthermore, many features of quantum mechanics can seem to be paradoxical and are counterintuitive because they describe behavior quite dissimilar to that seen at larger scales. The famous quantum physicist Richard Feynman describes quantum mechanics as a theory that deals with "nature as She is—absurd." One major

DOI: http://dx.doi.org/10.5772/intechopen.1001985

"paradox" is the apparent inconsistency between quantum mechanics and Newton's laws and which can be clarified using the theorem of Ehrenfest. In his theorem, the latter proves that the obtained quantum mechanics average values (like position and momentum) obey and respect classical laws. However, the theorem of Ehrenfest is far from being capable of explaining all the observed counterintuitive phenomena of quantum weirdness, but rather is a mathematical expression of the principle of correspondence.

Moreover, the quantum-mechanical analog of the classical harmonic oscillator is the quantum harmonic oscillator. It is one of the most important model systems in quantum mechanics because an arbitrary smooth potential can generally be estimated as a harmonic potential at the neighborhood of a stable equilibrium point. Furthermore, since an exact, analytical solution is known, then it is one of the few quantum-mechanical systems for which this kind of solution is provided. Consequently, I will relate my complex probability paradigm (*CPP*) to this well-known and important problem in quantum mechanics in order to express it completely deterministically.

At the end, and to conclude, this research work is organized as follows: In Section 1, we will present the introduction, and then in Section 2 we will explain the advantages and the purpose of the present work. Afterward, in Section 3, we will explain and summarize the extended Kolmogorov's axioms and hence present the original parameters and interpretation of the complex probability paradigm. Additionally, in Section 4, the new paradigm will be related to the quantum harmonic oscillators with Gaussian initial condition problem after applying *CPP* to the momentum wavefunction of the problem in this current second chapter, hence some corresponding simulations will be done, and afterward the characteristics of this stochastic distribution will be computed in the probabilities sets $\mathcal{R}$, $\mathcal{M}$, and $\mathcal{C}$. Furthermore, in Section 5, *CPP* will be used to extend and to verify the Quantum Uncertainty Principle in $\mathcal{R}$, $\mathcal{M}$, and $\mathcal{C}$. In addition, in Section 6, we will calculate and determine the position and the momentum wavefunctions entropies in $\mathcal{R}$, $\mathcal{M}$, and $\mathcal{C}$. Finally, we conclude the work by doing a comprehensive summary in Section 7 and then present the list of references cited in the current research work.

2. The purpose and the advantages of the current publication

Computing probabilities is all our work in the classical theory of probability [14–36]. Adding new dimensions to our stochastic experiment is the innovative idea in the current paradigm which will make the study absolutely deterministic. As a matter of fact, the theory of probability is a nondeterministic theory by essence that means that all the random events outcome is due to luck and chance. Hence, we make the study deterministic by adding new imaginary dimensions to the phenomenon occurring in the "real" laboratory which is $\mathcal{R}$, and therefore, a stochastic experiment will have a certain outcome in the complex probabilities set $\mathcal{C}$. It is of great significance that random systems become completely predictable since we will be perfectly knowledgeable to predict the outcome of all stochastic and chaotic phenomena that occur in nature like for example in all stochastic processes, in statistical mechanics, or in the well-established field of quantum mechanics. Consequently, the work that should be done is to add the contributions of $\mathcal{M}$ which is the set of imaginary probabilities to the set of real probabilities $\mathcal{R}$ that will make the random phenomenon in $\mathcal{C} = \mathcal{R} + \mathcal{M}$ completely deterministic. Since this paradigm is found to be fruitful, then a new theory in prognostic and stochastic sciences is established and this is to understand deterministically those events that used to be stochastic events in $\mathcal{R}$. This is what I coined by the term "The Complex Probability Paradigm" that was elaborated and initiated in my 23 previous papers.

To summarize, the advantages and the purposes of this current work and chapter are to:

1. Relate probability theory to the field of complex variables and analysis in mathematics and therefore to extend the theory of classical probability to the set of complex numbers. This task was elaborated and initiated in my 23 previous papers.

2. Apply the novel probability axioms and *CPP* paradigm to quantum mechanics, specifically to the quantum harmonic oscillators with Gaussian initial condition problem.

3. Demonstrate that any stochastic and random event and experiment can be expressed deterministically in the complex probabilities set $\mathcal{C}$.

4. Quantify both the chaos magnitude and the degree of our knowledge of the wavefunction momentum distribution and *CPP* in the sets $\mathcal{R}$, $\mathcal{M}$, and $\mathcal{C}$.

5. Represent graphically and illustrate the parameters and functions of the original paradigm related to this quantum mechanics problem.

6. Evaluate all the characteristics of the wavefunction momentum distribution.

7. Demonstrate that the classical concepts of stochastic system have a probability of occurring permanently equal to one in the complex set; consequently, no ignorance, no unpredictability, no stochasticity, no disorder, no randomness, no nondeterminism, and no chaos exist in:

$$\mathcal{C}\ (\text{complex set}) = \mathcal{R}\ (\text{real set}) + \mathcal{M}\ (\text{imaginary set}). \tag{1}$$

8. Verify and extend the Quantum Uncertainty Principle in $\mathcal{R}$ to $\mathcal{M}$ and $\mathcal{C}$.

9. Calculate the problem entropies in $\mathcal{R}$, $\mathcal{M}$, and $\mathcal{C}$ and show that there is no disorder and no information loss or gain in *CPP* but conservation of information.

10. Prepare to apply the novel paradigm to other topics in stochastic processes, in statistical mechanics, and to the field of prognostics in science and engineering and quantum mechanics. This will be the task in my following research work and publications.

Compared with existing literature, the major contribution of the current research work is to apply the novel paradigm of *CPP* to quantum mechanics and to express it completely deterministically. And concerning some applications of the novel developed paradigm and as a future work, it can be applied to any nondeterministic

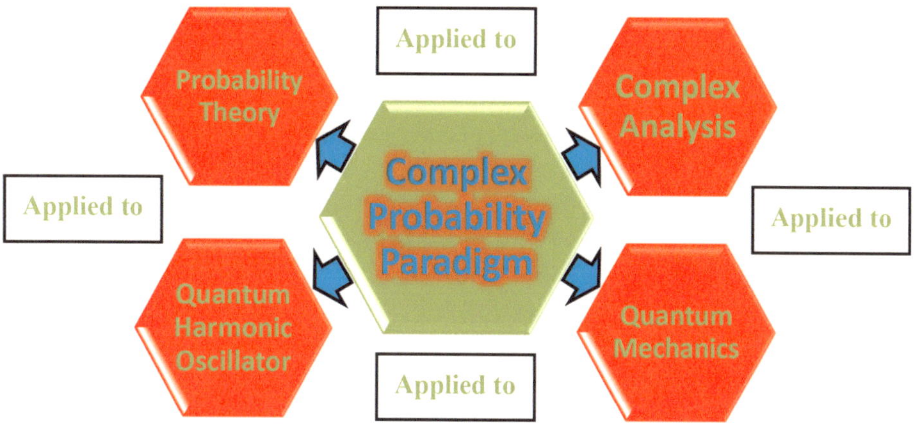

Figure 1.
The diagram of the Complex Probability Paradigm applied to Quantum Mechanics major purposes and goals.

phenomenon in quantum mechanics. The next figure displays the major purposes of the Complex Probability Paradigm (*CPP*) (**Figure 1**).

3. The complex probability paradigm

3.1 The original Andrey Nikolaevich Kolmogorov system of axioms

The simplicity of Kolmogorov's system of axioms may be surprising [14–36]. Let E be a collection of elements $\{E_1, E_2, \dots\}$ called elementary events and let F be a set of subsets of E called random events [37–40]. The five axioms for a finite set E are:

Axiom 1: F is a field of sets.

Axiom 2: F contains the set E.

Axiom 3: A non-negative real number $P_{rob}(A)$, called the probability of A, is assigned to each set A in F. We have always $0 \le P_{rob}(A) \le 1$.

Axiom 4: $P_{rob}(E)$ equals 1.

Axiom 5: If A and B have no elements in common, the number assigned to their union is:

$$P_{rob}(A \cup B) = P_{rob}(A) + P_{rob}(B) \tag{2}$$

hence, we say that A and B are disjoint; otherwise, we have:

$$P_{rob}(A \cup B) = P_{rob}(A) + P_{rob}(B) - P_{rob}(A \cap B) \tag{3}$$

And we say also that: $P_{rob}(A \cap B) = P_{rob}(A) \times P_{rob}(B/A) = P_{rob}(B) \times P_{rob}(A/B)$ which is the conditional probability. If both A and B are independent then: $P_{rob}(A \cap B) = P_{rob}(A) \times P_{rob}(B)$.

Moreover, we can generalize and say that for N disjoint (mutually exclusive) events $A_1, A_2, \dots, A_j, \dots, A_N$ (for $1 \le j \le N$), we have the following additivity rule:

$$P_{rob}\left(\bigcup_{j=1}^{N} A_j\right) = \sum_{j=1}^{N} P_{rob}(A_j) \tag{4}$$

And we say also that for N independent events $A_1, A_2, \dots, A_j, \dots, A_N$ (for $1 \le j \le N$), we have the following product rule:

$$P_{rob}\left(\bigcap_{j=1}^{N} A_j\right) = \prod_{j=1}^{N} P_{rob}(A_j) \tag{5}$$

3.2 Adding the imaginary part $\mathcal{M}$

Now, we can add to this system of axioms an imaginary part such that:
Axiom 6: Let $P_m = i \times (1 - P_r)$ be the probability of an associated complementary event in $\mathcal{M}$ (the imaginary part or universe) to the event A in $\mathcal{R}$ (the real part or universe). It follows that $P_r + P_m/i = 1$ where i is the imaginary number with $i = \sqrt{-1}$ or $i^2 = -1$.
Axiom 7: We construct the complex number or vector $Z = P_r + P_m = P_r + i(1 - P_r)$ having a norm $|Z|$ such that:

$$|Z|^2 = P_r^2 + (P_m/i)^2. \tag{6}$$

Axiom 8: Let Pc denote the probability of an event in the complex probability set and universe $\mathcal{C}$ where $\mathcal{C} = \mathcal{R} + \mathcal{M}$. We say that Pc is the probability of an event A in $\mathcal{R}$ with its associated and complementary event in $\mathcal{M}$ such that:

$$Pc^2 = (P_r + P_m/i)^2 = |Z|^2 - 2iP_rP_m \text{ and is always equal to } 1. \tag{7}$$

We can see that by taking into consideration the set of imaginary probabilities we added three new and original axioms and consequently the system of axioms defined by Kolmogorov was hence expanded to encompass the set of imaginary numbers and realm [41–68].

3.3 A concise interpretation of the original *CPP* paradigm

To conclude and to summarize, we state that our degree of our certain knowledge is undesirably incomplete and imperfect and thus unsatisfactory in the real probability universe $\mathcal{R}$. Hence, we extend our study to the set of complex numbers $\mathcal{C}$ which includes the contributions of both the set of real probabilities which is $\mathcal{R}$ and the set of complementary imaginary probabilities which is $\mathcal{M}$. Consequently, this will result to a perfect and absolute degree of our knowledge in the probability universe $\mathcal{C} = \mathcal{R} + \mathcal{M}$ because $Pc = 1$ continuously. In fact, the study in the complex universe $\mathcal{C}$ leads to a certain prediction of any stochastic and random event and experiment since in $\mathcal{C}$ we subtract and eliminate the measured chaotic factor from our computed degree of our knowledge. This will result to a probability permanently equal to 1 in the universe $\mathcal{C}$ as it is shown in the following equation deduced from *CPP*: $Pc^2 = DOK - Chf = DOK + MChf = 1 = Pc$. Many numerous discrete and continuous probability distributions were illustrated in my 23 previous research works and that confirm this hypothesis and original paradigm [14–36]. The Extended Kolmogorov Axioms (*EKA* for short) or the Complex Probability Paradigm (*CPP* for short) can be summarized and shown in the next illustration (**Figure 2**):

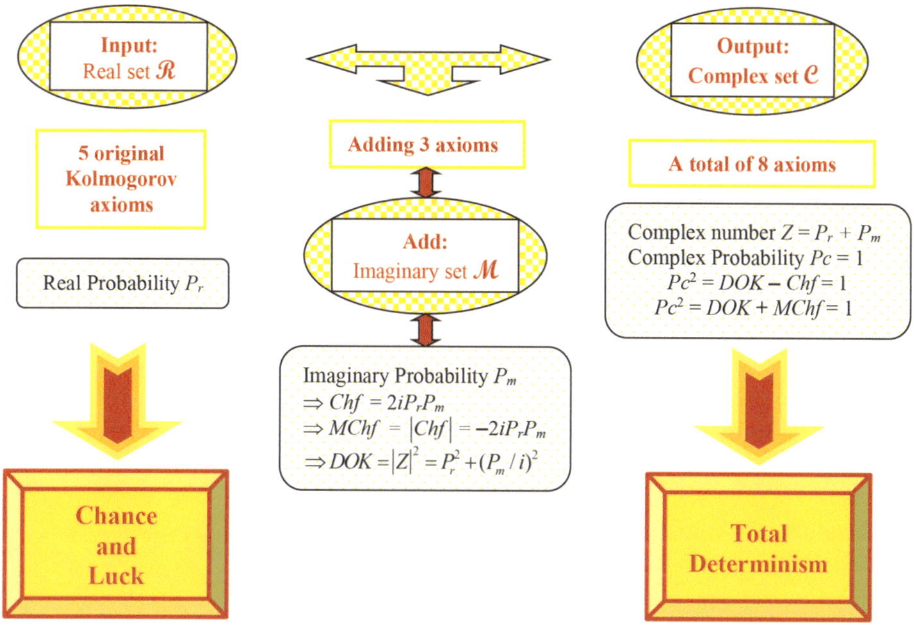

Figure 2.
The EKA *or the* CPP *diagram.*

4. The quantum harmonic oscillator with Gaussian initial condition and the complex probability paradigm (*CPP*) parameters: The momentum wavefunction and *CPP*

In this section, we will relate and link quantum mechanics to the complex probability paradigm with all its parameters by applying it to the quantum harmonic oscillators with Gaussian initial condition and this by using the four *CPP* concepts which are: the real probability P_r in the real probability set $\mathcal{R}$, the imaginary probability P_m in the imaginary probability set $\mathcal{M}$, the complex random vector or number Z in the complex probability set $\mathcal{C} = \mathcal{R} + \mathcal{M}$, and the deterministic real probability Pc also in the probability set $\mathcal{C}$ [1–36, 69–80].

4.1 The momentum wavefunction probability distribution and *CPP*

The probability momentum density for the quantum harmonic oscillators with Gaussian initial condition problem is derived from the wavefunction as $f(p) = |\phi(p,t)|^2$. Through integration over the propagator, we can solve for the full time-dependent solution. After many cancelations, and as with position, the wavefunction momentum probability density function (*PDF*) reduces to and is given by [1–3]:

$$f(p) = |\phi(p,t)|^2 = \mathbb{N}\left(-mx_0w\sin(wt), \sqrt{\frac{\hbar m\Omega}{2}\left(\cos^2(wt) + \frac{w^2}{\Omega^2}\sin^2(wt)\right)}\right) \quad (8)$$

Where the mean of the normal distribution $\mathbb{N}(\mu_P, \sigma_P)$ is = $\mu_p = -mx_0 w \sin(wt)$. and the standard deviation of this normal distribution is =

$\sigma_p = \sqrt{\frac{\hbar m\Omega}{2}\left(\cos^2(wt) + \frac{w^2}{\Omega^2}\sin^2(wt)\right)}$ and $\hbar = \frac{h}{2\pi}$ is the reduced Planck constant, and w is the characteristic angular frequency, and with Ω describing the width of the initial state but need not be the same as w. Knowing that, we have taken in this study $\Omega = n \times w$ where n is a simple multiplier and can be equal to ½, or 1, or 2, or 30, or 100, or 500, etc., as it will be shown afterward in the simulations section. Also, in a quantum harmonic oscillator of characteristic angular frequency w, we place a state that is offset from the bottom of the potential by some displacement x_0 as it is shown in the equation [1–3]:

$$\Psi(x, t = 0) = \left(\frac{m\Omega}{\pi\hbar}\right)^{1/4} \exp\left(-\frac{m\Omega(x - x_0)^2}{2\hbar}\right) \tag{9}$$

Therefore, the wavefunction momentum cumulative probability distribution function (*CDF*) which is equal to $P_r(T)$ in $\mathcal{R}$ is:

$$\begin{aligned} P_r(T) = F(t_j) = P_{rob}(T \leq t_j) &= \int_0^{t_j} |\phi(p,t)|^2 dt \\ &= \int_0^{t_j} \mathbb{N}\left(-mx_0 w \sin(wt), \sqrt{\frac{\hbar m\Omega}{2}\left(\cos^2(wt) + \frac{w^2}{\Omega^2}\sin^2(wt)\right)}\right) dt \end{aligned} \tag{10}$$

And the real complementary probability to $P_r(T)$ in $\mathcal{R}$ which is $P_m(T)/i$ is:

$$\begin{aligned} P_m(T)/i &= 1 - P_r(T) = 1 - F(t_j) = 1 - P_{rob}(T \leq t_j) = P_{rob}(T > t_j) = 1 - \int_0^{t_j} |\phi(p,t)|^2 dt \\ &= \int_{t_j}^{+\infty} |\phi(p,t)|^2 dt \\ &= 1 - \int_0^{t_j} \mathbb{N}\left(-mx_0 w \sin(wt), \sqrt{\frac{\hbar m\Omega}{2}\left(\cos^2(wt) + \frac{w^2}{\Omega^2}\sin^2(wt)\right)}\right) dt \\ &= \int_{t_j}^{+\infty} \mathbb{N}\left(-mx_0 w \sin(wt), \sqrt{\frac{\hbar m\Omega}{2}\left(\cos^2(wt) + \frac{w^2}{\Omega^2}\sin^2(wt)\right)}\right) dt \end{aligned} \tag{11}$$

Consequently, the imaginary complementary probability to $P_r(T)$ in $\mathcal{M}$ which is $P_m(T)$ is:

$$
\begin{aligned}
P_m(T) &= i[1 - P_r(T)] = i[1 - F(t_j)] = i[1 - P_{rob}(T \le t_j)] = iP_{rob}(T > t_j) \\
&= i\left[1 - \int_0^{t_j} |\phi(p,t)|^2 dt\right] = i\int_{t_j}^{+\infty} |\phi(p,t)|^2 dt \\
&= i\left[1 - \int_0^{t_j} \mathbb{N}\left(-mx_0 w \sin(wt), \sqrt{\frac{\hbar m \Omega}{2}\left(\cos^2(wt) + \frac{w^2}{\Omega^2}\sin^2(wt)\right)}\right) dt\right] \\
&= i\int_{t_j}^{+\infty} \mathbb{N}\left(-mx_0 w \sin(wt), \sqrt{\frac{\hbar m \Omega}{2}\left(\cos^2(wt) + \frac{w^2}{\Omega^2}\sin^2(wt)\right)}\right) dt
\end{aligned}
\tag{12}
$$

Furthermore, the complex random number or vector in $\mathcal{C} = \mathcal{R} + \mathcal{M}$ which is $Z(T)$ is:

$$
\begin{aligned}
Z(T) &= P_r(T) + P_m(T) = P_r(T) + i[1 - P_r(T)] = F(t_j) + i[1 - F(t_j)] \\
&= P_{rob}(T \le t_j) + i[1 - P_{rob}(T \le t_j)] = P_{rob}(T \le t_j) + iP_{rob}(T > t_j) \\
&= \int_0^{t_j} |\phi(p,t)|^2 dt + i\left[1 - \int_0^{t_j} |\phi(p,t)|^2 dt\right] = \int_0^{t_j} |\phi(p,t)|^2 dt + i\int_{t_j}^{+\infty} |\phi(p,t)|^2 dt \\
&= \left[\int_0^{t_j} \mathbb{N}\left(-mx_0 w \sin(wt), \sqrt{\frac{\hbar m \Omega}{2}\left(\cos^2(wt) + \frac{w^2}{\Omega^2}\sin^2(wt)\right)}\right) dt\right] \\
&\quad + i\left[1 - \int_0^{t_j} \mathbb{N}\left(-mx_0 w \sin(wt), \sqrt{\frac{\hbar m \Omega}{2}\left(\cos^2(wt) + \frac{w^2}{\Omega^2}\sin^2(wt)\right)}\right) dt\right] \\
&= \left[\int_0^{t_j} \mathbb{N}\left(-mx_0 w \sin(wt), \sqrt{\frac{\hbar m \Omega}{2}\left(\cos^2(wt) + \frac{w^2}{\Omega^2}\sin^2(wt)\right)}\right) dt\right] \\
&\quad + i\left[\int_{t_j}^{+\infty} \mathbb{N}\left(-mx_0 w \sin(wt), \sqrt{\frac{\hbar m \Omega}{2}\left(\cos^2(wt) + \frac{w^2}{\Omega^2}\sin^2(wt)\right)}\right) dt\right]
\end{aligned}
\tag{13}
$$

Additionally, the degree of our knowledge which is $DOK(T)$ is:

$$
\begin{aligned}
DOK(T) &= [P_r(T)]^2 + [P_m(T)/i]^2 = [P_r(T)]^2 + [1 - P_r(T)]^2 = [F(t_j)]^2 + [1 - F(t_j)]^2 \\
&= [P_{rob}(T \le t_j)]^2 + [1 - P_{rob}(T \le t_j)]^2 = [P_{rob}(T \le t_j)]^2 + [P_{rob}(T > t_j)]^2 \\
&= \left[\int_0^{t_j} |\phi(p,t)|^2 dt\right]^2 + \left[1 - \int_0^{t_j} |\phi(p,t)|^2 dt\right]^2 \\
&= \left[\int_0^{t_j} |\phi(p,t)|^2 dt\right]^2 + \left[\int_{t_j}^{+\infty} |\phi(p,t)|^2 dt\right]^2
\end{aligned}
$$

$$
\begin{aligned}
&= \left[\int_0^{t_j} \mathbb{N}\left(-mx_0 w \sin(wt), \sqrt{\frac{\hbar m\Omega}{2}\left(\cos^2(wt) + \frac{w^2}{\Omega^2}\sin^2(wt)\right)}\right)dt\right]^2 \\
&\quad + \left[1 - \int_0^{t_j} \mathbb{N}\left(-mx_0 w \sin(wt), \sqrt{\frac{\hbar m\Omega}{2}\left(\cos^2(wt) + \frac{w^2}{\Omega^2}\sin^2(wt)\right)}\right)dt\right]^2 \\
&= \left[\int_0^{t_j} \mathbb{N}\left(-mx_0 w \sin(wt), \sqrt{\frac{\hbar m\Omega}{2}\left(\cos^2(wt) + \frac{w^2}{\Omega^2}\sin^2(wt)\right)}\right)dt\right]^2 \\
&\quad + \left[\int_{t_j}^{+\infty} \mathbb{N}\left(-mx_0 w \sin(wt), \sqrt{\frac{\hbar m\Omega}{2}\left(\cos^2(wt) + \frac{w^2}{\Omega^2}\sin^2(wt)\right)}\right)dt\right]^2
\end{aligned}
\tag{14}
$$

Moreover, the chaotic factor which is $Chf(T)$ is:

$$
\begin{aligned}
Chf(T) &= 2iP_r(T)P_m(T) = 2iP_r(T) \times i[1 - P_r(T)] = -2P_r(T)[1 - P_r(T)] \\
&= -2F(t_j)\left[1 - F(t_j)\right] \\
&= -2P_{rob}(T \le t_j)\left[1 - P_{rob}(T \le t_j)\right] = -2P_{rob}(T \le t_j)P_{rob}(T > t_j) \\
&= -2\left[\int_0^{t_j} |\phi(p,t)|^2 dt\right] \times \left[1 - \int_0^{t_j} |\phi(p,t)|^2 dt\right] \\
&= -2\left[\int_0^{t_j} |\phi(p,t)|^2 dt\right] \times \left[\int_{t_j}^{+\infty} |\phi(p,t)|^2 dt\right] \\
&= -2\left[\int_0^{t_j} \mathbb{N}\left(-mx_0 w \sin(wt), \sqrt{\frac{\hbar m\Omega}{2}\left(\cos^2(wt) + \frac{w^2}{\Omega^2}\sin^2(wt)\right)}\right)dt\right] \\
&\quad \times \left[1 - \int_0^{t_j} \mathbb{N}\left(-mx_0 w \sin(wt), \sqrt{\frac{\hbar m\Omega}{2}\left(\cos^2(wt) + \frac{w^2}{\Omega^2}\sin^2(wt)\right)}\right)dt\right] \\
&= -2\left[\int_0^{t_j} \mathbb{N}\left(-mx_0 w \sin(wt), \sqrt{\frac{\hbar m\Omega}{2}\left(\cos^2(wt) + \frac{w^2}{\Omega^2}\sin^2(wt)\right)}\right)dt\right] \\
&\quad \times \left[\int_{t_j}^{+\infty} \mathbb{N}\left(-mx_0 w \sin(wt), \sqrt{\frac{\hbar m\Omega}{2}\left(\cos^2(wt) + \frac{w^2}{\Omega^2}\sin^2(wt)\right)}\right)dt\right]
\end{aligned}
\tag{15}
$$

In addition, the magnitude of the chaotic factor which is $MChf(T)$ is:

$$
\begin{aligned}
MChf(T) &= |Chf(T)| = -2iP_r(T)P_m(T) = -2iP_r(T) \times i[1-P_r(T)] = 2P_r(T)[1-P_r(T)] \\
&= 2F(t_j)\left[1-F(t_j)\right] \\
&= 2P_{rob}\left(T \le t_j\right)\left[1-P_{rob}\left(T \le t_j\right)\right] = 2P_{rob}\left(T \le t_j\right)P_{rob}\left(T > t_j\right) \\
&= 2\left[\int_0^{t_j} |\phi(p,t)|^2 dt\right] \times \left[1-\int_0^{t_j} |\phi(p,t)|^2 dt\right] \\
&= 2\left[\int_0^{t_j} |\phi(p,t)|^2 dt\right] \times \left[\int_{t_j}^{+\infty} |\phi(p,t)|^2 dt\right] \\
&= 2\left[\int_0^{t_j} \mathbb{N}\left(-mx_0 w\sin(wt), \sqrt{\frac{\hbar m\Omega}{2}\left(\cos^2(wt)+\frac{w^2}{\Omega^2}\sin^2(wt)\right)}\right)dt\right] \\
&\times \left[1-\int_0^{t_j} \mathbb{N}\left(-mx_0 w\sin(wt), \sqrt{\frac{\hbar m\Omega}{2}\left(\cos^2(wt)+\frac{w^2}{\Omega^2}\sin^2(wt)\right)}\right)dt\right] \\
&= 2\left[\int_0^{t_j} \mathbb{N}\left(-mx_0 w\sin(wt), \sqrt{\frac{\hbar m\Omega}{2}\left(\cos^2(wt)+\frac{w^2}{\Omega^2}\sin^2(wt)\right)}\right)dt\right] \\
&\times \left[\int_{t_j}^{+\infty} \mathbb{N}\left(-mx_0 w\sin(wt), \sqrt{\frac{\hbar m\Omega}{2}\left(\cos^2(wt)+\frac{w^2}{\Omega^2}\sin^2(wt)\right)}\right)dt\right]
\end{aligned}
\tag{16}
$$

Finally, the real probability in the complex probability universe $\mathcal{C} = \mathcal{R} + \mathcal{M}$ which is $Pc(T)$ is:

$$
\begin{aligned}
Pc^2(T) &= \{[P_r(T)] + [P_m(T)/i]\}^2 = \{[P_r(T)] + [1-P_r(T)]\}^2 \\
&= \left\{\left[F(t_j)\right] + \left[1-F(t_j)\right]\right\}^2 = \left\{P_{rob}\left(T \le t_j\right) + \left[1-P_{rob}\left(T \le t_j\right)\right]\right\}^2 \\
&= \left\{P_{rob}\left(T \le t_j\right) + P_{rob}\left(T > t_j\right)\right\}^2 \\
&= \left\{\int_0^{t_j} |\phi(p,t)|^2 dt + \left[1-\int_0^{t_j} |\phi(p,t)|^2 dt\right]\right\}^2 \\
&= \left\{\int_0^{t_j} |\phi(p,t)|^2 dt + \int_{t_j}^{+\infty} |\phi(p,t)|^2 dt\right\}^2 = \left\{\int_0^{+\infty} |\phi(p,t)|^2 dt\right\}^2 \\
&= \left\{\begin{aligned}&\left[\int_0^{t_j} \mathbb{N}\left(-mx_0 w\sin(wt), \sqrt{\frac{\hbar m\Omega}{2}\left(\cos^2(wt)+\frac{w^2}{\Omega^2}\sin^2(wt)\right)}\right)dt\right] \\ &+\left[1-\int_0^{t_j} \mathbb{N}\left(-mx_0 w\sin(wt), \sqrt{\frac{\hbar m\Omega}{2}\left(\cos^2(wt)+\frac{w^2}{\Omega^2}\sin^2(wt)\right)}\right)dt\right]\end{aligned}\right\}^2 \\
&= \left\{\begin{aligned}&\left[\int_0^{t_j} \mathbb{N}\left(-mx_0 w\sin(wt), \sqrt{\frac{\hbar m\Omega}{2}\left(\cos^2(wt)+\frac{w^2}{\Omega^2}\sin^2(wt)\right)}\right)dt\right] \\ &+\left[\int_{t_j}^{+\infty} \mathbb{N}\left(-mx_0 w\sin(wt), \sqrt{\frac{\hbar m\Omega}{2}\left(\cos^2(wt)+\frac{w^2}{\Omega^2}\sin^2(wt)\right)}\right)dt\right]\end{aligned}\right\}^2
\end{aligned}
\tag{17}
$$

$$= \left\{ \int_0^{+\infty} \mathbb{N}\left(-mx_0 w \sin(wt), \sqrt{\frac{\hbar m \Omega}{2}\left(\cos^2(wt) + \frac{w^2}{\Omega^2}\sin^2(wt)\right)}\right) dt \right\}^2$$
$$= 1^2 = 1$$
$$= Pc(T) \tag{18}$$

And, $Pc(T)$ can be computed using *CPP* as follows:

$$\begin{aligned} Pc^2(T) &= DOK(T) - Chf(T) = [P_r(T)]^2 + [P_m(T)/i]^2 - 2iP_r(T)P_m(T) \\ &= [P_r(T)]^2 + [1 - P_r(T)]^2 + 2P_r(T)[1 - P_r(T)] = \{P_r(T) + [1 - P_r(T)]\}^2 \end{aligned}$$

$$\begin{aligned} &= \left\{ \int_0^{t_j} |\phi(p,t)|^2 dt + \left[1 - \int_0^{t_j} |\phi(p,t)|^2 dt \right] \right\}^2 = \left\{ \int_0^{t_j} |\phi(p,t)|^2 dt + \int_{t_j}^{+\infty} |\phi(p,t)|^2 dt \right\}^2 = \left\{ \int_0^{+\infty} |\phi(p,t)|^2 dt \right\}^2 \\ &= 1^2 = 1 \\ &= Pc(T) \end{aligned} \tag{19}$$

And, $Pc(T)$ can be computed using always *CPP* as follows:

$$\begin{aligned} Pc^2(T) &= DOK(T) + MChf(T) = [P_r(T)]^2 + [P_m(T)/i]^2 + [-2iP_r(T)P_m(T)] \\ &= [P_r(T)]^2 + [1 - P_r(T)]^2 + 2P_r(T)[1 - P_r(T)] = \{P_r(T) + [1 - P_r(T)]\}^2 \\ &= \left\{ \int_0^{t_j} |\phi(p,t)|^2 dt + \left[1 - \int_0^{t_j} |\phi(p,t)|^2 dt \right] \right\}^2 \\ &= \left\{ \int_0^{t_j} |\phi(p,t)|^2 dt + \int_{t_j}^{+\infty} |\phi(p,t)|^2 dt \right\}^2 = \left\{ \int_0^{+\infty} |\phi(p,t)|^2 dt \right\}^2 = 1^2 = 1 = Pc(T) \end{aligned} \tag{20}$$

Hence, the prediction of all the wavefunction momentum probabilities of the quantum harmonic oscillators with Gaussian initial condition problem in the universe $\mathcal{C} = \mathcal{R} + \mathcal{M}$ is permanently certain and perfectly deterministic.

4.2 The new model simulations

The following figures (**Figures 3–17**) illustrate all the calculations done above.

4.2.1 Simulations interpretation

In **Figures 3, 8**, and **13**, we can see the graphs of the probability density functions (*PDF*) of the wavefunction momentum probability distribution for this problem as functions of the time random variable $T : 0 \leq T \leq 4.17$ for n = 30, 100, 500.

In **Figures 4, 9**, and **14**, we can see also the graphs and the simulations of all the *CPP* parameters (*Chf*, *MChf*, *DOK*, P_r, P_m/i, *Pc*) as functions of the time random variable *T* for the wavefunction momentum probability distribution of the quantum

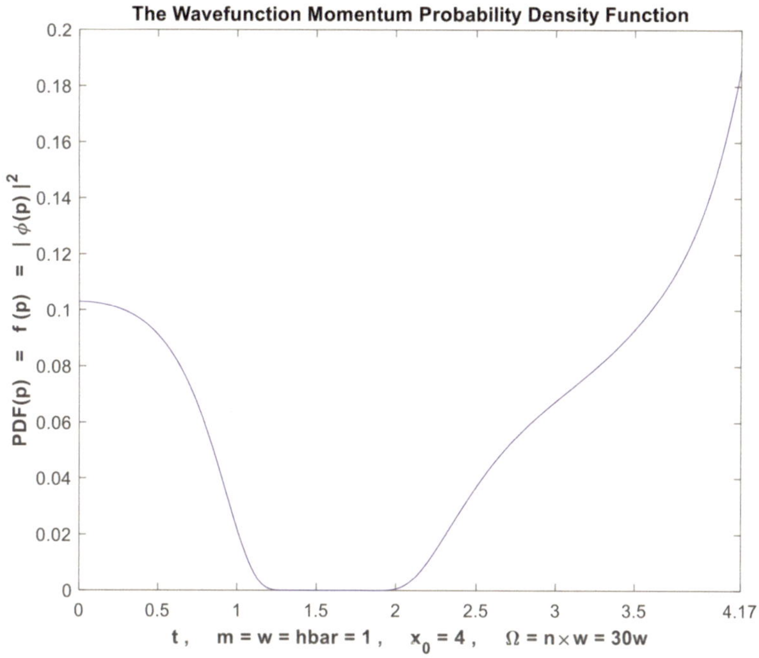

Figure 3.
The graph of the PDF *as a function of the random variable* T *of the wavefunction momentum probability density for* n = *30.*

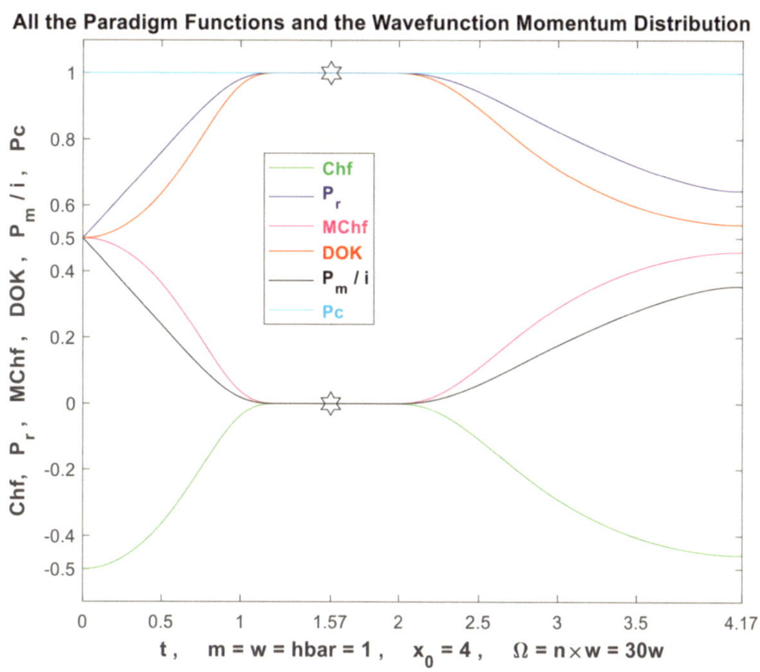

Figure 4.
The graphs of all the CPP *parameters for the wavefunction momentum probability distribution as functions of the random variable* T *for* n = *30.*

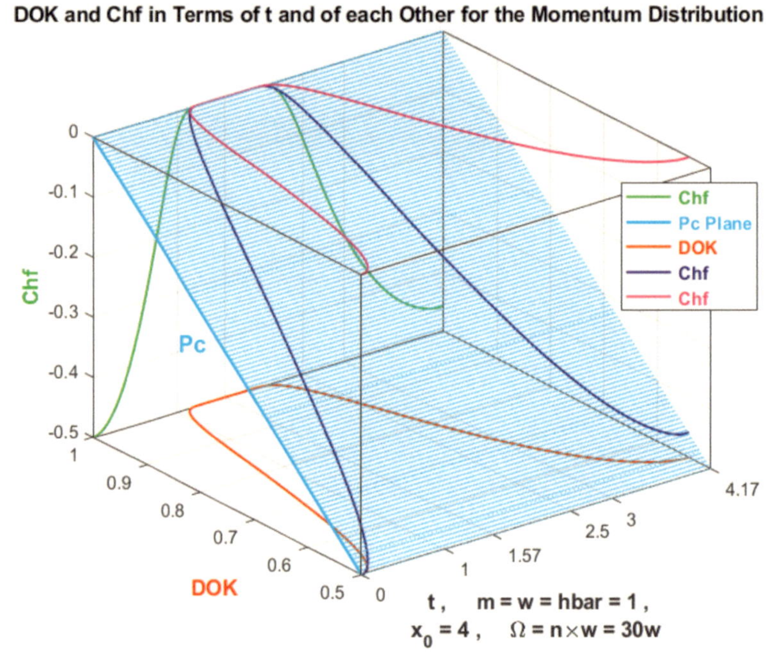

Figure 5.
The graphs of DOK *and* Chf *and the deterministic probability* Pc *for the wavefunction momentum probability distribution in terms of* T *and of each other for* n = 30.

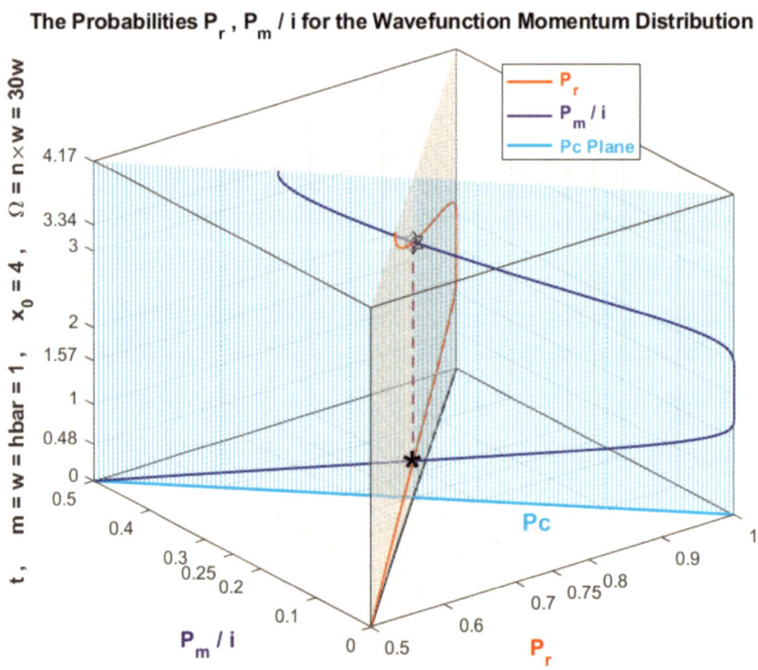

Figure 6.
The graphs of P_r *and* P_m/i *and* Pc *for the wavefunction momentum probability distribution in terms of* T *and of each other for* n = 30.

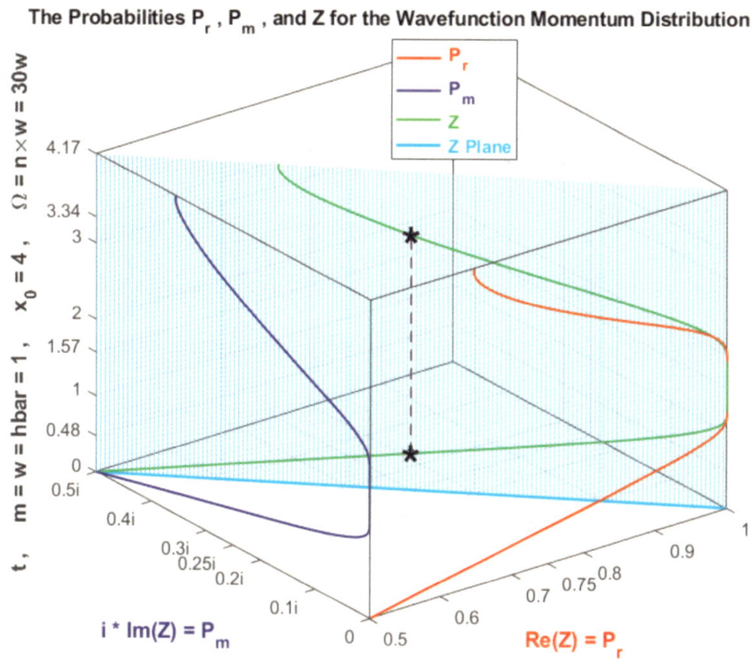

Figure 7.
The graphs of the probabilities P_r *and* P_m *and* Z *for the wavefunction momentum probability distribution in terms of* T *for* n = 30.

harmonic oscillators with Gaussian initial condition problem for $n = 30, 100, 500$. Hence, we can visualize all the new paradigm functions for this problem.

In the cubes (**Figures 5, 10**, and **15**), the simulation of *DOK* and *Chf* as functions of each other and of the time random variable *T* for the quantum harmonic oscillators with Gaussian initial condition problem wavefunction momentum probability distribution can be seen. The thick line in cyan is the projection of the plane $Pc^2(T) = DOK(T) - Chf(T) = 1 = Pc(T)$ on the plane $T = L_b$ = lower bound of $T = 0$. This thick line starts at the point $(DOK = 0.5, Chf = -0.5)$ when $T = L_b = 0$, reaches the point $(DOK = 1\ Chf = 0)$ when $T = 1.57$, and returns at the end to $(DOK = 0.5, Chf = -0.5)$ when $T = U_b$ = upper bound of $T = 4.17$. The other curves are the graphs of $DOK(T)$ (red) and $Chf(T)$ (green, blue, pink) in different simulation planes. Notice that they all have a maximum at the point $(DOK = 1, Chf = 0, T = 1.57)$. The last simulation point corresponds to $(DOK = 0.5, Chf = -0.5, T = U_b = 4.17)$.

In the cubes (**Figures 6, 11**, and **16**), we can notice the simulation of the real probability $P_r(T)$ in $\mathcal{R}$ and its complementary real probability $P_m(T)/i$ in $\mathcal{R}$ also in terms of the time random variable *T* for the quantum harmonic oscillators with Gaussian initial condition problem wavefunction momentum probability distribution. The thick line in cyan is the projection of the plane $Pc^2(T) = P_r(T) + P_m(T)/i = 1 = Pc(T)$ on the plane $T = L_b$ = lower bound of $T = 0$. This thick line starts at the point ($P_r = 0.5$, $P_m/i = 0.5$) and ends at the point ($P_r = 1$, $P_m/i = 0$). The red curve represents $P_r(T)$ in the plane $P_r(T) = P_m(T)/i + 0.5$ in light gray. This curve starts at the point ($P_r = 0.5$, $P_m/i = 0, T = L_b$ = lower bound of $T = 0$), reaches the point ($P_r = 1, P_m/i = 0.5, T = 1.57$), and gets at the end to ($P_r = 0.5, P_m/i = 0, T = U_b$ = upper bound of $T = 4.17$). The blue curve represents $P_m(T)/i$ in the plane in cyan $Pc^2(T) = P_r(T) + P_m(T)/i = 1 = Pc(T)$.

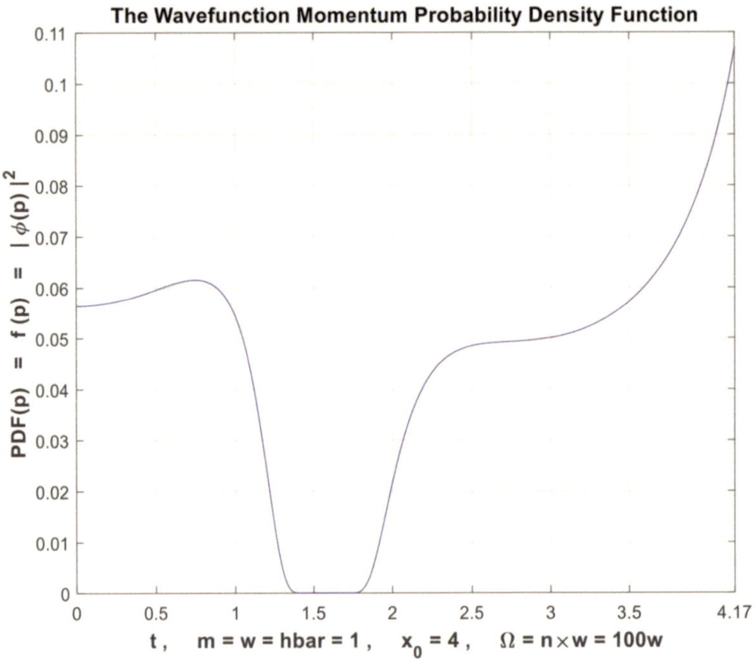

Figure 8.
The graph of the PDF *of the random variable* T *of the wavefunction momentum probability density as a function for* n = 100.

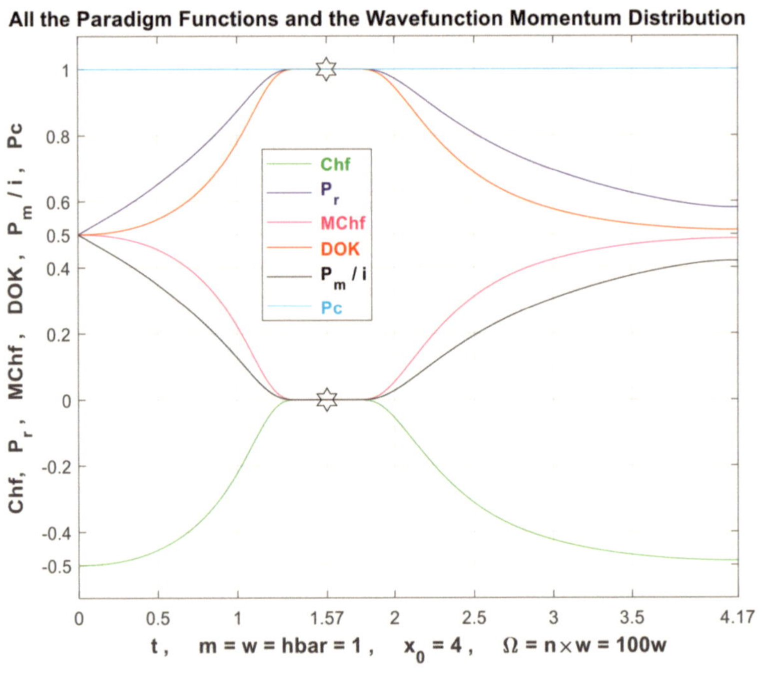

Figure 9.
The graphs of all the CPP *parameters for the wavefunction momentum probability distribution as functions of the random variable* T *for* n = 100.

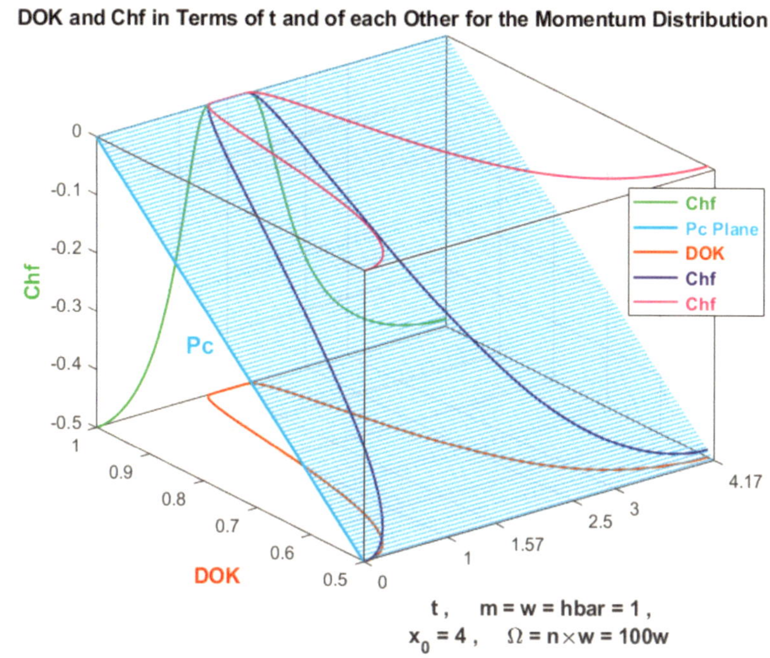

Figure 10.
The graphs of DOK *and* Chf *and the deterministic probability* Pc *for the wavefunction momentum probability distribution in terms of* T *and of each other for* n = 100.

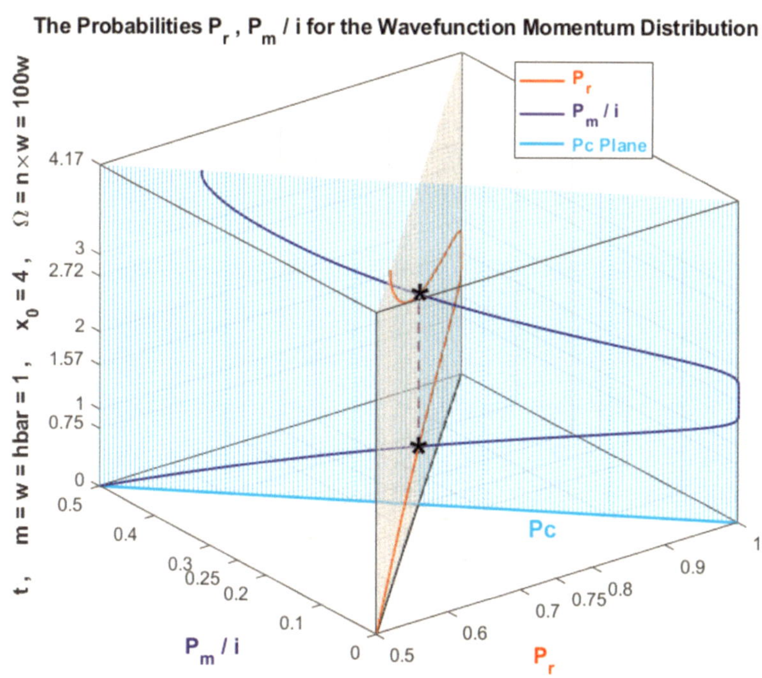

Figure 11.
The graphs of P_r *and* P_m/i *and* Pc *for the wavefunction momentum probability distribution in terms of* T *and of each other for* n = 100.

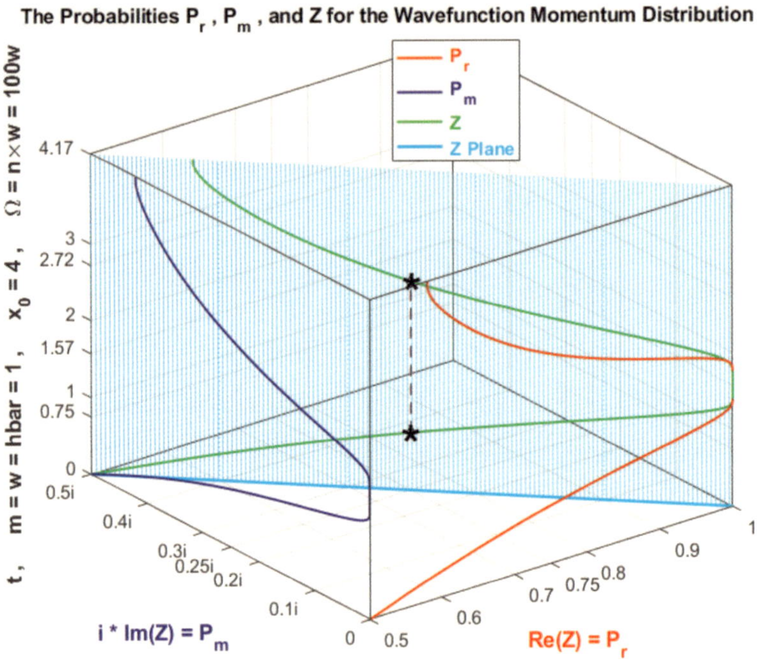

Figure 12.
The graphs of the probabilities P_r *and* P_m *and* Z *for the wavefunction momentum probability distribution in terms of* T *for* n = 100.

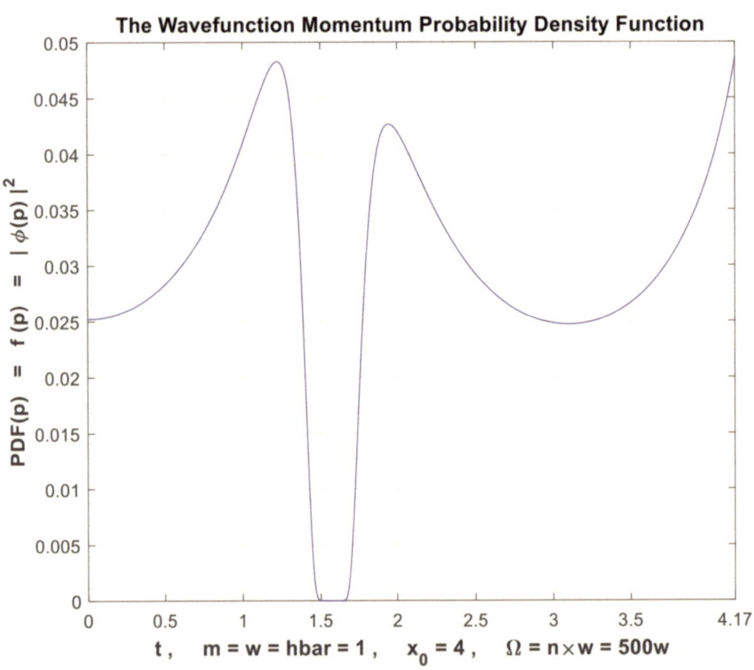

Figure 13.
The graph of the PDF *of the random variable* T *of the wavefunction momentum probability density as a function for* n = 500.

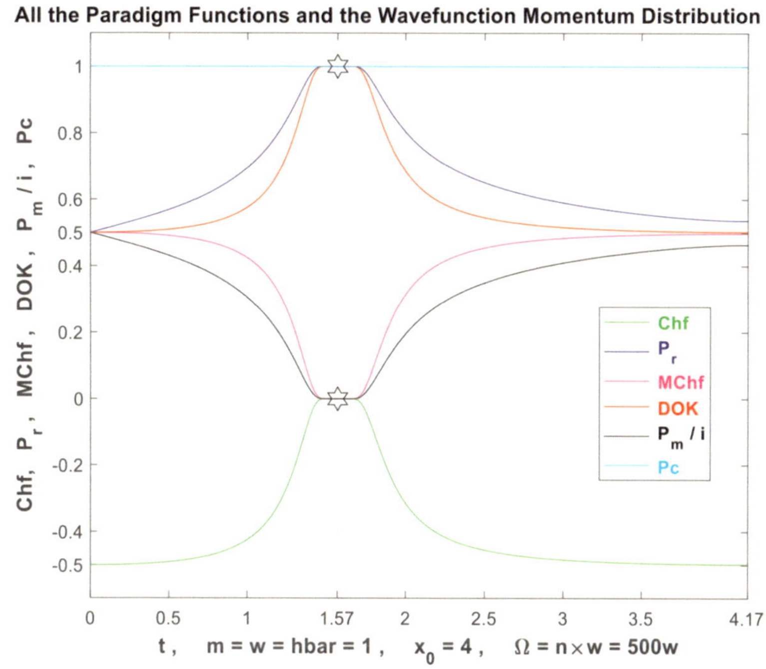

Figure 14.
The graphs of all the CPP *parameters for the wavefunction momentum probability distribution as functions of the random variable* T *for* n = 500.

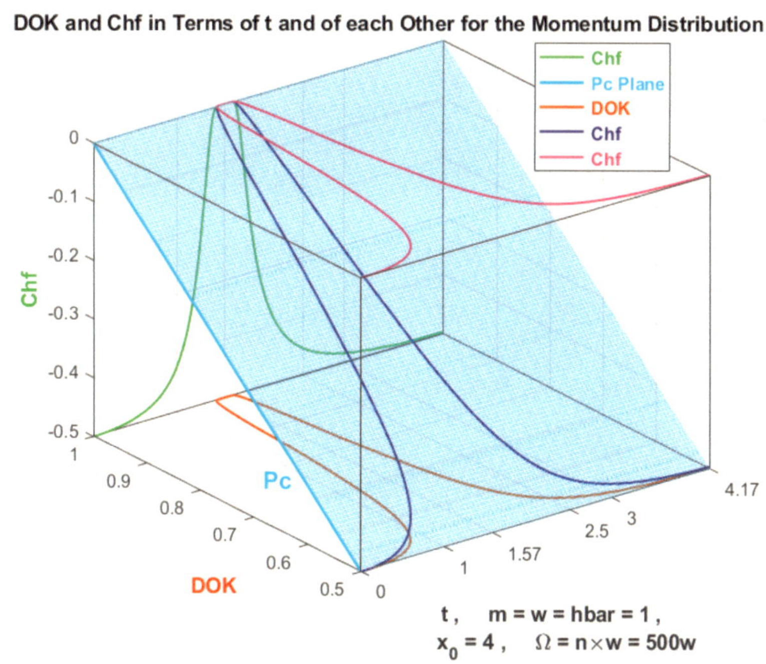

Figure 15.
The graphs of DOK *and* Chf *and the deterministic probability* Pc *for the wavefunction momentum probability distribution in terms of* T *and of each other for* n = 500.

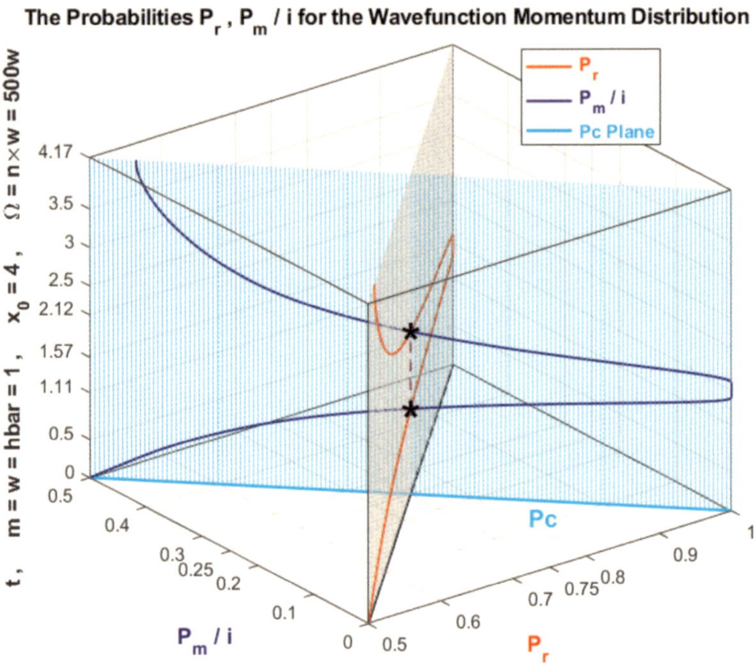

Figure 16.
The graphs of P_r *and* P_m/i *and* Pc *for the wavefunction momentum probability distribution in terms of* T *and of each other for* n = 500.

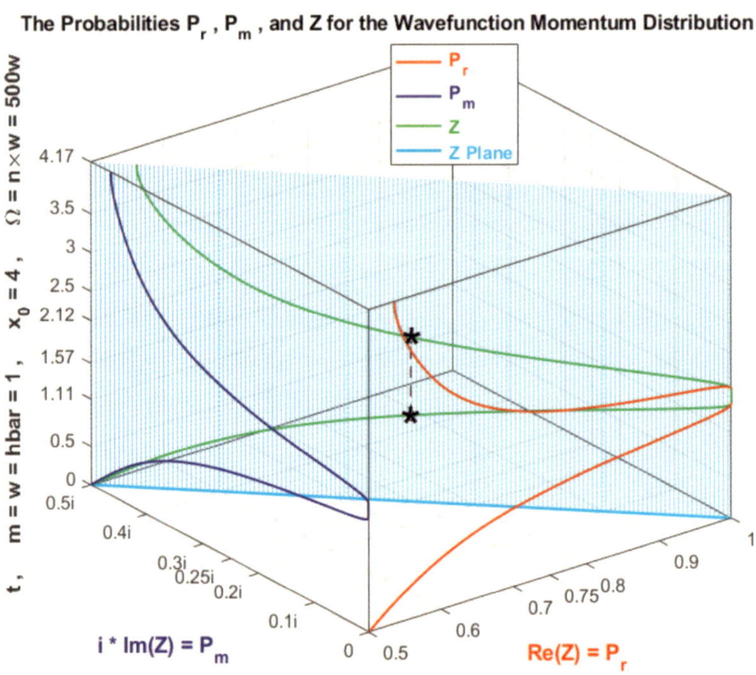

Figure 17.
The graphs of the probabilities P_r *and* P_m *and* Z *for the wavefunction momentum probability distribution in terms of* T *for* n = 500.

This curve starts at the point ($P_r = 0.5, P_m/i = 0.5, T = L_b =$ lower bound of $T = 0$), reaches the point ($P_r = 1, P_m/i = 0, T = 1.57$), and gets at the end to ($P_r = 0.5, P_m/i = 0.5$, $T = U_b =$ upper bound of $T = 4.17$). Notice the importance of the point which is on the intersection of the gray and cyan planes at $T = 1.57$ and when $P_r(T) = 0.75$ and $P_m(T)/i = 0.25$.

In the cubes (**Figures 7, 12**, and **17**), we can notice the simulation of the complex probability $Z(T)$ in $\mathcal{C} = \mathcal{R} + \mathcal{M}$ as a function of the real probability $P_r(T) = \text{Re}(Z)$ in $\mathcal{R}$ and of its complementary imaginary probability $P_m(T) = i \times \text{Im}(Z)$ in $\mathcal{M}$, and this is in terms of the time random variable T for the quantum harmonic oscillators with Gaussian initial condition problem wavefunction momentum probability distribution. The red curve represents $P_r(T)$ in the plane $P_m(T) = 0$ and the blue curve represents $P_m(T)$ in the plane $P_r(T) = 0.5$. The green curve represents the complex probability $Z(T) = P_r(T) + P_m(T) = \text{Re}(Z) + i \times \text{Im}(Z)$ in the plane $P_r(T) = iP_m(T) + 1$ or $Z(T)$ plane in cyan. The curve of $Z(T)$ starts at the point ($P_r = 0.5, P_m = 0.5i, T = L_b =$ lower bound of $T = 0$), reaches the point ($P_r = 1, P_m/i = 0, T = 1.57$), and ends at the point ($P_r = 0.5, P_m = 0.5i, T = U_b =$ upper bound of $T = 4.17$). The thick line in cyan is $P_r(T = L_b = 0) = iP_m(T = L_b = 0) + 1$ and it is the projection of the $Z(T)$ curve on the complex probability plane whose equation is: $T = L_b = 0$. This projected thick line starts at the point ($P_r = 0.5, P_m = 0.5i, T = L_b = 0$) and ends at the point ($P_r = 1, P_m = 0$, $T = L_b = 0$). Notice the importance of the point corresponding to $T = 1.57$ and $Z = 0.75 + 0.25i$ when $P_r = 0.75$ and $P_m = 0.25i$.

Furthermore, as it was verified and proved and shown in this original paradigm (*CPP*) simulations, before the beginning of the simulation of the random event and at its end we have the chaotic factor (*Chf* and *MChf*) is 0 and the degree of our knowledge (*DOK*) is 1 since the stochastic and probabilistic effects and fluctuations have either not started yet or they have finished and terminated their task on the random phenomenon. During the execution of the nondeterministic experiment and process we also have: $-0.5 \leq Chf < 0$, $0 < MChf \leq 0.5$, and $0.5 \leq DOK < 1$. We can see that during the whole and entire process we have constantly and incessantly $Pc^2 = DOK - Chf = DOK + MChf = 1 = Pc$ that shows that the simulation which behaved probabilistically and randomly in the real universe and set $\mathcal{R}$ is now deterministic and certain in the complex probability universe and set $\mathcal{C} = \mathcal{R} + \mathcal{M}$ of *CPP*, and this after adding to the stochastic phenomenon performed in the real universe and set $\mathcal{R}$ the contributions of the imaginary universe and set $\mathcal{M}$ and thus after subtracting and eliminating from the degree of our knowledge the chaotic factor.

Finally, we can conclude that the quantum harmonic oscillator is the quantum-mechanical analog of the classical harmonic oscillator. Because an arbitrary smooth potential can usually be approximated as a harmonic potential at the vicinity of a stable equilibrium point, it is one of the most important model systems in quantum mechanics. Additionally, it is one of the few quantum-mechanical systems for which an exact and analytical solution is known [1–3]. Hence, we can see directly from all the simulations done and achieved that its relation to *CPP* is very fruitful, fascinating, and wonderful and which leads to delightful results and successful consequences.

4.3 The characteristics of the momentum probability distribution

In this quantum mechanics problem [20], the average, or expectation value of the momentum of a particle is given by:

$$\langle p\rangle = \int_0^{+\infty} t|\phi(p,t)|^2 dt = \int_0^{+\infty} t\mathbb{N}\left(-mx_0w\sin(wt), \sqrt{\frac{\hbar m\Omega}{2}\left(\cos^2(wt)+\frac{w^2}{\Omega^2}\sin^2(wt)\right)}\right)dt \quad (21)$$

For this problem, it can be shown that the average momentum is always $\langle p\rangle = -mx_0w\sin(wt)$.

So, in the probability set and universe $\mathcal{R}$, we have:

$$\langle p\rangle_R = \langle p\rangle = -mx_0w\sin(wt) \quad (22)$$

The variance in the momentum is a measure of the uncertainty in position of the particle, so in the probability set and universe $\mathcal{R}$, we have:

$$\mathrm{Var}_{p,R} = \mathrm{Var}(p) = \langle p^2\rangle_R - \langle p\rangle_R^2 \quad (23)$$

$$= \left\{\int_0^{+\infty} t^2|\phi(p,t)|^2 dt\right\} - \left\{\int_0^{+\infty} t|\phi(p,t)|^2 dt\right\}^2 \quad (24)$$

$$= \left\{\int_0^{+\infty} t^2\mathbb{N}\left(-mx_0w\sin(wt), \sqrt{\frac{\hbar m\Omega}{2}\left(\cos^2(wt)+\frac{w^2}{\Omega^2}\sin^2(wt)\right)}\right)dt\right\} - [-mx_0w\sin(wt)]^2$$
$$= \frac{\hbar m\Omega}{2}\left(\cos^2(wt)+\frac{w^2}{\Omega^2}\sin^2(wt)\right) \quad (25)$$

In the probability set and universe $\mathcal{M}$, we have:

$$\langle p\rangle_M = \int_0^{+\infty} t\left\{i\left[1-|\phi(p,t)|^2\right]\right\}dt$$
$$= i\int_0^{+\infty} t\left\{1-\mathbb{N}\left(-mx_0w\sin(wt), \sqrt{\frac{\hbar m\Omega}{2}\left(\cos^2(wt)+\frac{w^2}{\Omega^2}\sin^2(wt)\right)}\right)\right\}dt$$
$$= i\left\{\int_0^{U_b} t\,dt - \int_0^{+\infty} t\mathbb{N}\left(-mx_0w\sin(wt), \sqrt{\frac{\hbar m\Omega}{2}\left(\cos^2(wt)+\frac{w^2}{\Omega^2}\sin^2(wt)\right)}\right)dt\right\} = i\left\{\left[\frac{t^2}{2}\right]_0^{U_b} - \langle p\rangle_R\right\}$$
$$= i\left\{\frac{U_b^2}{2} - [-mx_0w\sin(wt)]\right\} = i\left\{\frac{U_b^2}{2} + mx_0w\sin(wt)\right\} \quad (26)$$

Where U_b is the upper bound of the definite integral above. Practically, the standard normal distribution probability is very nearly equal to 1.0000 (0.99997 exactly) for $U_b = 4$.

Note that, If $U_b \to +\infty \Leftrightarrow \langle p\rangle_M \to +\infty$. Moreover,

$$\mathrm{Var}_{p,M} = \langle p^2\rangle_M - \langle p\rangle_M^2 = \left\{\int_0^{+\infty} t^2\left\{i\left[1-|\phi(p,t)|^2\right]\right\}dt\right\} - \left\{\int_0^{+\infty} t\left\{i\left[1-|\phi(p,t)|^2\right]\right\}dt\right\}^2$$
$$= i\int_0^{+\infty} t^2\left\{1-\mathbb{N}\left(-mx_0w\sin(wt), \sqrt{\frac{\hbar m\Omega}{2}\left(\cos^2(wt)+\frac{w^2}{\Omega^2}\sin^2(wt)\right)}\right)\right\}dt - \left\{i\left[\frac{U_b^2}{2} + mx_0w\sin(wt)\right]\right\}^2 \quad (27)$$

$$= i\left\{\int_0^{U_b} t^2 dt - \int_0^{+\infty} t^2\left\{\mathbb{N}\left(-mx_0w\sin(wt), \sqrt{\frac{\hbar m\Omega}{2}\left(\cos^2(wt) + \frac{w^2}{\Omega^2}\sin^2(wt)\right)}\right)\right\}dt\right\} + \left\{\frac{U_b^2}{2} + mx_0w\sin(wt)\right\}^2 \quad (28)$$

$$= i\left\{\int_0^{U_b} t^2 dt - \mathrm{Var}_{p,R}\right\} + \left\{\frac{U_b^2}{2} + mx_0w\sin(wt)\right\}^2 = i\left\{\left[\frac{t^3}{3}\right]_0^{U_b} - \mathrm{Var}_{p,R}\right\} + \left\{\frac{U_b^2}{2} + mx_0w\sin(wt)\right\}^2 \quad (29)$$

$$= i\left\{\frac{U_b^3}{3} - \frac{\hbar m\Omega}{2}\left(\cos^2(wt) + \frac{w^2}{\Omega^2}\sin^2(wt)\right)\right\} + \left\{\frac{U_b^2}{2} + mx_0w\sin(wt)\right\}^2 \quad (30)$$

Note that, If $U_b \to +\infty \Leftrightarrow \mathrm{Var}_{p,M} \to +\infty$.

In the probability set and the universe $\mathcal{C} = \mathcal{R} + \mathcal{M}$, we have from *CPP*:

$$\begin{aligned}
\langle p\rangle_C &= \int_0^{+\infty} t[z(p,t)]dt = \int_0^{+\infty} t\left\{|\phi(p,t)|^2 + i\left[1 - |\phi(p,t)|^2\right]\right\}dt \\
&= \int_0^{+\infty} t|\phi(p,t)|^2 dt + \int_0^{+\infty} t\left\{i\left[1 - |\phi(p,t)|^2\right]\right\}dt \\
&= \int_0^{+\infty} t\mathbb{N}\left(-mx_0w\sin(wt), \sqrt{\frac{\hbar m\Omega}{2}\left(\cos^2(wt) + \frac{w^2}{\Omega^2}\sin^2(wt)\right)}\right)dt \\
&\quad + i\int_0^{+\infty} t\left\{1 - \mathbb{N}\left(-mx_0w\sin(wt), \sqrt{\frac{\hbar m\Omega}{2}\left(\cos^2(wt) + \frac{w^2}{\Omega^2}\sin^2(wt)\right)}\right)\right\}dt \\
&= \langle p\rangle_R + \langle p\rangle_M = -mx_0w\sin(wt) + i\left\{\frac{U_b^2}{2} + mx_0w\sin(wt)\right\}
\end{aligned} \quad (31)$$

Note that, If $U_b \to +\infty \Leftrightarrow \langle p\rangle_C \to +\infty$.

$$\begin{aligned}
\mathrm{Var}_{p,C} &= \langle p^2\rangle_C - \langle p\rangle_C^2 = \left[\int_0^{+\infty} t^2[z(p,t)]dt\right] - \left[\langle p\rangle_R + \langle p\rangle_M\right]^2 \\
&= \left[\int_0^{+\infty} t^2\left\{|\phi(p,t)|^2 + i\left[1 - |\phi(p,t)|^2\right]\right\}dt\right] - \left[\langle p\rangle_R + \langle p\rangle_M\right]^2 \\
&= \left[\int_0^{+\infty} t^2|\phi(p,t)|^2 dt + \int_0^{+\infty} t^2\left\{i\left[1 - |\phi(p,t)|^2\right]\right\}dt\right] - \left[\langle p\rangle_R + \langle p\rangle_M\right]^2 \\
&= \left[\langle p^2\rangle_R + \langle p^2\rangle_M\right] - \left[\langle p\rangle_R + \langle p\rangle_M\right]^2 = \left[\langle p^2\rangle_R + \langle p^2\rangle_M\right] - \left[\langle p\rangle_R^2 + \langle p\rangle_M^2 + 2\langle p\rangle_R\langle p\rangle_M\right] \\
&= \left[\langle p^2\rangle_R - \langle p\rangle_R^2\right] + \left[\langle p^2\rangle_M - \langle p\rangle_M^2\right] - 2\langle p\rangle_R\langle p\rangle_M = \mathrm{Var}_{p,R} + \mathrm{Var}_{p,M} - 2\langle p\rangle_R\langle p\rangle_M \\
&= \frac{\hbar m\Omega}{2}\left(\cos^2(wt) + \frac{w^2}{\Omega^2}\sin^2(wt)\right) + i\left\{\frac{U_b^3}{3} - \frac{\hbar m\Omega}{2}\left(\cos^2(wt) + \frac{w^2}{\Omega^2}\sin^2(wt)\right)\right\} \\
&\quad + \left\{\frac{U_b^2}{2} + mx_0w\sin(wt)\right\}^2 - 2\{-mx_0w\sin(wt)\} \times \left\{i\left\{\frac{U_b^2}{2} + mx_0w\sin(wt)\right\}\right\}
\end{aligned} \quad (32)$$

Note that, If $U_b \to +\infty \Leftrightarrow \text{Var}_{p,C} \to +\infty$.

The following tables (**Tables 1–4**) compute the momentum distribution characteristics for $x_0 = 4, m = w = \hbar = 1, n = 1/4$ where $\Omega = n \times w$, and $U_b = 6, 20, 50, 1000$.

Momentum distribution characteristics	$x_0 = 4, m = w = \hbar = 1, \Omega = w/4, U_b = 6$
$\langle p \rangle_R$	1.1177
$\text{Var}_{p,R}$	0.2714
$\langle p \rangle_M$	$i \times$ 16.8823
$\text{Var}_{p,M}$	285.0133 + $i \times$ 71.7286
$\langle p \rangle_C = \langle p \rangle_R + \langle p \rangle_M$	1.1177 + $i \times$ 16.8823
$\text{Var}_{p,C} = \text{Var}_{p,R} + \text{Var}_{p,M} - 2\langle p \rangle_R \langle p \rangle_M$	285.2847 + $i \times$ 33.9911

Table 1.
The momentum distribution characteristics for $U_b = 6$ and $n = 1/4$.

Momentum distribution characteristics	$x_0 = 4, m = w = \hbar = 1, \Omega = w/4, U_b = 20$
$\langle p \rangle_R$	−3.6518
$\text{Var}_{p,R}$	1.6878
$\langle p \rangle_M$	$i \times$ 203.6518
$\text{Var}_{p,M}$	4.1474e+04 + $i \times$ 2.6650e+03
$\langle p \rangle_C = \langle p \rangle_R + \langle p \rangle_M$	−3.6518 + $i \times$ 203.6518
$\text{Var}_{p,C} = \text{Var}_{p,R} + \text{Var}_{p,M} - 2\langle p \rangle_R \langle p \rangle_M$	4.1476e+04 + $i \times$ 4.1524e+03

Table 2.
The momentum distribution characteristics for $U_b = 20$ and $n = 1/4$.

Momentum distribution characteristics	$x_0 = 4, m = w = \hbar = 1, \Omega = w/4, U_b = 50$
$\langle p \rangle_R$	1.0495
$\text{Var}_{p,R}$	0.2541
$\langle p \rangle_M$	$i \times$ 1.2490e+03
$\text{Var}_{p,M}$	1.5599e+06 + $i \times$ 4.1666e+04
$\langle p \rangle_C = \langle p \rangle_R + \langle p \rangle_M$	1.0495 + $i \times$ 1.2490e+03
$\text{Var}_{p,C} = \text{Var}_{p,R} + \text{Var}_{p,M} - 2\langle p \rangle_R \langle p \rangle_M$	1.5599e+06 + $i \times$ 3.9045 + 04

Table 3.
The momentum distribution characteristics for $U_b = 50$ and $n = 1/4$.

Momentum distribution characteristics	$x_0 = 4, m = w = \hbar = 1, \Omega = w/4, U_b = 1000$
$\langle p \rangle_R$	−3.3075
$\text{Var}_{p,R}$	1.4070
$\langle p \rangle_M$	$i \times$ 5.0000e+05

Momentum distribution characteristics	$x_0 = 4, m = w = \hbar = 1, \Omega = w/4, U_b = 1000$
$\text{Var}_{p,M}$	2.5000e+11 + $i \times$ 3.3333e+08
$\langle p \rangle_C = \langle p \rangle_R + \langle p \rangle_M$	−3.3075 + $i \times$ 5.0000e+05
$\text{Var}_{p,C} = \text{Var}_{p,R} + \text{Var}_{p,M} - 2\langle p \rangle_R \langle p \rangle_M$	2.5000e+11 + $i \times$ 3.3664 + 08

Table 4.
The momentum distribution characteristics for $U_b = 1000$ and $n = 1/4$.

5. The quantum uncertainty principle in $\mathcal{R}$, $\mathcal{M}$, and $\mathcal{C}$

Copying the variances above and applying trigonometric identities [1–3], we can write the product of the standard deviations as:

$$\begin{aligned}\sigma_{x_R} \times \sigma_{p_R} &= \sqrt{\text{Var}_{x,R}} \times \sqrt{\text{Var}_{p,R}} \\ &= \sqrt{\frac{\hbar}{2m\Omega}\left(\cos^2(wt) + \frac{\Omega^2}{w^2}\sin^2(wt)\right)} \times \sqrt{\frac{\hbar m\Omega}{2}\left(\cos^2(wt) + \frac{w^2}{\Omega^2}\sin^2(wt)\right)} \\ &= \frac{\hbar}{2}\sqrt{\left(\cos^2(wt) + \frac{\Omega^2}{w^2}\sin^2(wt)\right)\left(\cos^2(wt) + \frac{w^2}{\Omega^2}\sin^2(wt)\right)} \\ &= \frac{\hbar}{4}\sqrt{3 + \frac{1}{2}\left(\frac{\Omega^2}{w^2} + \frac{w^2}{\Omega^2}\right) - \left(\frac{1}{2}\left(\frac{\Omega^2}{w^2} + \frac{w^2}{\Omega^2}\right) - 1\right)\cos(4wt)}\end{aligned} \tag{33}$$

Since $\forall (w,t) \in \mathbb{R}^2 : \cos^2(wt) + \sin^2(wt) = 1$.
and $\cos(4wt) = 2\cos^2(2wt) - 1 = 1 - 2\sin^2(2wt)$.
and $\sin(2wt) = 2\sin(wt)\cos(wt)$.
which are the half angles formulas.
From the relations:
$\frac{\Omega^2}{w^2} + \frac{w^2}{\Omega^2} \geq 2$ and $\forall (w,t) \in \mathbb{R}^2 : -1 \leq \cos(4wt) \leq 1 \Leftrightarrow 0 \leq |\cos(4wt)| \leq 1$.
Then:

$$\sigma_{x_R} \times \sigma_{p_R} \geq \frac{\hbar}{4}\sqrt{3 + \frac{1}{2}\left(\frac{\Omega^2}{w^2} + \frac{w^2}{\Omega^2}\right) - \left(\frac{1}{2}\left(\frac{\Omega^2}{w^2} + \frac{w^2}{\Omega^2}\right) - 1\right)} = \frac{\hbar}{2} \tag{34}$$

which obeys the quantum uncertainty principle in the probability set and universe $\mathcal{R}$.

The uncertainties in the probability set and universe $\mathcal{M}$ in position and momentum (σ_{x_M} and σ_{p_M}) are defined as being equal to the square root of their respective variances in $\mathcal{M}$, so that:

$$\sigma_{x_M} \times \sigma_{p_M} = \sqrt{\text{Var}_{x,M}} \times \sqrt{\text{Var}_{p,M}} \to \sqrt{+\infty} \times \sqrt{+\infty} \to +\infty \tag{35}$$

$\Leftrightarrow \sigma_{x_M} \times \sigma_{p_M} \geq \frac{\hbar}{2}$, in accordance with the quantum uncertainty principle.

The uncertainties in the probability set and universe $\mathcal{C} = \mathcal{R} + \mathcal{M}$ in position and momentum (σ_{x_C} and σ_{p_C}) are defined as being equal to the square root of their respective variances in $\mathcal{C}$, so that:

$$\sigma_{x_C} \times \sigma_{p_C} = \sqrt{\text{Var}_{x,C}} \times \sqrt{\text{Var}_{p,C}} \rightarrow \sqrt{+\infty} \times \sqrt{+\infty} \rightarrow +\infty \tag{36}$$

$\Leftrightarrow \sigma_{x_C} \times \sigma_{p_C} \geq \frac{\hbar}{2}$, in accordance with the quantum uncertainty principle.

Consequently, the quantum uncertainty principle is verified in the universe $\mathcal{R}$, in the universe $\mathcal{M}$, and in the complex universe $\mathcal{C} = \mathcal{R} + \mathcal{M}$.

6. The wavefunction entropies in $\mathcal{R}$, $\mathcal{M}$, and $\mathcal{C}$

Another measure of uncertainty in position is the Shannon information entropy of the probability distribution H_x which is the entropy in $\mathcal{R}$ and is equal to [1–36]:

$$H_x = -\int_0^{+\infty} |\psi(x,t)|^2 Ln\left[|\psi(x,t)|^2 x'_0\right] dt = -\int_0^{U_b} |\psi(x,t)|^2 Ln\left[|\psi(x,t)|^2 x'_0\right] dt \tag{37}$$

Where U_b is the upper bound of the definite integral above. Practically, the standard normal distribution probability is very nearly equal to 1.0000 (0.99997 exactly) for $U_b = 4$.

$= -\sum_{t=0}^{t=U_b} |\psi(x,t)|^2 Ln\left[|\psi(x,t)|^2 x'_0\right] = H_x^R$ with $dt = 0.01$ and where x'_0 is an arbitrary reference length.

Take $x'_0 = 1 \Leftrightarrow H_x^R = -\sum_{t=0}^{t=U_b} |\psi(x,t)|^2 Ln\left[|\psi(x,t)|^2\right]$.

$\Leftrightarrow \forall t : 0 \leq t < +\infty$, we have : $d\left[H_x^R\right] \geq 0$, that means that H_x^R is a nondecreasing series with time t and converging and that also in $\mathcal{R}$, chaos and disorder are increasing with time t.

The negative real entropy corresponding to H_x^R in $\mathcal{R}$ is $NegH_x^R$ and is the following:

$$NegH_x^R = -H_x^R = \sum_{t=0}^{t=U_b} |\psi(x,t)|^2 Ln\left[|\psi(x,t)|^2\right] \tag{38}$$

$\Leftrightarrow \forall t : 0 \leq t < +\infty$, we have : $d\left[NegH_x^R\right] \leq 0$, that means that $NegH_x^R$ is a nonincreasing series with time t and converging. Therefore, if H_x^R measures in $\mathcal{R}$ the amount of disorder, of uncertainty, of chaos, of ignorance, of unpredictability, and of information gain in a random system then since $NegH_x^R = -H_x^R$, that means the opposite of H_x^R, $NegH_x^R$ measures in $\mathcal{R}$ the amount of order, of certainty, of predictability, and of information loss in a stochastic system.

The complementary real entropy to H_x^R in $\mathcal{R}$ is $\overline{H}_x^R$ and is the following:

$$\overline{H}_x^R = -\sum_{t=0}^{t=U_b} \left[1 - |\psi(x,t)|^2\right] Ln\left[1 - |\psi(x,t)|^2\right] \tag{39}$$

In the complementary real probability set to $\mathcal{R}$, we denote the corresponding real entropy by $\overline{H}_x^R$.

The meaning of $\overline{H}_x^R$ is the following: It is the real entropy in the real set $\mathcal{R}$ and which is related to the complementary real probability $P_m/i = 1 - P_r$.

$\Leftrightarrow \forall t : 0 \leq t < +\infty$, we have : $d\left[\overline{H}_x^R\right] \geq 0$, that means that $\overline{H}_x^R$ is a nondecreasing series with time t and converging and that also means that in the complementary real probability set to $\mathcal{R}$, chaos and disorder are increasing with time t.

In the complementary imaginary probability set $\mathcal{M}$ to the set $\mathcal{R}$, we denote the corresponding imaginary entropy by H_x^M. The meaning of H_x^M is the following: It is the imaginary entropy in the imaginary set $\mathcal{M}$ and which is related to the complementary imaginary probability $P_m = i(1 - P_r)$. The complementary entropy to H_x^R in $\mathcal{M}$ is H_x^M and is computed as follows:

$$\begin{aligned}
H_x^M &= -\sum_{t=0}^{t=U_b} i\left[1-|\psi(x,t)|^2\right] Ln\left\{i\left[1-|\psi(x,t)|^2\right]\right\} \\
&= -\sum_{t=0}^{t=U_b} i\left[1-|\psi(x,t)|^2\right]\left\{Lni + Ln\left[1-|\psi(x,t)|^2\right]\right\} \\
&= -\sum_{t=0}^{t=U_b} i\left\{Lni + Ln\left[1-|\psi(x,t)|^2\right] - \left[|\psi(x,t)|^2\right]Lni - \left[|\psi(x,t)|^2\right]Ln\left[1-|\psi(x,t)|^2\right]\right\} \\
&= -\sum_{t=0}^{t=U_b} iLni + iLn\left[1-|\psi(x,t)|^2\right] - i\left[|\psi(x,t)|^2\right]Lni - i\left[|\psi(x,t)|^2\right]Ln\left[1-|\psi(x,t)|^2\right] \\
&= -\sum_{t=0}^{t=U_b} iLni\left[1-|\psi(x,t)|^2\right] + i\left[1-|\psi(x,t)|^2\right]Ln\left[1-|\psi(x,t)|^2\right] \\
&= -\sum_{t=0}^{t=U_b} iLni\left[1-|\psi(x,t)|^2\right] - i\sum_{t=0}^{t=U_b}\left[1-|\psi(x,t)|^2\right]Ln\left[1-|\psi(x,t)|^2\right] \\
&= -\sum_{t=0}^{t=U_b} iLni\left[1-|\psi(x,t)|^2\right] + i\overline{H}_x^R = -iLni\sum_{t=0}^{t=U_b}\left[1-|\psi(x,t)|^2\right] + i\overline{H}_x^R \\
&= -iLni\left\{\sum_{t=0}^{t=U_b} 1 - \sum_{t=0}^{t=U_b}|\psi(x,t)|^2\right\} + i\overline{H}_x^R \\
&= -iLni\{[(U_b - 0) + 1] - 1\} + i\overline{H}_x^R \quad \text{since} \quad \sum_{t=0}^{t=U_b}|\psi(x,t)|^2 = 1 \\
&= -(iLni)U_b + i\overline{H}_x^R
\end{aligned} \tag{40}$$

From the properties of logarithms, we have: $\theta Lnx = Ln\left(x^\theta\right)$ then $iLni = Lni^i$.
Moreover, Leonhard Euler's formula for complex numbers gives:

$$e^{i\theta} = \cos\theta + i\sin\theta \tag{41}$$

Take $\theta = \pi/2 + 2k\pi \Leftrightarrow e^{i(\pi/2+2k\pi)} = \cos(\pi/2 + 2k\pi) + i\sin(\pi/2 + 2k\pi) = 0 + i(1) = i$, then:

$i^i = \left(e^{i(\pi/2+2k\pi)}\right)^i = e^{i^2(\pi/2+2k\pi)} = e^{-(\pi/2+2k\pi)}$ since $i^2 = -1$, therefore:

$-iLni = -Lni^i = -Ln\left[e^{-(\pi/2+2k\pi)}\right] = \pi/2 + 2k\pi$ since $Ln[e] = 1$ and where k belongs to the set of integer numbers $\mathbb{Z}$.

Consequently,

$$H_x^M = -(iLni)U_b + i\overline{H}_x^R = (\pi/2 + 2k\pi)U_b + i\overline{H}_x^R \tag{42}$$

That means that H_x^M is a complex number where:
the real part is: $\text{Re}(H_x^M) = (\pi/2 + 2k\pi)U_b$.
and the imaginary part is: $\text{Im}(H_x^M) = \overline{H}_x^R$.

For $k = -1$ then $\text{Re}(H_x^M) = (-3\pi/2)U_b = -4.71238898U_b = -18.84955592\ldots$ for $U_b = 4$.

For $k = 0$ then $\text{Re}(H_x^M) = (\pi/2)U_b = 1.570796327U_b = 6.283185307\ldots$ for $U_b = 4$.

For $k = 1$ then $\text{Re}(H_x^M) = (5\pi/2)U_b = 7.853981634U_b = 31.41592654\ldots$ for $U_b = 4$.

etc. ...

Finally, the entropy H_x^C in $\mathcal{C} = \mathcal{R} + \mathcal{M}$ is the following:

$$\begin{aligned} H_x^C &= -\sum_{t=0}^{t=U_b} Pc(t)Ln[Pc(t)] \\ &= -\sum_{t=0}^{t=U_b} 1 \times Ln[1] = -\sum_{t=0}^{t=U_b} (1 \times 0) = 0 \\ &= H_x^R + NegH_x^R \end{aligned} \tag{43}$$

$\Leftrightarrow \forall t : 0 \leq t < +\infty$, we have : $d[H_x^C] = 0$, that means that H_x^C is a constant series with time t and is always equal to 0. That means also and most importantly, for the wavefunction position distribution and in the probability set and universe $\mathcal{C} = \mathcal{R} + \mathcal{M}$, we have complete order, no chaos, no ignorance, no uncertainty, no disorder, no randomness, no information loss or gain but a conservation of information, and no unpredictability since all measurements are completely and perfectly deterministic ($Pc(t) = 1$ and $H_x^C = 0$).

Similarly, we can determine another measure of uncertainty in momentum which is the information entropy of the probability distribution H_p and which is:

$$H_p = -\int_0^{+\infty} |\phi(p,t)|^2 Ln\left[|\phi(p,t)|^2 p_0'\right] dt = -\int_0^{U_b} |\phi(p,t)|^2 Ln\left[|\phi(p,t)|^2 p_0'\right] dt \tag{44}$$

where U_b is the upper bound of the definite integral above. Practically, the standard normal distribution probability is very nearly equal to 1.0000 (0.99997 exactly) for $U_b = 4$.

$= -\sum_{t=0}^{t=U_b} |\phi(p,t)|^2 Ln\left[|\phi(p,t)|^2 p_0'\right]$ with $dt = 0.01$ and where p_0' is an arbitrary reference momentum.

For $p_0' = 1$ we can compute similarly all the defined entropies in $\mathcal{R}$, $\mathcal{M}$, and $\mathcal{C}$ and which are:

$$H_p^R = -\sum_{t=0}^{t=U_b} |\phi(p,t)|^2 Ln\left[|\phi(p,t)|^2\right] \tag{45}$$

$$NegH_p^R = \sum_{t=0}^{t=U_b} |\phi(p,t)|^2 Ln\left[|\phi(p,t)|^2\right] \tag{46}$$

$$\overline{H}_p^R = -\sum_{t=0}^{t=U_b} \left[1 - |\phi(p,t)|^2\right] Ln\left[1 - |\phi(p,t)|^2\right] \tag{47}$$

$$H_p^M = -\sum_{t=0}^{t=U_b} i\left[1 - |\phi(p,t)|^2\right] Ln\left\{i\left[1 - |\phi(p,t)|^2\right]\right\} = (\pi/2 + 2k\pi)U_b + i\overline{H}_p^R \tag{48}$$

$$\begin{aligned} H_p^C &= -\sum_{t=0}^{t=U_b} Pc(t)Ln[Pc(t)] = -\sum_{t=0}^{t=U_b} 1 \times Ln[1] = -\sum_{t=0}^{t=U_b} (1 \times 0) = 0 \\ &= H_p^R + NegH_p^R \end{aligned} \tag{49}$$

That means also and most importantly, for the wavefunction momentum distribution and in the probability set and universe $\mathcal{C} = \mathcal{R} + \mathcal{M}$, we have complete order, no chaos, no ignorance, no uncertainty, no disorder, no randomness, no information loss or gain but a conservation of information, and no unpredictability since all measurements are completely and perfectly deterministic ($Pc(t) = 1$ and $H_p^C = 0$).

Due to the Fourier transform relation between the position wavefunction $\psi(x)$ and the momentum wavefunction $\phi(p)$, the above constraint can be written for the corresponding entropies as:

$H_x + H_p \geq Ln\left(\frac{eh}{2x'_0 p'_0}\right)$, where h is Planck's constant.

Depending on one's choice of the $x'_0 p'_0$ product, the expression may be written in many ways. If $x'_0 p'_0$ is chosen to be h, then:

$$H_x + H_p \geq Ln\left(\frac{e}{2}\right) \tag{50}$$

If, instead, $x'_0 p'_0$ is chosen to be $\hbar$, then:

$$H_x + H_p \geq Ln(e\pi) \tag{51}$$

If x'_0 and p'_0 are chosen to be unity in whatever system of units are being used, then:

$$H_x + H_p \geq Ln\left(\frac{eh}{2}\right) \tag{52}$$

where h is interpreted as a dimensionless number equal to the value of Planck's constant in the chosen system of units.

The following figures (**Figures 18–29**) illustrate all the computations done above.

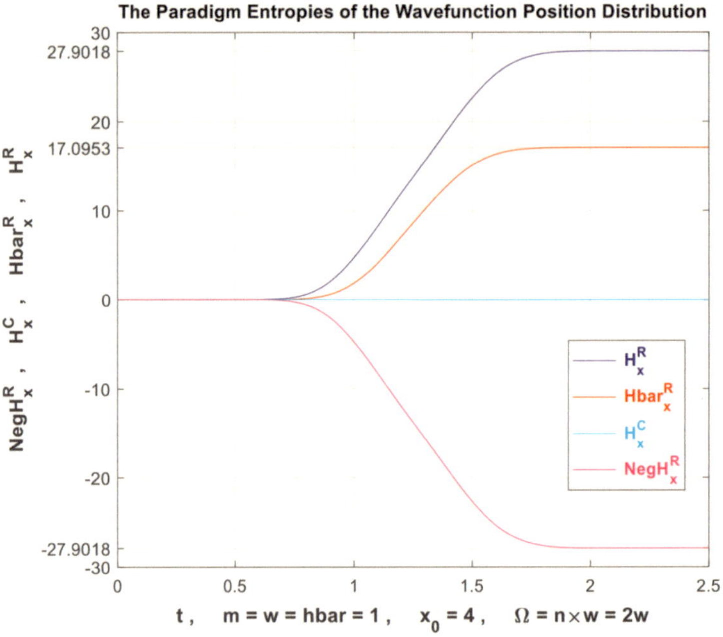

Figure 18.
The graphs of $H_x^R, \overline{H}_x^R, H_x^C, NegH_x^R$ *as functions of* T *for* $n = 2$.

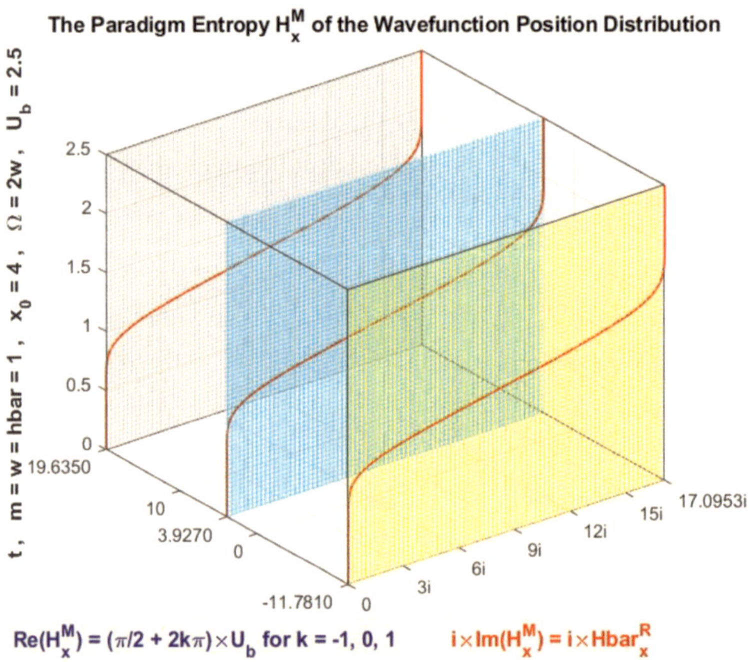

Figure 19.
The graph of $H_x^M = \mathrm{Re}(H_x^M) + i\mathrm{Im}(H_x^M)$ *in red as functions of* T *for* $n = 2$ *and for* $k = -1, 0, 1$ *in the planes in yellow, in cyan, and in light gray, respectively.*

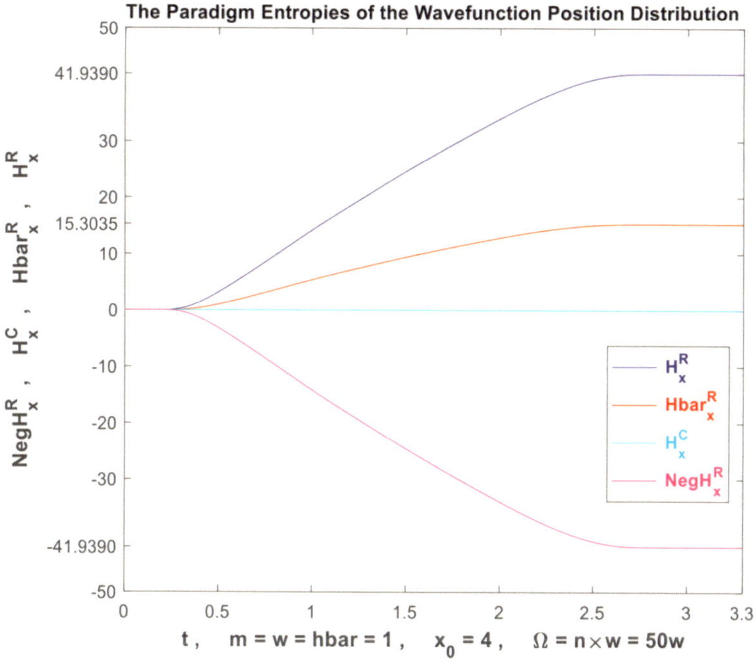

Figure 20.
The graphs of $H_x^R, \overline{H}_x^R, H_x^C, NegH_x^R$ *as functions of* T *for* $n = 50$.

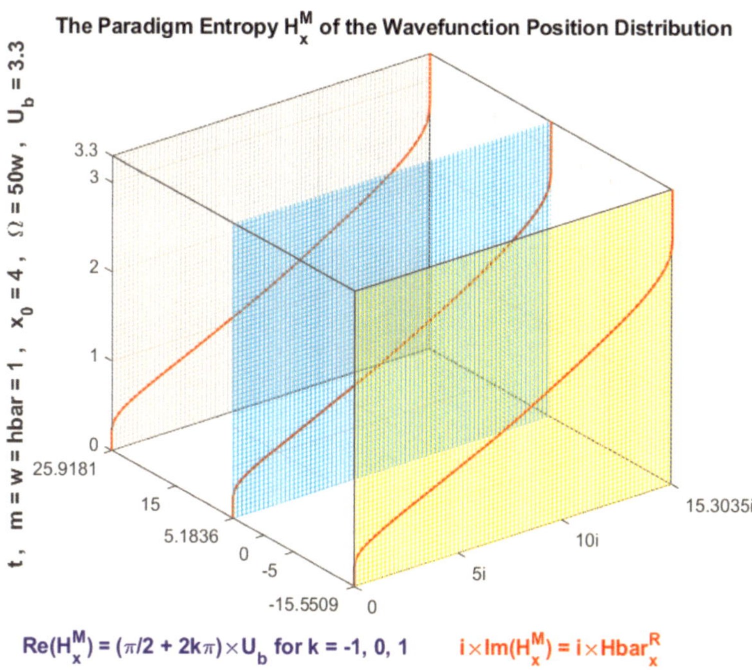

Figure 21.
The graph of $H_x^M = \text{Re}(H_x^M) + i\text{Im}(H_x^M)$ *in red as functions of* T *for* $n = 50$ *and for* $k = -1, 0, 1$ *in the planes in yellow, in cyan, and in light gray, respectively.*

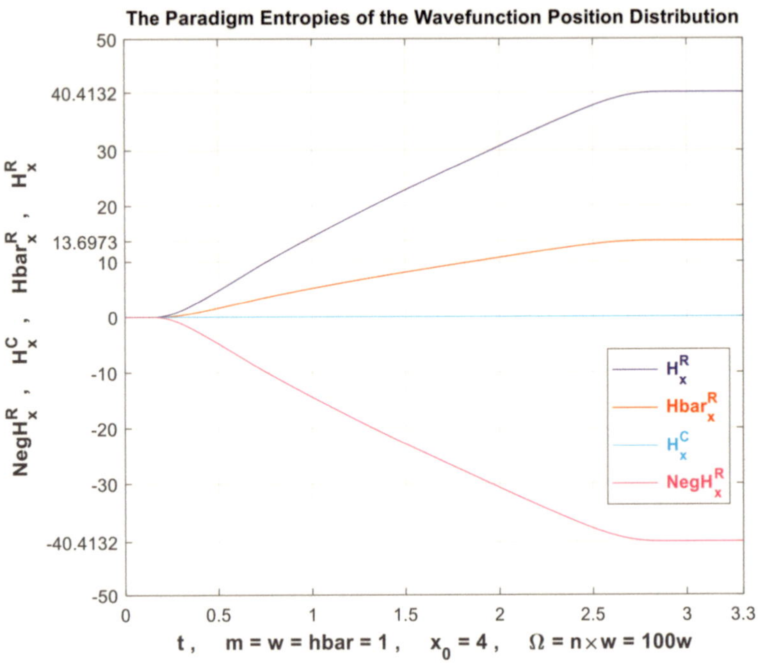

Figure 22.
The graphs of $H_x^R, \overline{H}_x^R, H_x^C, NegH_x^R$ *as functions of* T *for* $n = 100$.

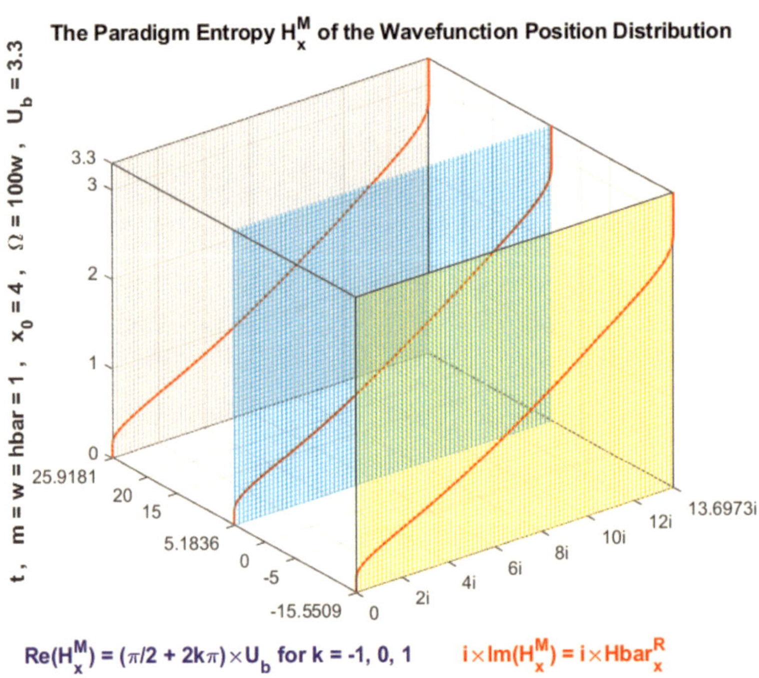

Figure 23.
The graph of $H_x^M = \mathrm{Re}(H_x^M) + i\mathrm{Im}(H_x^M)$ *in red as functions of* T *for* $n = 100$ *and for* $k = -1, 0, 1$ *in the planes in yellow, in cyan, and in light gray, respectively.*

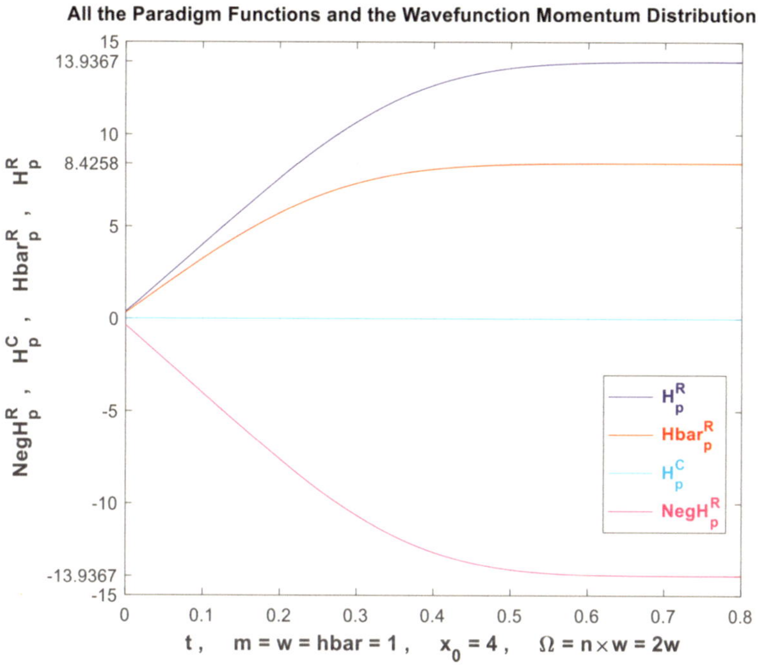

Figure 24.
The graphs of $H_p^R, \overline{H}_p^R, H_p^C, NegH_p^R$ *as functions of* T *for* $n = 2$.

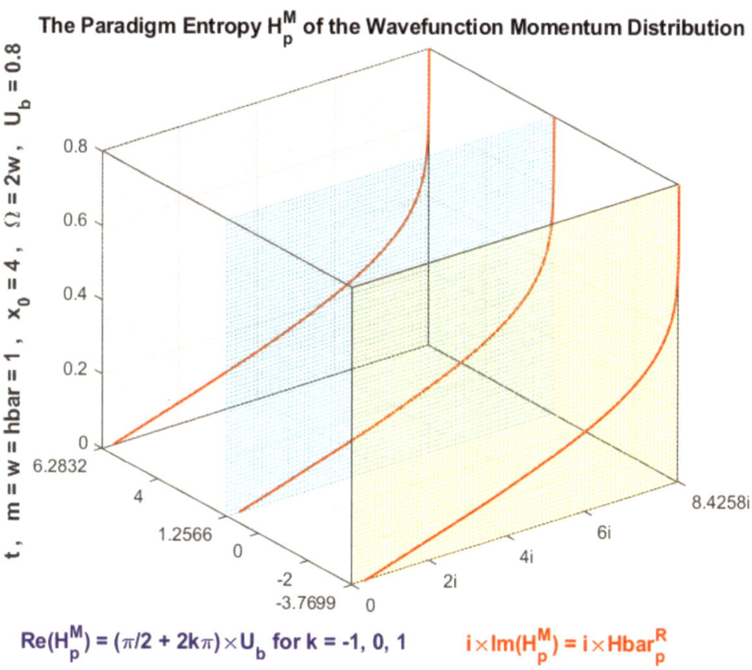

Figure 25.
The graph of $H_p^M = \mathrm{Re}\left(H_p^M\right) + i\mathrm{Im}\left(H_p^M\right)$ *in red as functions of* T *for* $n = 2$ *and for* $k = -1, 0, 1$ *in the planes in yellow, in cyan, and in light gray, respectively.*

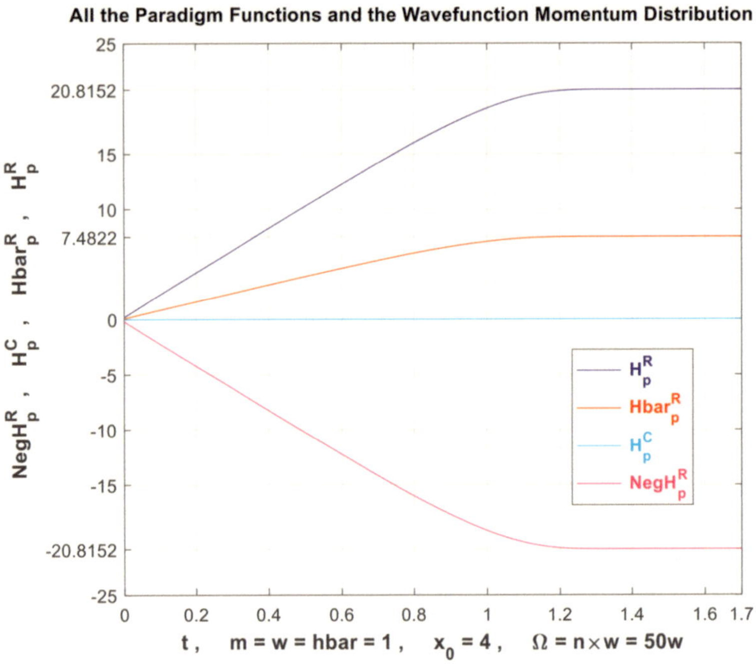

Figure 26.
The graphs of $H_p^R, \overline{H}_p^R, H_p^C, NegH_p^R$ *as functions of* T *for* $n = 50$.

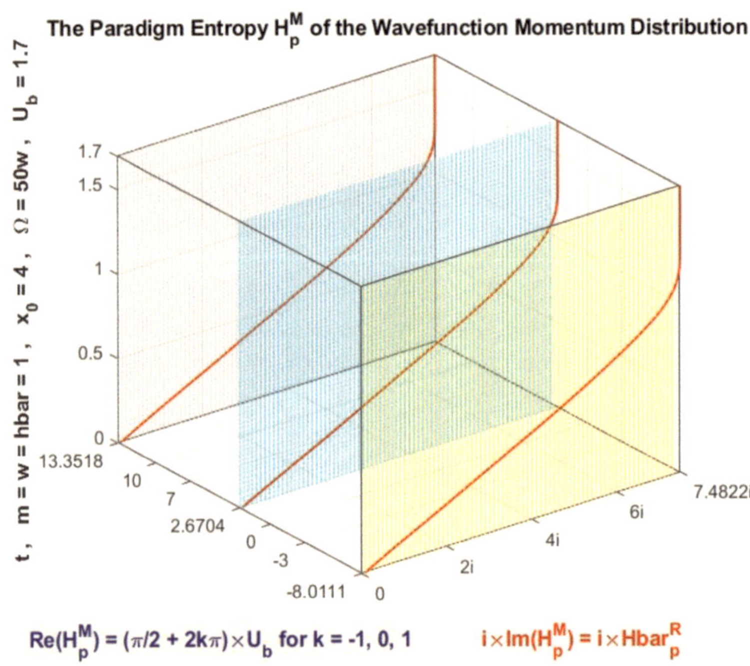

Figure 27.
The graph of $H_p^M = \text{Re}\left(H_p^M\right) + i\text{Im}\left(H_p^M\right)$ *in red as functions of* T *for* $n = 50$ *and for* $k = -1, 0, 1$ *in the planes in yellow, in cyan, and in light gray, respectively.*

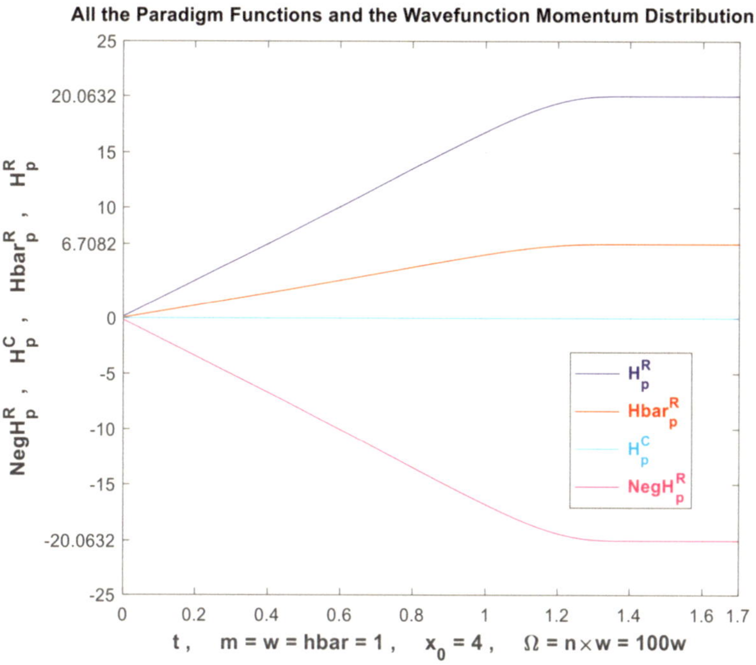

Figure 28.
The graphs of $H_p^R, \overline{H}_p^R, H_p^C, NegH_p^R$ as functions of T *for $n = 100$.*

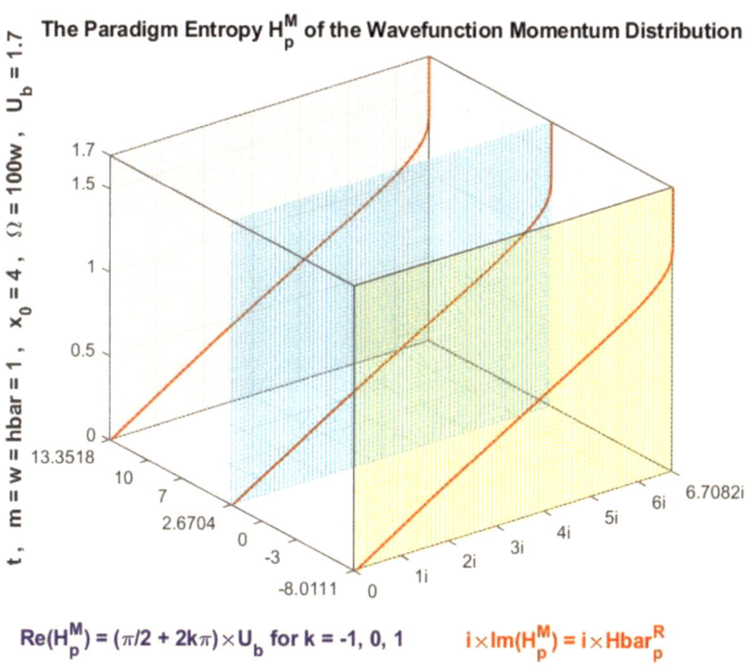

Figure 29.
The graph of $H_p^M = \mathrm{Re}\left(H_p^M\right) + i\mathrm{Im}\left(H_p^M\right)$ in red as functions of T *for $n = 100$ and for $k = -1, 0, 1$ in the planes in yellow, in cyan, and in light gray, respectively.*

7. Conclusion and perspectives

In the current research work, the original extended model of eight axioms (*EKA*) of A. N. Kolmogorov was connected and applied to the quantum harmonic oscillators with Gaussian initial condition problem in quantum mechanics theory. Thus, a tight link between quantum mechanics and the novel paradigm (*CPP*) was achieved. Consequently, the model of "Complex Probability" was more developed beyond the scope of my 23 previous research works on this topic. We can realize that during this whole process we have continually and incessantly $Pc^2 = DOK - Chf = DOK + MChf = 1 = Pc$, that means that the simulation which behaved stochastically and randomly in the real universe and probability set $\mathcal{R}$ is now deterministic and certain in the complex probability universe and set $\mathcal{C} = \mathcal{R} + \mathcal{M}$. This is accomplished by adding to the stochastic phenomenon occurring in the real probability universe $\mathcal{R}$ the contributions of the imaginary probability universe and set $\mathcal{M}$ and hence after subtracting and eliminating from the degree of our knowledge the chaotic factor. Additionally, the real, imaginary, complex, and deterministic probabilities that correspond to each value of the momentum random variable have been evaluated in the three probabilities universes and sets which are $\mathcal{R}$, $\mathcal{M}$, and $\mathcal{C}$ by P_r, P_m, Z and Pc, respectively. Consequently, at each value of time t, the novel quantum mechanics and *CPP* parameters P_r, P_m, P_m/i, DOK, Chf, $MChf$, Pc, and Z are perfectly and surely evaluated and predicted in the complex probabilities universe and set $\mathcal{C}$ with Pc maintained equal to 1 repeatedly and permanently. Furthermore, we are successful to visualize and to quantify both the system certain knowledge (expressed and materialized by *DOK* and *Pc*) and the chaos and stochastic influences and effects (expressed and materialized by *Chf* and *MChf*) of the new paradigm when referring to all these obtained graphs and executed simulations throughout the whole research work. This is certainly very wonderful, fascinating, and fruitful and shows and proves once again the rewards of extending the five axioms of probability of A. N. Kolmogorov and thus the benefits and novelty of my original and novel model in the fields of applied mathematics, prognostics, and quantum mechanics that can be called verily: "The Complex Probability Paradigm." As a prospective and future research and challenges, we aim to more elaborate the novel prognostic paradigm developed and to apply it to a wide set of nondeterministic and stochastic phenomena in quantum mechanics theory.

Conflicts of interest

The author declares that there are no conflict of interests regarding the publication of this work.

Data availability

The data used to support the findings of this study are available from the author upon request.

Nomenclature

$\mathcal{R}$	probabilities and events real set
$\mathcal{M}$	probabilities and events imaginary set

DOI: http://dx.doi.org/10.5772/intechopen.1001985

$\mathcal{C}$	probabilities and events complex set
i	the imaginary number where $i^2 = -1$ or $i = \sqrt{-1}$
EKA	Extended Kolmogorov's Axioms
CPP	Complex Probability Paradigm
P_{rob}	any event probability
P_r	probability of an event in the real set $\mathcal{R}$
P_m	probability of an event in the imaginary set $\mathcal{M}$ associated with the real probability in $\mathcal{R}$
Pc	probability of an event in $\mathcal{R}$ with its corresponding complementary event in $\mathcal{M}$ = probability in the complex probability set $\mathcal{C} = \mathcal{R} + \mathcal{M}$
Z	complex probability number = sum of P_r and P_m = complex random vector
DOK	$= \lvert Z\rvert^2$= the random experiment or system degree of our knowledge, it is the square of the norm of Z
Chf	the chaotic factor of Z
$MChf$	the magnitude of the chaotic factor of Z
$\lvert\psi(x,t)\rvert^2$	probability density function of the position wavefunction
$\lvert\phi(p,t)\rvert^2$	probability density function of the momentum wavefunction
$\langle x\rangle_R, \langle x\rangle_M, \langle x\rangle_C$	averages, or expectations, or means of the position wavefunction probability density function in $\mathcal{R}$, $\mathcal{M}$, and $\mathcal{C}$, respectively
$\text{Var}_{x,R}, \text{Var}_{x,M}, \text{Var}_{x,C}$	variances of the position wavefunction probability density function in $\mathcal{R}$, $\mathcal{M}$, and $\mathcal{C}$, respectively
$\langle p\rangle_R, \langle p\rangle_M, \langle p\rangle_C$	averages, or expectations, or means of the momentum wavefunction probability density function in $\mathcal{R}$, $\mathcal{M}$, and $\mathcal{C}$, respectively
$\text{Var}_{p,R}, \text{Var}_{p,M}, \text{Var}_{p,C}$	variances of the momentum wavefunction probability density function in $\mathcal{R}$, $\mathcal{M}$, and $\mathcal{C}$, respectively
H_x^R	entropy in the real universe $\mathcal{R}$ of the particle position
$NegH_x^R$	negative entropy in the real universe $\mathcal{R}$ of the particle position
$\overline{H}_x^R$	complementary entropy in the real universe $\mathcal{R}$ of the particle position
H_x^M	entropy in the imaginary universe $\mathcal{M}$ of the particle position
H_x^C	entropy in the complex universe $\mathcal{C}$ of the particle position
H_p^R	entropy in the real universe $\mathcal{R}$ of the particle momentum
$NegH_p^R$	negative entropy in the real universe $\mathcal{R}$ of the particle momentum
$\overline{H}_p^R$	complementary entropy in the real universe $\mathcal{R}$ of the particle momentum
H_p^M	entropy in the imaginary universe $\mathcal{M}$ of the particle momentum
H_p^C	entropy in the complex universe $\mathcal{C}$ of the particle momentum

Author details

Abdo Abou Jaoudé
Faculty of Natural and Applied Sciences, Department of Mathematics and Statistics, Notre Dame University-Louaize, Lebanon

*Address all correspondence to: abdoaj@idm.net.lb

References

[1] Wikipedia, the free encyclopedia, Quantum Mechanics. Available from: https://en.wikipedia.org/

[2] Wikipedia, the free encyclopedia, Uncertainty Principle. Available from: https://en.wikipedia.org/

[3] Wikipedia, the free encyclopedia, Quantum Harmonic Oscillator. Available from: https://en.wikipedia.org/

[4] Griffiths DJ. Introduction to Quantum Mechanics. 2nd ed. New Jersey, United States: Prentice Hall; 2004

[5] Liboff RL. Introductory Quantum Mechanics. Boston, United States: Addison–Wesley; 2002

[6] Rashid MA. "Transition Amplitude for Time-Dependent Linear Harmonic Oscillator with Linear Time-Dependent Terms Added to the Hamiltonian" (PDF-Microsoft Power Point). M.A. Rashid – Center for Advanced Mathematics and Physics. Islamabad, Islamabad Capital Territory, Pakistan: National Center for Physics; 2006

[7] Hall BC. Quantum theory for mathematicians. In: Graduate Texts in Mathematics. Vol. 267. New York City, United States: Springer; 2013

[8] Pauli W. Wave Mechanics: Volume 5 of Pauli Lectures on Physics. New York, United States: Dover Books on Physics; 2000

[9] Condon EU. Immersion of the Fourier transform in a continuous Group of Functional Transformations. Proceedings of the National Academy of Sciences of the United States of America. 1937;**23**:158-164

[10] Albert M. Chapter XII, paragraph 15. In: Quantum Mechanics. Mineola, New York, United States: Dover Publication Inc.; 1967. p. 456

[11] Fradkin DM. Three-dimensional isotropic harmonic oscillator and SU3. American Journal of Physics. 1965;**33**(3): 207-211

[12] Mahan GD. Many Particles Physics. New York: Springer; 1981

[13] Arthur B. Quantum Harmonic Oscillator. Hyperphysics. 5th Ed. New York, United States: McGraw-Hill; 2009

[14] Abou Jaoude A, El-Tawil K, Kadry S. Prediction in complex dimension using Kolmogorov's set of axioms. Journal of Mathematics and Statistics, Science Publications. 2010;**6**(2):116-124

[15] Abou Jaoude A. "The Complex statistics paradigm and the law of large numbers", Journal of Mathematics and Statistics, Science Publications. 2013;**9** (4):289-304

[16] Abou Jaoude A. The theory of complex probability and the first order reliability method. Journal of Mathematics and Statistics, Science Publications. 2013;**9**(4):310-324

[17] Abou Jaoude A. Complex probability theory and prognostic. Journal of Mathematics and Statistics, Science Publications. 2014;**10**(1):1-24

[18] Abou Jaoude A. The complex probability paradigm and analytic linear prognostic for vehicle suspension systems. American Journal of Engineering and Applied Sciences, Science Publications. 2015;**8**(1):147-175

[19] Abou Jaoude A. The paradigm of complex probability and the Brownian motion. Systems Science and Control

Engineering, Taylor and Francis Publishers. 2015;**3**(1):478-503

[20] Abou Jaoude A. "The paradigm of complex probability and Chebyshev's inequality", Systems Science and Control Engineering. London, United Kingdom: Taylor and Francis Publishers; 2016;**4**(1):99-137

[21] Abou Jaoude A. "The paradigm of complex probability and analytic nonlinear prognostic for vehicle suspension systems", Systems Science and Control Engineering. London, United Kingdom: Taylor and Francis Publishers; 2016;**4**(1):99-137

[22] Abou Jaoude A. "The paradigm of complex probability and analytic linear prognostic for unburied petrochemical pipelines", Systems Science and Control Engineering. London, United Kingdom: Taylor and Francis Publishers; 2017;**5**(1): 178-214

[23] Abou Jaoude A. "The paradigm of complex probability and Claude Shannon's information theory", Systems Science and Control Engineering. London, United Kingdom: Taylor and Francis Publishers; 2017;**5**(1):380-425

[24] Abou Jaoude A. "The paradigm of complex probability and analytic nonlinear prognostic for unburied petrochemical pipelines", Systems Science and Control Engineering. London, United Kingdom: Taylor and Francis Publishers; 2017;**5**(1):495-534

[25] Abou Jaoude A. "The paradigm of complex probability and Ludwig Boltzmann's entropy", Systems Science and Control Engineering. London, United Kingdom: Taylor and Francis Publishers; 2018;**6**(1):108-149

[26] Abou Jaoude A. "The paradigm of complex probability and Monte Carlo methods", Systems Science and Control Engineering. London, United Kingdom: Taylor and Francis Publishers; 2019;**7**(1): 407-451

[27] Abou, Jaoude A. Analytic prognostic in the linear damage case applied to buried petrochemical pipelines and the complex probability paradigm. In: Fault Detection, Diagnosis and Prognosis. Vol. 1, Chap. 5. London, United Kingdom: IntechOpen; 2020. pp. 65-103. DOI: 10.5772/intechopen.90157

[28] Abou, Jaoude A. The Monte Carlo techniques and the complex probability paradigm. In: Forecasting in Mathematics - Recent Advances, New Perspectives and Applications. Vol. 1. Chap. 1. London, United Kingdom: IntechOpen; 2020. pp. 1-29. DOI: 10.5772/intechopen.93048

[29] Abou Jaoude A. The paradigm of complex probability and prognostic using FORM. London Journal of Research in Science: Natural and Formal (LJRS), London, United Kingdom. Chapter 1. 2020;**20**(4):1-65. DOI: 10.17472/LJRS, 2020

[30] Abou Jaoude A. The paradigm of complex probability and the central limit theorem. London Journal of Research in Science: Natural and Formal (LJRS), London, United Kingdom. Chapter 1. 2020;**20**(5):1-57. DOI: 10.17472/LJRS, 2020

[31] Abou, Jaoude A. The paradigm of complex probability and Thomas Bayes' theorem. In: The Monte Carlo Methods - Recent Advances, New Perspectives and Applications. Vol. 1. Chap. 1. London, United Kingdom: IntechOpen; 2021. pp. 1-44. DOI: 10.5772/intechopen.98340

[32] Abou, Jaoude A. The paradigm of complex probability and Isaac Newton's classical mechanics: On the Foundation of Statistical Physics. In: The Monte

Carlo Methods - Recent Advances, New Perspectives and Applications. Vol. 1, Chap. 2. London, United Kingdom: IntechOpen; 2021. pp. 45-116. DOI: 10.5772/intechopen.98341

[33] Abou, Jaoude A. The paradigm of complex probability and quantum mechanics: The infinite potential well problem - the position wave function. In: Applied Probability Theory - New Perspectives, Recent Advances and Trends. Vol. 1. Chap. 1. London, United Kingdom: IntechOpen; 2022. pp. 1-44. DOI: 10.5772/intechopen.107300

[34] Abou, Jaoude A. The paradigm of complex probability and quantum mechanics: The infinite potential well problem - the momentum Wavefunction and the Wavefunction entropies. In: Applied Probability Theory - New Perspectives, Recent Advances and Trends. Vol. 1. Chap. 2. London, United Kingdom: IntechOpen; 2022. pp. 45-88. DOI: 10.5772/intechopen.107665

[35] Abou Jaoude A. The paradigm of complex probability and the theory of Metarelativity – A simplified model of MCPP. In: Operator Theory – Recent Advances, New Perspectives and Applications. Vol. 1. Chap. 1. London, United Kingdom: IntechOpen; 2023. pp. 1-36. DOI: 10.5772/intechopen.110378

[36] Abou Jaoude A. The paradigm of complex probability and the theory of Metarelativity – The general model and some consequences of MCPP. In: Operator Theory – Recent Advances, New Perspectives and Applications. Vol. 1. Chap. 2. London, United Kingdom: IntechOpen; 2023. pp. 37-78. DOI: 10.5772/intechopen.110377

[37] Benton W. Probability, Encyclopedia Britannica. Vol. 18. Chicago: Encyclopedia Britannica Inc.; 1966. pp. 570-574

[38] Benton W. Mathematical Probability, Encyclopedia Britannica. Vol. 18. Chicago: Encyclopedia Britannica Inc.; 1966. pp. 574-579

[39] Feller W. An Introduction to Probability Theory and its Applications. 3rd ed. New York: Wiley; 1968

[40] Walpole R, Myers R, Myers S, Ye K. Probability and Statistics for Engineers and Scientists. 7th ed. New Jersey: Prentice Hall; 2002

[41] De Broglie L. La Physique Nouvelle et les Quanta. Paris, France: Flammarion; 1937

[42] Feynmann R. Traduction Française: La Nature de la Physique. In: Hélène Isaac, Jean-Marc Lévy-Leblond, Françoise Balibar. Paris: Le Seuil; 1980

[43] Balibar F. Albert Einstein: Physique, Philosophie, Politique. Paris: Le Seuil; 2002

[44] Greene B. The Elegant Universe. New York City, New York, United States: Vintage; 2003

[45] Gribbin J. In: Cassé M, editor. Traduction Française: A la Poursuite du Big Bang. Paris, France: Flammarion; 1994

[46] Gubser SS. The Little Book of String Theory. Princeton, New Jersey, United States: Princeton; 2010

[47] Luminet J-P. Les Trous Noirs. Paris, France: Le Seuil; 1992

[48] Penrose R. The Road to Reality. New York City, New York, United States: Vintage; 2004

[49] Planck M. Traduction Française: Initiations à la Physique. Paris, France: J.

du Plessis de Grenédan, Flammarion; 1993

[50] Poincaré H. La Science et L'Hypothèse. Paris: Flammarion; 1968

[51] Reeves H. Patience dans L'Azur. In: L'évolution cosmique. Paris, France: Le Seuil; 1988

[52] Ronan C. Traduction Française: Histoire Mondiale des Sciences. Paris, France: Claude Bonnafont, Le Seuil; 1988

[53] Sagan C. Traduction Française: Cosmic Connection ou L'appel des étoiles. Paris, France: Vincent Bardet, Le Seuil; 1975

[54] Weinberg S. Dreams of a Final Theory. New York City, New York, United States: Vintage; 1993

[55] Stewart I. Does God Play Dice? 2nd ed. Oxford: Blackwell Publishing; 2002

[56] Barrow J. Pi in the Sky. Oxford: Oxford University Press; 1992

[57] Bogdanov I, Bogdanov G. Au Commencement du Temps. Paris: Flammarion; 2009

[58] Bogdanov I, Bogdanov G. Le Visage de Dieu. Paris: Editions Grasset et Fasquelle; 2010

[59] Bogdanov I, Bogdanov G. La Pensée de Dieu. Paris: Editions Grasset et Fasquelle; 2012

[60] Bogdanov I, Bogdanov G. La Fin du Hasard. Paris: Editions Grasset et Fasquelle; 2013

[61] Bell ET. The Development of Mathematics. New York, United States of America: Dover Publications, Inc; 1992

[62] Boursin J-L. Les Structures du Hasard. Paris: Editions du Seuil; 1986

[63] Dacunha-Castelle D. Chemins de l'Aléatoire. Paris: Flammarion; 1996

[64] Dalmedico-Dahan A, Chabert J-L, Chemla K. Chaos Et Déterminisme. Paris: Edition du Seuil; 1992

[65] Ekeland I. Au Hasard. La Chance, la Science et le Monde. Paris: Editions du Seuil; 1991

[66] Gleick J. Chaos, Making a New Science. New York: Penguin Books; 1997

[67] Davies P. The Mind of God. London: Penguin Books; 1993

[68] Gillies D. Philosophical Theories of Probability. London: Routledge; 2000

[69] Hawking S. On the Shoulders of Giants. London: Running Press; 2002

[70] Abou Jaoude A. The Computer Simulation of Monté Carlo Methods and Random Phenomena. United Kingdom: Cambridge Scholars Publishing; 2019

[71] Abou Jaoude A. The Analysis of Selected Algorithms for the Stochastic Paradigm. United Kingdom: Cambridge Scholars Publishing; 2019

[72] Abou Jaoude A. The Analysis of Selected Algorithms for the Statistical Paradigm, Volume 1. The Republic of Moldova: Generis Publishing; 2021

[73] Abou Jaoude A. The Analysis of Selected Algorithms for the Statistical Paradigm, Volume 2. The Republic of Moldova: Generis Publishing; 2021

[74] Abou Jaoude A. Forecasting in Mathematics - Recent Advances, New

Perspectives and Applications. London, UK, London: United Kingdom: IntechOpen; 2021

[75] Abou Jaoude A. The Monte Carlo Methods - Recent Advances, New Perspectives and Applications. London, UK, London: United Kingdom: IntechOpen; 2022

[76] Abou Jaoude A. Applied Probability Theory – New Perspectives, Recent Advances and Trends. London, UK, London: United Kingdom: IntechOpen; 2023

[77] Abou Jaoude A. Operator Theory – Recent Advances. London, UK, United Kingdom: New Perspectives and Applications. IntechOpen. London; 2023 In Press

[78] Abou Jaoude A. Numerical Methods and Algorithms for Applied Mathematicians [Thesis]. Spain: Bircham International University; 2004

[79] Abou Jaoude A. Computer Simulation of Monté Carlo Methods and Random Phenomena [Thesis]. Spain: Bircham International University; 2005

[80] Abou Jaoude A. Analysis and Algorithms for the Statistical and Stochastic Paradigm [Thesis]. Spain: Bircham International University; 2007

Section 2

Simulation and Physics

Chapter 3

Gravity in the Early Universe

Seddigheh Rezapour

Abstract

According to scientific evidence, the universe began after the "Big Bang" with a very high temperature and energy density. At early times, the temperature was certainly so high that all particles such as quarks, leptons, gluons, and so forth were strongly relativistic. Even strongly interacting quark and gluon particles interact relatively weakly due to asymptotic freedom. Therefore, it was only a system of hot and weak particles that can be called Quark-Gluon Plasma. As the universe cooled during the next phase of expansion, quarks, antiquarks, and gluons combined to form hadrons, leading to the formation of baryonic matter, the phase transition of quantum color dynamics (QCD), during which quarks and gluons are confined. In the first-order cosmic phase transition with the formation of bubbles, the transition to a true vacuum occurs. The bubbles expand and collide with each other, and part of the energy stored in the walls of the bubbles turns into gravitational waves.

Keywords: early universe, Quark-Gluon Plasma, gravity, gravitational waves, cosmic phase transition, QCD phase transition

1. Introduction

Discovering the mystery of the universe from the beginning and how it evolved and expanded is still interesting to people. The Big Bang model provides a general framework for describing the evolution of the universe and has been able to provide several successful fundamental predictions, including the existence of cosmic microwave background radiation and the abundance of light elements in the universe. It is also possible to obtain other indirect information from the early universe through the phase transition that occurred in it and their effects.

In the evolution of the early universe, there were at least two phase transitions. The electroweak theory predicts that at about 100 GeV, a transition from the symmetric high-temperature phase to the broken symmetry phase has occurred, in which the SU (2) $\times$ U(1) dimensional symmetry is spontaneously broken and the particle mass, which is proportional to the vacuum expectation of the Higgs, becomes nonzero. Also, quantum color dynamics (QCD) predicts that at an energy of 100 MeV, there is a phase transition from quark-gluon plasma to a confinement phase without free quarks and gluons. At approximately the same energy, we expect that the global chiral symmetry of the QCD with massless fermions is spontaneously broken by the formation of dense quark pairs. Cosmic phase transitions, especially QCD and electroweak phase transitions, may be first-order phase transitions under some conditions [1].

First-order phase transitions in the early universe can be powerful sources of gravitational wave radiation that can suggest a new way to explore the early universe. In a first-order phase transition, the universe starts in a quasi-stable high-temperature phase (symmetric phase) and turns into a stable low-temperature phase (broken symmetry phase). This transition continues through formation of low-temperature phase bubbles in the high-temperature phase. These bubbles then expand and merge. Finally, the universe remains in the phase of broken symmetry.

In a cosmological first-order phase transition that occurs in a thermal bath, bubbles form and expand due to the released vacuum energy. In the hydrodynamic description of the bubble evolution, the bubble velocity, v_b, is an important parameter that affects the production of gravity waves caused by this process. The realistic background is actually a thermal plasma full of relativistic particles that collide with the bubble wall and produce friction. The main problem of bubble expansion in plasma is to understand the relationship between the following quantities, by which the characteristics of the resulting gravitational waves can be known: The velocity of the bubble wall v_b, the friction on the wall from the plasma side; the phase transition intensity α, which measures the energy density of the released vacuum compared to the radiation energy density; and the factors κ and κ_v, which calculate the ability to transfer the energy of the released vacuum to bubble wall expansion, and the fraction of energy converted to the plasma motion, respectively. All parameters are measured at transition temperature. These quantities are used as an interface between theoretical models of particle physics and bubble collision simulations. The bubble collision is one of the sources of gravitational waves [2].

Gravitational waves from first-order phase transitions propagate gravitational field fluctuations and move at the speed of light. Any object in the path of the wave receives the tidal force that is applied perpendicular to the direction of wave propagation. Gravitational waves, after being produced, propagate without any restrictions. It has been proven that these waves are harder to stop than neutrinos. An important change that occurs as a result of their propagation is the decrease in amplitude when moving away from the source. These waves can be redshifted like electromagnetic waves.

In the QCD phase transition, leptonic asymmetry may lead to the generation of a finite bariochemical potential. In the strong baryon generation mechanism where a high baryochemical potential is produced, a short period of supercooling may occur during the QCD phase transition and lead to a dilution of the baryon density. Because in these conditions, the vacuum energy may be much greater than the thermal energy of the universe, and in this case, the plasma becomes very diluted. In this scenario, the phase transition occurs through the expansion and cooling of the universe, which results in the generation of gravitational waves. The phase transition proceeds through the formation of stable and low-energy phase bubbles inside the supercooled phase. These bubbles eventually expand and collide, causing gravitational waves [3].

2. Big bang and the early universe

In 1926, a Belgian priest named Georges Lemaitre proposed the theory that the universe started from a primordial atom. In 1929, the astronomer Edwin Hubble, by analyzing the redshift of the light of the galaxies, observed that they are moving away from each other in all directions. The expansion of the universe reinforces the idea that galaxies must have been closer together in the past when they are increasing in distance today. This theory of the expanding universe eventually led to what is now called the Big

Bang model. The framework of this model is based on Einstein's theory of general relativity and has been able to successfully explain many phenomena of the universe [4].

The evolution of the world is explained based on the Friedmann-Robertson-Walker (FRW) cosmological model or the Big Bang model, which includes the expansion of the world and homogeneous and isotropic approximations of the distribution of matter and energy on a large scale. This model has been so successful that it is called the Standard Model of Cosmology.

Assume a uniform distribution of matter. An observer who looks at the world from all sides and in all directions will find the same world. These two characteristics are called homogeneity and isotropy. According to Einstein's equations of general relativity, the space-time that includes the mentioned features is given by the Friedman-Robertson-Walker metric:

$$ds^2 = dt^2 - a^2(t)\left[\frac{dr^2}{1-kr^2} + r^2 d\theta^2 + r^2 \sin^2\theta \, d\varphi^2\right] \tag{1}$$

where $k = 0$ is space with zero or Euclidean curvature, $k = 1$ is closed space with positive curvature, and $k = -1$ is open space with negative curvature. At $k = 0$, the Robertson-Walker time space is not flat, but only its t = const subspaces are flat. The variable $a(t)$, which is called cosmic scale factor or expansion factor, is dimensionless.

The general relativity field equations published in 1915 were as follows:

$$G_{\mu\nu} \equiv \mathcal{R}_{\mu\nu} - \frac{1}{2} g_{\mu\nu}\mathcal{R} = -\kappa T_{\mu\nu} \tag{2}$$

where $G_{\mu\nu}$ is the Einstein tensor, $\mathcal{R}_{\mu\nu}$ is the Ricci curvature tensor, $g_{\mu\nu}$ is the metric tensor and $\mathcal{R}$ is the scalar curvature, κ is the Einstein gravitational constant, and $T_{\mu\nu}$ is the stress-energy tensor that describes the matter-energy distribution on a large scale. Stress-energy tensor is a diagonal tensor whose elements are relativistic mass density and isotropic pressure. From Eqs. (1) and (2), the relation $p = \omega\rho$ is obtained, where w is the state parameter.

When $\omega = -1$, the energy of the universe was made up of vacuum energy. $\omega = 1/3$ corresponds to the condition that the energy of the universe is determined by relativistic particles. It means that radiation dominates the universe. In this case, the factor of 1/3 is due to curving in all directions. $\omega = 0$ describes the condition where nonrelativistic particles or matter is dominant in the universe. In this case, the matter of the universe behaves as dust and the pressure is zero. This condition has been true for recent era.

From the combination of Einstein and FRW equations, the following relations can be reached:

$$\frac{d}{da}\left(\rho a^3\right) + 3pa^2 = 0 \Longrightarrow \rho \propto a^{-3(1+\omega)} \tag{3}$$

The Big Bang era starts with $a = 0$. At this point the cosmic time is zero. Since the singularity means that some terms become infinite and others become zero. As a result, the concept of space-time geometry is violated. Therefore, physicists cannot explain what happened at that moment or before. In other words, the theory of general relativity is not complete, and maybe there is a more complete theory that provides a better explanation, such as hypotheses based on string theory and avoid this space-time singularity [5].

Some theories and hypotheses that try to avoid space-time singularity are:

A theoretical framework of ring quantum gravity that quantizes space and time, leading to the idea that spacetime is granular at the smallest scales, potentially avoiding singularities.

String theory proposes that the fundamental building blocks of the universe are not particles, but tiny strings that can potentially resolve singularities.

Modified theories of gravity that modify general relativity on large scales and aim to avoid singularities without invoking quantum effects.

About 10^{-43} seconds after the Big Bang, the dynamics of the universe is described by the theory of quantum gravity, where the quantum effects of gravity are important. Experimental studies have shown that with increasing energy, the intensity of electromagnetic and weak interactions becomes comparable and the electroweak theory becomes dominant. This prediction was made for energies around 10^{15} GeV, approximately 10^{-36} seconds after the Big Bang, which is called the Grand Unification Epoch. In this era, electroweak and strong interactions are unified.

Quarks and leptons become members of a common irreducible group; for example, quarks are converted into a lepton by boson exchange. If this theory is correct, the difference between these three interactions is due to the fact that we are working at low energies, in which case the unification of forces is lost.

Today's observational evidence confirms the exponential growth at this stage of the world's expansion. As the temperature decreases around 100 GeV, approximately 10^{-10} s after the Big Bang, the weak and electromagnetic forces are separated. At times around 10^{-5} after the big bang, when the temperature was around 100 MeV, quarks could exist in their bound states, i.e. baryons and mesons, and then protons and neutrons were able to come together and in energy around 1 MeV formed light nuclei [4].

3. Quark-gluon plasma

The universe began with extremely high temperature and energy density. At early times, the temperature was about 100 GeV. At such a temperature, all particles are relativistic so that quarks and gluons, which are strongly interacting particles, interact relatively weakly due to their asymptotic freedom. Thus, a system of weakly interacting hot particles called quark-gluon plasma (QGP) was in equilibrium with other species.

The interaction between quarks becomes vanishingly weakened as they get closer to each other at high density; these quarks are no longer confined inside the hadrons and become free. This process is called asymptotic freedom.

At high temperature, due to the asymptotic freedom, the quark-gluon plasma can be described as the simplest strongly interacting particle system in the context of QCD. As the universe cooled during the next phase of expansion, quarks, antiquarks, and gluons combined to form hadrons, the result of which is baryonic matter. The cooling scenario of the universe presents first- or second-order phase transitions associated with spontaneous symmetries breaking [6–8].

Phase transition at temperatures of more than 100 GeV causes the spontaneous breaking of the electroweak symmetry, providing masses to elementary particles. It is also related to the electroweak baryon-number-violating processes, which had a major influence on the observed baryon asymmetry of the universe [9].

The next phase transition is the transition from QGP to hadronic matter, which occurs at $T < 200$ MeV. This transition is related to the spontaneous breaking of the

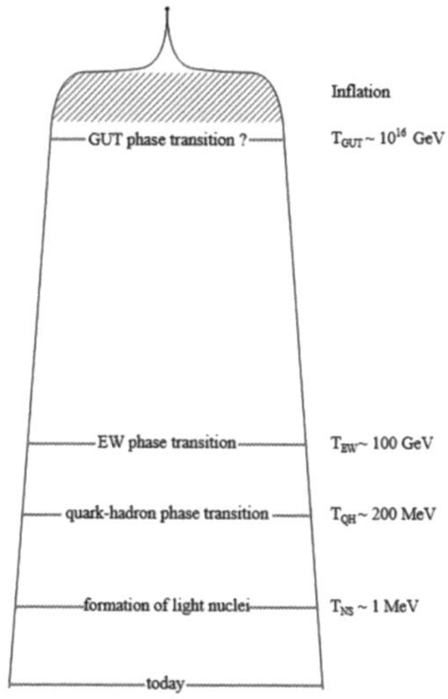

Figure 1.
Schematic diagram the history of the universe [10].

chiral symmetry strong interactions [8]. In a strong first-order phase transition, the QGP supercools before bubbles of hadron gas are formed. Since the hadronic phase is the initial condition for nucleosynthesis, the inhomogeneities in this phase could have a strong effect on the nucleosynthesis epoch [8]. **Figure 1** shows in schematic diagram the history of the universe, beginning with the Big Bang and cooling as the universe expanded. The increase in time is accompanied by a decrease in temperature, which is given by the relation $t = \left(\frac{3}{16\,\pi G g_* A}\right) T^{-2}$, where A is the radiation constant and g_* is the total number of relativistic degrees of freedom. But in the quark-hadron epoch, the equation of state of matter becomes complicated.

4. Cosmic first-order phase transition

The transition between physical states of matter is called phase transition. If the phase transition is accompanied by sudden changes in entropy and definite volume, then it is a first-order phase transition, which usually occurs in the change of state of matter and fundamental alteration in the crystal structure of solids. The first-order phase transition equations are given in Eq. (4):

$$T1 = T2, P1 = P2, G1(P, T) = G2(P, T) \tag{4}$$

where T, P, and G are temperature, pressure, and Gibbs thermodynamic potential, respectively. Using Eq. (4), one can obtain the Clausius-Clapeyron equation, which determines the slope of the phase equilibrium curves:

$$\frac{\Delta P}{\Delta T} = \frac{\Delta S}{\Delta V} \tag{5}$$

In this equation, ΔS and ΔV are entropy changes and volume changes in phase transition, respectively. It is possible to consider the phase transition as a function of volume using the Helmholtz free energy.

Standard cosmology predicts several phase transitions for the evolution of the universe. One of the results of these transitions is the production of gravitational waves. But only if the phase transition is of the first-order type, gravitational waves will be produced. In the cosmic first-order phase transition, tunneling to the new phase is done by creating bubbles. The expansion of the bubbles and their collision with each other produce gravitational waves [11].

If the system starts from the highest Helmholtz free energy level at high temperatures and cools rapidly to below T_c, it is trapped in the high free energy phase. In this case, the system is in a quasi-stable phase. The transition to the lowest free energy phase is discontinuous and occurs through the formation of low free energy phase bubbles in the quasi-stable phase, which is an example of a discontinuous or first-order phase transition. The bubble structure shows a coherent oscillation of the field, which is induced by thermal or quantum effects.

If σ is the surface density and ΔF is the free energy difference between the two phases, the critical radius is $R_c = 2\sigma/\Delta F$, which is suitable for configuration or bubble growth. If $R_c < R$, surface tension prevails and the bubble shrinks. If $\Delta F = 0$, R_c tends to infinity. R_c also determines the free energy barrier for critical bubble formation. If bubbles form with $R_c > R$, they will grow and coalesce.

The expansion rate of the universe compared to the time scale of particle interaction is the factor that determines the rate of temperature drop. The expansion rate of the universe for a period of dominant radiation can be obtained from Friedman's equations, and the curvature constant is considered $k = 0$:

$$\frac{\dot{R}}{R} = H \equiv 1.66\, g_*^{\ 1/2} \frac{T^2}{m_{Pl}}, \tag{6}$$

In cosmic expansion, all types of particles are in thermal equilibrium until the rate of interaction exceeds the rate of expansion, $\Gamma_{int} > H$. At very high energies, until $k_B T \leq 10^{16-17}$, GeV thermal equilibrium is established.

As the universe expands and cools, the scalar field, which is the cause of symmetry breaking, is trapped in a quasi-steady state. In this case, the energy density of matter at high temperatures includes two terms, one of which comes from relativistic radiation and the other from vacuum energy:

$$\rho(T) = \frac{\pi^2}{30} g_* T^4 + V_0 \tag{7}$$

V_0 is the vacuum energy density that is trapped in the quasi-stable minimum. If the temperature is $T < (30\, V_0/g_*\, \pi^2)^{1/4}$, the constant vacuum energy dominates the energy density and the scale factor increases exponentially and rapidly. That is why it is called inflation. The universe is expanding at a speed close to the speed of light, and supercooling occurs. A sudden drop in temperature will make quantum effects dominate over thermal effects. Finally, the quantum bubble formation mechanism collapses to the lowest energy state, and the phase transition is complete by merging the bubbles [12].

5. Cosmic phase transition QCD

Quantum chromodynamics (QCD) is a suitable theoretical framework for describing strong interactions. This theory is a part of the standard model of particle physics that describes the physics of quarks and gluons. It also explains mechanisms such as confinement, chiral symmetry breaking, and asymptotic freedom [13].

QCD phase transition usually occurs in time $t \approx 10^{-5}$ s and temperature $T \approx 150$ MeV. This transition results from the temperature evolution of the strong coupling constant gs. At a temperature higher than the transition temperature, the coupling constant is so small that the behavior of the system is disordered and the system is in the phase of quark-gluon plasma.

As the universe cools, the strong coupling constant grows and quarks and gluons confined into colorless hadrons. All multicolors are enclosed in monochromatic baryons. The QCD phase diagram in terms of temperature and baryonic chemical potential is shown in **Figure 2**.

The QCD phase diagram in **Figure 2** is predicted for different phases of nuclear matter, which shows the temperature value for different barochemical potential. The bariochemical potential describes the energy required to add or remove a baryon at constant pressure and temperature. At small chemical potential and high temperatures, matter is in the quark-gluon plasma phase. The early universe went through this phase in the first few microseconds after the Big Bbang. At low temperatures and high baryon densities, such as in neutron star cores, color superconducting phases are predicted. The phase transition between quark-gluon plasma and ordinary hadron gas appears continuous for small chemical potentials (dashed line). However, a critical point appears at higher potential values [15].

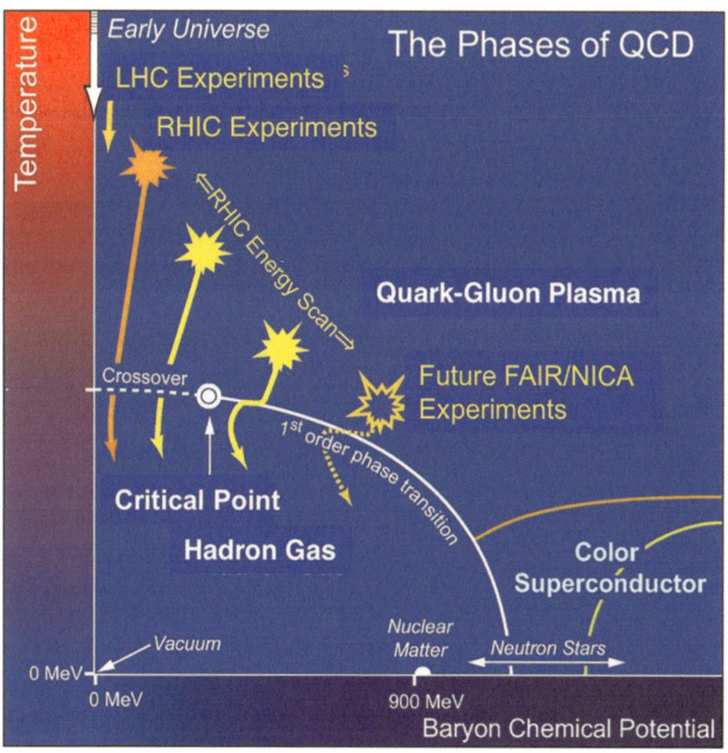

Figure 2.
A schematic QCD phase diagram in the temperature (T) and chemical potential (μ). The solid lines show the phase boundaries for the indicated phases. The solid circle depicts the critical point [14].

Numerical calculations show that the first-order QCD phase transition occurs at finite temperature for very small and large quark masses [16]. However, for the average mass of quarks with zero chemical potential, especially for three light quarks and small chemical potential, the first-order phase transition does not occur [17]. QCD has approximate chiral and central symmetries for very small and large quark masses, respectively. The remainder of these symmetries are spontaneously broken, and the corresponding order parameters become nonzero at all temperatures.

The confinement-deconfinement phase transition at the critical temperature T_c is a first order phase transition in pure Yang–Mills theory. While at $\mu = 0$ above the temperature T_c, suggest a crossover in full quantum chromodynamics. At temperatures below T_c, quarks and gluons are confined into hadronic bound states, which are the effective degrees of freedom of low-energy QCD.

In the QCD phase diagram, a little inflation starts when the baryon to photon ratio has a larger initial value. For massless particles, this value is proportional to the ratio of barochemical potential to temperature so that $\mu/T \sim 1$. Such a large quantity is produced by some primary disequilibrium processes. Then, in the early universe, the conditions are the same as the first-order phase transition line in QCD large barochemical potentials, where the universe is trapped in a pseudo-vacuum state. The constant and nonzero vacuum energy density leads to inflationary expansion, which is accompanied by exponential supercooling and dilution of matter.

During this evolution, the universe is not in equilibrium, while the temperature and barochemical potential decay exponentially with μ/T = const. At low temperatures, the true vacuum barrier becomes so low that the universe spins into a true vacuum. The latent heat released is used to produce particles and eventually causes the world to reheat to a temperature of $T \sim Tc$.

During the reheating, the baryon number is constant so that the net baryon density still decreases compared to the initial state before inflation, Today, the observed value of μ/T is approximately 10^{-9}. After that, standard cosmological evolution continues the Big Bang nucleation process [18].

According to **Figure 2**, parts of the phase diagram have been investigated by several accelerator-based experimental programs. Experience has shown that T and μ change as a function of the energy of the center of mass [19]. This plan will be followed by experimental programs such as LHC, RHIC, FAIR, and NICA. Among these experiments, the RHIC results provide evidence of QGP formation.

6. Bubble nucleation rate

The bubble nucleation rate per unit volume, Γ, becomes zero at temperatures $T \geq T_c$. In other words, the phase transition does not occur exactly at $T = T_c$. At temperatures $T < T_c$, the nucleation rate grows continuously from $\Gamma = 0$ and may reach a value of $\Gamma \sim {T_c}^4$. As a result, the entire universe becomes a stable phase. The peak bubble nucleation rate determines the nucleation temperature T_n, $T_n < T_c$. At this temperature, the nucleation rate is extremely large.

For weak first-order phase transition, although $T_n < T_c$, the temperature T_n is very close to T_c. However, for a stronger phase transition, we have a lower temperature or even in $T = 0$. In such a case, there is no temperature T_n and the phase transition can take longer.

The time taken to complete the transition can be expressed as the transition rate β. The value of β is usually positive. Hence, the nucleation rate grows monotonously with time. The parameter β determines the time scale of the phase transition with the approximation $\Gamma = \Gamma_0 e^{\beta(t-t_0)}$. Therefore, the duration is given by $\Delta t \sim \beta^{-1}$, and the typical bubble size is given by $R_b \sim v_w \beta^{-1}$, where v_w is the bubble wall velocity. If the duration of the phase transition is short enough, the bubble radius can be assumed to be constant.

In fact, the progress of the phase transition depends on the relationship between Γ and the expansion rate H. In other words, bubble nucleation becomes important when Γ is comparable to the Hubble rate [11], since the number of bubbles, N, nucleated in the volume $V \sim H^{-3}$ in a cosmic time $t \sim H^{-1}$ will be $N \sim \Gamma H^{-4}$ and H is approximately given by $H \sim T^2/M_{Pl}$, where M_{Pl} is the Planck mass. Therefore, for T close enough to T_c, the bubble nucleation will be very slow, that is $\Gamma \ll H^4$. The phase transition generally occurs at an intermediate temperature between T_n and T_c so that $\Gamma H^{-4} \sim 1$, and due to the rapid growth of Γ, ends at a time $\Delta t \ll t$ [20].

The nucleation process transfers part of the vacuum energy to the bubble wall and leads to an increase in the wall velocity [21]. For a very strong phase transition, the wall velocity is $v_w \simeq 1$, but for a weak phase transition, the wall velocity is obtained as a function of T_n. Of course, these calculations may be different in different models and regimes. For example, in Jouguet detonation regime, the wall speed is calculated from Eq. (8) [22]:

$$v_w = \frac{\sqrt{\alpha^2 + \frac{2\alpha}{3}} + \sqrt{\frac{1}{3}}}{1+\alpha} \tag{8}$$

In Eq. (8), the quantity α is the ratio of the vacuum energy density to the radiation energy density, which is calculated from relations Eq. (9):

$$\alpha = \frac{\varepsilon}{\rho_R(T_n)}, \rho_R(T_n) = \frac{\pi^2}{30} g_* {T_n}^4 \tag{9}$$

The quantity ε represents the vacuum energy density, which is calculated using Eq. (10) for ΔF, where ΔF is the free energy difference.

$$\varepsilon = \left(\Delta F(T) - T\frac{d\Delta F(T)}{dT}\right)_{T=T_n} \tag{10}$$

Since gravitational waves are generated during the first-order phase transition, it is very important to calculate the values that play a key role in calculating the gravitational wave spectrum of the early universe. These values are more than the difference of free energy in two phases, vacuum energy density, ratio of vacuum energy to radiant energy, core temperature and wall velocity. In the next section, we will discuss the spectrum of gravitational waves.

7. Gravitational wave spectrum

The discovery of gravitational waves is a promising way to investigate the events of the early universe. One of the sources of the production of these gravitational waves is the cosmic first-order phase transition. When a first-order phase transition occurs in a thermal bath, bubbles are formed. Then, due to the release of the initial phase vacuum energy, the bubbles in the plasma expand and collide with each other. During the collision of the bubbles, their spherical symmetry is broken and part of the stored energy leads to the generation of gravitational waves [13]. For the transition of the first-order phases, during the formation of the bubble, three gravitational wave sources have been proposed: bubble collision, sound waves, and turbulent motion of the fluid. To calculate the energy density spectrum of gravitational waves, the sum of these three sources can be considered:

$$h^2\Omega(f) \simeq h^2\Omega_{Col}(f) + h^2\Omega_{Sw}(f) + h^2\Omega_{Tu}(f) \tag{11}$$

where the coefficient h is the current Hubble parameter in unit of 100 km/(s MPc). In the following, the contribution of three sources in the energy density of gravitational waves produced during a first-order phase transition is expressed.

The contribution of the collision of bubbles in the frequency spectrum of gravity waves using the envelope approximation has been calculated by numerical simulations as follows [23]:

$$h^2\Omega_{Col}(f) = 1.67 \times 10^{-5}\left(\frac{0.11{v_w}^3}{0.42 + {v_w}^2}\right)\left(\frac{H_*}{\beta}\right)^2\left(\frac{\kappa\alpha}{1+\alpha}\right)^2\left(\frac{100}{g_*}\right)^{\frac{1}{3}} S_{en}(f) \tag{12}$$

The κ parameter is a fraction of the vacuum energy that has been converted into the kinetic energy of the bubbles. In other words, for runaway bubbles, the efficiency factor of the bubble collision source is given by $\kappa = 1 - \frac{\alpha_\infty}{\alpha}$, α_∞ that is a critical value of α. The Hubble parameter at T_* is donated by $H_* = 1.66\sqrt{g_*}{T_*}^2/M_{Pl}$ where T_* is the temperature at which gravitational waves are produced. The spectral shape of the gravitational wave is analytically fitted as:

$$S_{en}(f) = \frac{3.8\left(\frac{f}{f_{en}}\right)^{2.8}}{1 + 2.8\left(\frac{f}{f_{en}}\right)^{3.8}} \tag{13}$$

The peak frequency of redshift is given by Eq. (14):

$$f_{en} = 16.5 \times 10^{-6}(Hz)\left(\frac{0.62}{1.8 - 0.1v_w + {v_w}^2}\right)\left(\frac{\beta}{H_*}\right)\left(\frac{T_*}{100\ GeV}\right)\left(\frac{g_*}{100}\right)^{\frac{1}{6}} \tag{14}$$

After the collision of the bubbles and before the expansion in the plasma, sound waves are formed and receive the kinetic energy in the plasma. Due to their distinct peak frequency distribution and the fact that they are potentially a long-period gravitational wave source, they have attracted special attention. In general, these sound waves can lead to a significant gravitational wave signal [24]. Gravitational wave spectrum resulting from sound waves has been calculated based on numerical methods as follows:

$$h^2\Omega_{Sw}(f) = 2.65 \times 10^{-6}\left(\frac{H_*}{\beta}\right)\left(\frac{\kappa_{sw}\alpha}{1+\alpha}\right)^{100}\left(\frac{100}{g_*}\right)^{\frac{1}{3}} v_w S_{sw}(f) \tag{15}$$

The value of $\kappa_{sw} = (1-\delta)\kappa_v$ is a fraction of the latent heat that is converted into the movement of the fluid and depends on the state of expansion of the bubble and δ can be of the order of δ = 0.1. Also, κ_v is a part of energy and has been transformed into plasma motion that is given by $\kappa_v = \frac{\alpha_\infty}{\alpha}\frac{\alpha_\infty}{0.73+0.083\sqrt{\alpha_\infty}+\alpha_\infty}$ and for non-runaway bubbles is expressed as $\kappa_v = \frac{\alpha}{0.73+0.083\sqrt{\alpha}+\alpha}$.

The shape of the spectrum S_{sw} and the peak frequency of the redshift f_{sw} are given by [25]:

$$S_{Sw}(f) = \left(\frac{f}{f_{sw}}\right)^3 \left(\frac{7}{4+3\left(\frac{f}{f_{sw}}\right)^2}\right)^{\frac{7}{2}}, f_{sw} = 1.9 \times 10^{-5}(\mathrm{Hz})\left(\frac{1}{v_w}\right)\left(\frac{\beta}{H_*}\right)\left(\frac{T_*}{100\ \mathrm{GeV}}\right)\left(\frac{g_*}{100}\right)^{\frac{1}{6}} \tag{16}$$

Magnetohydrodynamic (MHD) occurs in plasma formation after bubbles collide. In a first-order phase transition, bubbles may generate magnetohydrodynamic turbulence. The resulting seed magnetic fields are converted into magnetic fields, and the created MHD perturbation produces gravitational waves that may be detected by gravitational wave detectors. Gravitational waves resulting from the first-order QCD phase transition are placed in the frequency range of pulsar timing arrays, that is, nanohertz frequencies [26]. The contribution of the turbulent motion of the fluid is given by:

$$h^2\Omega_{Tu}(f) = 3.35 \times 10^{-4}\left(\frac{H_*}{\beta}\right)^2\left(\frac{\kappa_{tu}\alpha}{1+\alpha}\right)^{3/2}\left(\frac{100}{g_*}\right)^{\frac{1}{3}} v_w S_{tu}(f), \delta = \kappa_{tu}/\kappa_v \tag{17}$$

$$S_{tu}(f) = \frac{\left(\frac{f}{f_{tu}}\right)^3}{\left(1+\frac{f}{f_{en}}\right)^{11/3}\left(1+\frac{8\pi f}{h_*}\right)}, h_* = 16.5 \times 10^{-6}(\mathrm{Hz})\left(\frac{T_*}{100\ \mathrm{GeV}}\right)\left(\frac{g_*}{100}\right)^{\frac{1}{6}} \tag{18}$$

In Eq. (18), h_* is the redshifted Hubble parameter. The explicit dependence of the spectral shape on h_* shows that the disturbance acts as a gravitational wave source, several times the Hubble parameter. The peak frequency in the form of the spectrum is given by the following relationship:

$$f_{tu} = 2.7 \times 10^{-5}(\mathrm{Hz})\left(\frac{1}{v_w}\right)\left(\frac{\beta}{H_*}\right)\left(\frac{T_*}{100\ \mathrm{GeV}}\right)\left(\frac{g_*}{100}\right)^{\frac{1}{6}} \tag{19}$$

8. Summary

This chapter is a brief overview of the effects of gravity in the early universe. Because the Big Bang is regarded as the beginning of our universe. It is predicted that in the first moments, the universe was a hot bath of fundamental particles that, although relativistic, had a weak interaction. In fact, in that era, the universe only consisted of quark-gluon plasma, which underwent several phase transitions and entered the confined phase from the deconfined state. This phase transition was accompanied by the nucleation of bubbles that grew and expanded. As a result, some of these bubbles collided with each other and created sound waves and magnetic turbulences, leading to the production of gravitational waves. In fact, during bubble formation, we may have three sources of gravity waves: bubble collisions, sound waves, and turbulent fluid motion. In order to reveal and study the energy density spectrum of gravitational waves, the total of these three sources should be considered and it can be checked which of these three sources has the most contribution. The study of the early universe and the discovery of gravitational waves resulting from it are very popular among physicists, and the results of these studies are consistent with the frequencies of some detectors and arrays.

Author details

Seddigheh Rezapour
Astronomical Pole of Student Research Institutes, Mashhad, Iran

*Address all correspondence to: ch.r.s1979@gmail.com

References

[1] Pasechnik R, Sumbera M. Phenomenological review on quark-gluon plasma: Concepts vs. observations. Universe. 2017;**3**:7. DOI: 10.3390/universe3010007 [arXiv: 1611.01533[hep-ph]]

[2] Zhong H, Gong B, Qiu T. Gravitational waves from bubble collisions in FLRW spacetime, 2107.01845. DOI: 10.1007/JHEP02%282022%29077 [arXiv:2107.01845 [gr-qc]]

[3] Bea Y et al. Bubble wall velocity from holography. Physical Review D. 2021; **104**:L121903. DOI: 10.1103/PhysRevD.104.L121903 [arXiv: 2104.05708 [hep-ph]]

[4] Ahmadvand M. Baryogenesis within gauge/gravity correspondence [thesis]. Shahrood: Shahrood University of Technology; 2018

[5] Rubakov VA. Cosmology. 2015. pp. 151-196. arXiv: 1504.03587 [astroph. CO]. DOI: 10.5170/CERN-2014-003.151. Available from: https://doi.org/10.48550/arXiv.1504.03587, CERN Yellow Report CERN-2014-003

[6] Linde AD. Phase transitions in gauge theories and cosmology. Reports on Progress in Physics. 1979;**42**:389-437. DOI: 10.1088/0034-4885/42/3/001

[7] Bailin D, Love A. Cosmology in Gauge Field Theory and String Theory. Bristol, UK: IOP; 2004. ISBN 9780750304924

[8] Boyanovsky D, de Vega HJ, Schwarz DJ. Phase transitions in the early and the present universe. Annual Review of Nuclear and Particle Science. 2006;**56**:441-500. DOI: 10.1146/annurev.nucl.56.080805.140539

[9] Trodden M. Electroweak baryogenesis. Reviews of Modern Physics. 1999;**71**:1463-1500. DOI: 10.1088/1367-2630/14/12/125003

[10] Bernreuther W. CP violation and baryogenesis. Lecture Notes in Physics. 2002;**591**:237. DOI: 10.48550/arXiv.hep-ph/0205279

[11] Kosowsky A, Turner MS, Watkins R. Gravitational waves from first order cosmological phase transitions. 5 Oct 1992;**69**(14):2026-2029. DOI: 10.1103/PhysRevLett.69.2026. PMID: 10046380. PMID: 10046380

[12] Gleiser M. Phase transitions in the universe. Contemporary Physics. 1998; **39**:239-253. DOI: 10.1080/001075198181937. arXiv:hep-ph/9803291

[13] Zhu ZR, Chen J, Hou D. Gravitational waves from holographic QCD phase transition with gluon condensate. [arXiv:2109.09933[hep-ph]]. DOI: 10.1140/epja/s10050-022-00754-2. Available from: https://doi.org/10.48550/arXiv.2109.09933

[14] Nayak TK. Probing the QCD phase structure using event-by-event fluctuations. arXiv:2008.04643. DOI: 10.1088/1742-6596/1602/1/012003

[15] The DOE/NSF Nuclear Science Advisory Committee. The Frontiers of nuclear science, a long range plan. Nuclex. 2008. DOI: 10.48550/arXiv.0809.3137

[16] Petreczky P. Lattice QCD at non-zero temperature. Journal of Physics G. 2012;**39**:093002. DOI: 10.1088/0954-3899/39/9/093002 [arXiv: 1203.5320 [hep-lat]]

[17] Bhattacharya T et al. QCD phase transition with chiral quarks and physical quark masses. Physical Review Letters. 2014;**113**(8):082001. DOI: 10.1103/PhysRevLett.113.082001 [arXiv:1402.5175 [hep-lat]]. Available from: https://doi.org/10.48550/arXiv.1402.5175

[18] Boeckel T, Schettler S, Schaffner-Bielich J. Progress in Particle and Nuclear Physics. 2011;**66**:266-270. DOI: 10.1016/j.ppnp.2011.01.017 [arXiv:1012.3342 [astro-ph.CO]]

[19] Braun-Munzinger P, Stachel J. The quest for the quark-gluon plasma. Nature. 2007;**448**:302. DOI: 10.1038/nature06080

[20] Megevand A, Ramirez S. Bubble nucleation and growth in very strong cosmological phase transitions. Nuclear Physics B. 2017;**919**:74. DOI: 10.1016/j.nuclphysb.2017.03.009

[21] Hindmarsh M, Huber SJ, Rummukainen K, Weir DJ. Gravitational waves from the sound of a first order phase transition. Physical Review Letters. 2014;**112**:041301. DOI: 10.1103/PhysRevLett.112.041301 arXiv:1304.2433 [hep-ph]

[22] Espinosa JR, Konstandin T, No JM, Servant G. Energy budget of cosmological first-order phase transitions. Journal of Cosmology and Astroparticle Physics. 2010;**06**:028, arXiv:1004.4187 [hep-ph]. DOI: 10.1088/1475-7516/2010/06/028

[23] Kamionkowski M, Kosowsky A, Turner MS. Gravitational radiation from first-order phase transitions. Physical Review D. 1994;**49**:2837 [astro-ph/9310044]. DOI: 10.1103/PhysRevD.49.2837

[24] Hindmarsh M, Hijazi M. Gravitational waves from first order cosmological phase transitions in the sound shell model. JCAP. 2019;**12**:062. DOI: 10.1088/1475-7516/2019/12/062 [arXiv:1909.10040 [astro-ph.CO]]

[25] Caprini C et al. Science with the space-based interferometer eLISA II: Gravitational waves from cosmological phase transitions. Journal of Cosmology and Astroparticle Physics. 2016;**1604**: 001, arXiv:1512.06239 [astro-ph.CO]. DOI: 10.1088/1475-7516/2016/04/001

[26] Niksa P, Schlederer M, Sigl G. Gravitational waves produced by compressible MHD turbulence from cosmological phase transitions. [arXiv: 1803.02271 [astro-ph.CO]]. DOI: 10.1088/1361-6382/aac89c. Available from: https://doi.org/10.48550/arXiv.1803.02271

Chapter 4

The Energy-Momentum Tensor of the Gravitational Field as a Correction to the Einstein Equation

Valery Borisovich Morozov

Abstract

The energy-momentum-stress tensor of the Riemannian space is found and defined in the general theory of relativity as a function of the metric tensor. In inertial space, this tensor is equal to zero, but in the Newtonian limit, its energy density value is calculated. The general theory of relativity makes it possible to include the field energy tensor in the gravitational field equation. The new gravitational field equation is asymptotically equal to the Einstein equation when applied in weak fields. The region $0 < r < r_g$ is occupied by a potential well, which has an extremely low potential, this fact suggests that neutron stars and heavier objects have the same structure. The introduction of the new equation of the gravitational field and the energy-momentum tensor of the gravitational field into the classical Einsteinian relativistic theory of gravitation led us to a consistent continuation of the general theory of relativity.

Keywords: energy-momentum conservation law, Ricci tensor, gravitational field equation, gravitational field momentum energy tensor, black holes, neutron stars

1. Introduction

The proposed model is not an alternative gravitational model [1]. Criticism and comparison with these models are beyond the scope of this work. The work provides a response to the critical remarks of V.A. Fock [2] on the inadmissibility of solutions to the Einstein equation with special (singular) points.

Of particular importance were the results of Einstein himself when creating the gravitational field equation [3] and the work that followed immediately after this event. Our model is built in strict accordance with the principles obtained by A. Einstein [4] in 1913.

Einstein introduced the Riemannian space into physics as a physical object. This made it possible to consistently describe gravity and the phenomena accompanying it, at least in the region of fields which were not too large. The current material being presented does not go beyond the principles of the general theory of relativity, on the contrary, the theory has managed to include a full-fledged principle of conservation of energy, which is absent in Einstein's theory of gravity.

In this work, the energy-momentum tensor of the gravitational field and the equation of the gravitational field are successively introduced. The principles of

Einstein's general theory of relativity were established in 1913 [4]. One of the main principles: the law of conservation of energy-momentum, provided for the conservation of energy, not only the conservation of the energy-momentum of matter, but also the energy of complete systems of matter and their gravitational fields. In modern terms, the covariant equation of the gravitational field, which satisfies the law of the conservation of energy, should have looked like this:

$$G^{\mu}_{\nu} = \kappa\left(T^{\mu}_{\nu} + f^{\mu}_{\nu}\right), \tag{1}$$

where $f_{\mu\nu}$ is the energy-momentum tensor of the gravitational field. A conservation law could be derived from this field equation. However, Einstein failed to solve this problem. As a result, the current, simplified Einstein equation was obtained (December 1915 [5]):

$$R_{\mu\nu} = \frac{8\pi G}{c^4}\left(T_{\mu\nu} - \frac{1}{2}g_{\mu\nu}\,T\right), \tag{2}$$

the following shows the Ricci tensor

$$R_{\mu\nu} = \frac{\partial \Gamma^{\alpha}_{\mu\nu}}{\partial x^{\alpha}} - \Gamma^{\alpha}_{\mu\beta}\Gamma^{\beta}_{\nu\alpha} - \frac{\partial \Gamma^{\alpha}_{\mu\alpha}}{\partial x^{\nu}} + \Gamma^{\alpha}_{\mu\nu}\Gamma^{\beta}_{\alpha\beta}.$$

Einstein's equation is now more commonly used in the form

$$R_{\mu\nu} - \frac{1}{2}g_{\mu\nu}R = \frac{8\pi G}{c^4}T_{\mu\nu}. \tag{3}$$

In addition, Einstein showed that choosing coordinates such that $\sqrt{-g} = 1$ simplifies the equation. This brings the Einstein eq. (2) to the following form:

$$\begin{cases} \dfrac{\partial \Gamma^{\alpha}_{\mu\nu}}{\partial x_{\alpha}} - \Gamma^{\alpha}_{\mu\beta}\Gamma^{\beta}_{\nu\alpha} = \dfrac{8\pi G}{c^4}\left(T_{\mu\nu} - \dfrac{1}{2}g_{\mu\nu}\,T\right). \\ \sqrt{-g} = 1. \end{cases} \tag{4}$$

This form of the equation was solved by Schwarzschild [6]. Due to the complexity of this solution, this version of Einstein's equation has practically disappeared from the literature.

Landau and Lifshitz [7][1] noticed, that there is no gravitational field energy density as a source on the right-hand side of Einstein's field equations. In § 95, it is written, that.

> Gravitational interaction plays a role only for bodies with a sufficiently large mass (due to the small gravitational constant),. . .

and in § 96 regarding the vanishing covariant derivative of the energy-momentum density tensor of matter, $T_{ki;k} = 0$, (that)

[1] We can estimate the magnitude of the correction to the Newtonian law of gravity, if we take into account the gravitational density of the field. This correction is on the same order as the correction of the Newtonian gravitational theory caused by the general theory of relativity [5].

> In this form, however, this (Einstein) equation does not generally express any conservation law whatever. This is related to the fact that in a gravitational field the four-momentum of the matter alone must not be conserved, but rather the four-momentum of matter plus gravitational field;
> the latter is not included in the expression for T_i^k.

Indeed, Einstein initially introduced the energy-momentum density tensor of the gravitational field into his field eqs. [1]. However, it turned out, that this quantity, which is calculated from the conservation law of energy and momentum, is not a tensor. In this form, the expression for the gravitational field energy and momentum density was not covariant, which became a serious obstacle for creating the complete field equations. Within only two years, a way out of this dilemma was found, after Einstein simply had removed the gravitational field energy and momentum density from his theory. Then, the field equations became covariant and correctly described the motion of mercury.

2. Division of the ricci tensor

Following Einstein [6], we represent the Ricci tensor as the sum of two parts:

$$R_{\mu\nu} = A_{\mu\nu} + B_{\mu\nu};$$

$$A_{\mu\nu} = \frac{\partial \Gamma^{\alpha}_{\mu\nu}}{\partial x_{\alpha}} - \Gamma^{\alpha}_{\mu\beta}\Gamma^{\beta}_{\nu\alpha};$$

$$B_{\mu\nu} = -\frac{\partial \Gamma^{\alpha}_{\mu\alpha}}{\partial x^{\nu}} + \Gamma^{\alpha}_{\mu\nu}\Gamma^{\beta}_{\alpha\beta}. \tag{5}$$

However, we do not yet know whether parts of the Ricci tensor are tensor quantities. The following allows us thus to prove so.

Theorem 1. The magnitude of the simplified Christoffel symbol $\Gamma^{\alpha}_{\mu\alpha}$ is a 4-vector, $A_{\mu\nu}$ and $B_{\mu\nu}$ are tensors.

This is Proof.

We used the formula for the coordinate transformation of Christoffel symbols (Eq. (85.15), [4]).

$$\Gamma^{\iota}_{\mu\alpha} = \Gamma'^{\gamma}_{\nu\xi}\frac{\partial x^{\iota}}{\partial x'^{\gamma}}\frac{\partial x'^{\nu}}{\partial x^{\mu}}\frac{\partial x'^{\xi}}{\partial x^{\alpha}} + \frac{\partial^2 x'^{\gamma}}{\partial x^{\alpha}\partial x^{\mu}}\frac{\partial x^{\iota}}{\partial x'^{\gamma}}$$

If we are to add the following to this expression, $\iota = \alpha$ and $\gamma = \xi$, then the formula will be simplified as seen below:

$$\Gamma^{\alpha}_{\mu\alpha} = \Gamma'^{\xi}_{\nu\xi}\frac{\partial x^{\alpha}}{\partial x'^{\xi}}\frac{\partial x'^{\nu}}{\partial x^{\mu}}\frac{\partial x'^{\xi}}{\partial x^{\alpha}} + \frac{\partial^2 x'^{\xi}}{\partial x^{\alpha}\partial x^{\mu}}\frac{\partial x^{\alpha}}{\partial x'^{\xi}} = \Gamma'^{\xi}_{\nu\xi}\frac{\partial x'^{\nu}}{\partial x^{\mu}} + \frac{\partial^2}{\partial x^{\mu}\partial x'^{\xi}}x'^{\xi}$$

The last term is transformed according to the composite function differentiation theorem. According to the theorem on the equality of mixed derivatives, the order of differentiation can be changed. Then it turns out that the last term is equal to zero, and

$$\Gamma^{\alpha}_{\mu\alpha} = \Gamma^{\prime\xi}_{\nu\xi}\frac{\partial x^{\prime\nu}}{\partial x^{\kappa}}.$$

Thus, $\Gamma^{\alpha}_{\mu\alpha}$ transforms as a 4-vector and is therefore a 4-vector.

The covariant differentiation $\Gamma^{\alpha}_{\mu\alpha}$ generates the tensor

$$\frac{D\Gamma^{\alpha}_{\mu\alpha}}{\partial x_{\nu}} = \frac{\partial\Gamma^{\alpha}_{\mu\alpha}}{\partial x_{\nu}} - \Gamma^{\alpha}_{\mu\nu}\Gamma^{\beta}_{\alpha\beta},$$

the resulting tensor possesses and exactness up to the sign and is $B_{\mu\nu}$.

It follows from the proven theorem that since $R_{\mu\nu}$ is a tensor, it follows from $R_{\mu\nu} = A_{\mu\nu} + B_{\mu\nu}$ that $A_{\mu\nu}$ is also in fact a tensor quantity.

We'll need a mixed tensor soon.

$$B^{\nu}_{\mu} = -\frac{D\,\Gamma^{\alpha}_{\mu\alpha}}{\partial x^{\nu}} = -\frac{\partial\Gamma^{\alpha}_{\mu\alpha}}{\partial x^{\nu}} - \Gamma^{\alpha}_{\mu\nu}\Gamma^{\beta}_{\alpha\beta}. \tag{6}$$

3. Connection of the part of the tensor $B_{\mu\nu}$ with the gravitational field

Consider the quantity $B_{\mu\nu}$ in the Newtonian approximation, which is described by the stationary metric:

$$ds^2 = \left(1 + \frac{2\varphi}{c^2}\right)dt^2 - dx^2 - dy^2 - dz^2,$$

here $\varphi(x)$ is the Newtonian potential. Taking into account that the spatial components of the metric tensor and its determinant slightly differ from unity, we obtain:

$$B^{0}_{0} = -\Gamma^{1}_{00}\Gamma^{0}_{10} = -\frac{1}{c^2}\frac{\partial\varphi}{\partial x}\bullet\frac{1}{c^2 + 2\varphi}\frac{\partial\varphi}{\partial x} \approx -\frac{1}{c^4}\left(\frac{\partial\varphi}{\partial x}\right)^2.$$

This value of the energy density of the gravitational field, up to a constant factor, coincides with the known field density in the Newtonian approximation ([7] § 106, problem 1):

$$f = -\frac{(\nabla\varphi)^2}{8\pi G}$$

or

$$B^{0}_{0} = \frac{8\pi G}{c^4}f^{0}_{0} = \kappa f^{0}_{0}. \tag{7}$$

On the other hand, the tensor $B_{\mu\nu}$ vanishes in the Minkowski space in which the fundamental tensor consists of ones with a positive or negative sign, or zeros:

$$B_{\mu\nu} = -\frac{\partial\Gamma^{\alpha}_{\mu\alpha}}{\partial x^{\nu}} + \Gamma^{\alpha}_{\mu\nu}\Gamma^{\beta}_{\alpha\beta} \to 0.$$

On the other hand, the tensor B^{ν}_{μ} vanishes in Minkowski space, which does not contain gravitational fields. Relation (6) allows us to relate the tensor B^{ν}_{μ} with the energy-momentum-stress tensor of the gravitational field f^{ν}_{μ}:

$$B_{\mu}^{\nu} = \kappa f^{\nu}_{\ \mu}. \tag{8}$$

4. Gravitational field equation

The introduction of Eq. (8) into the Einstein Eq. (1) leads to a new equation for the gravitational field:

$$\frac{\partial \Gamma^{\alpha}_{\ \mu\nu}}{\partial x_{\alpha}} - \Gamma^{\alpha}_{\ \mu\beta}\Gamma^{\beta}_{\ \nu\alpha} - \frac{1}{2}g_{\mu\nu}R = \frac{8\pi G}{c^4}T_{\mu\nu}. \tag{9}$$

In this case, additional unknowns do not appear, since the quantities entering Eq. (9) do not depend on the components of the tensor $f_{\mu\nu}$.

Theorem 2. As the determinant of the metric tensor g approaches the value of the Minkowski tensor, which already differs little from the value of the Minkowski tensor, the Eq. (9) comes to an indeterminate approach of Einstein's Eq. (4).

Proof

As shown above the tensor $B_{\mu\nu} \to 0$ as the metric $g_{\mu\nu}$ approaches the Minkowski metric, we can write the asymptotic equality:

$$R_{\mu\nu} = \frac{\partial \Gamma^{\alpha}_{\ \mu\nu}}{\partial x_{\alpha}} - \Gamma^{\alpha}_{\ \mu\beta}\Gamma^{\beta}_{\ \nu\alpha} + B_{\mu\nu} \approx \frac{\partial \Gamma^{\alpha}_{\ \mu\nu}}{\partial x_{\alpha}} - \Gamma^{\alpha}_{\ \mu\beta}\Gamma^{\beta}_{\ \nu\alpha}.$$

Whence follows the asymptotic equivalence of Eq. (9) and the Einstein equation.

Theorem 3. From the equation of the gravitational field Eq. (9) follows the complete law of conservation of matter and the gravitational field.

Proof

We know that the covariant derivative of the Einstein tensor is:

$$G^{\mu}_{\nu;\mu} = 0.$$

Hence from Eq. (1) follows the general conservation law:

$$T^{\mu}_{\nu;\mu} + f^{\mu}_{\nu;\mu} = 0,$$

which includes not only matter but also the gravitational field. In fact, this is the conservation law of the $B_{\mu\nu}$ tensor, see Eq. (8). Thus, the equation of the gravitational field is obtained, which satisfies the postulates of Einstein + the law of conservation of energy-momentum.[2]

5. The Schwarzschild problem

Is solved in the same way as the Einstein equation. Looking for solutions in the form:

[2] Einstein failed to obtain an equation of the form (1). Therefore, from the Einstein equation, only the law of conservation of energy of matter $T^{\mu}_{\nu;\mu} \equiv 0$ (follows)

$$ds^2 = s(r)dt^2 - p(r)dr^2 - r^2 d\theta^2 - r^2 \sin^2\theta d\varphi^2. \tag{10}$$

The Christoffel symbols of the metric Eq. (10) have the meanings:

$$\Gamma^1_{11} = \frac{p'}{2p}; \Gamma^1_{22} = \frac{-r}{p}; \Gamma^1_{33} = -\frac{r\sin(\theta)^2}{p}; \Gamma^0_{10} = \Gamma^0_{01} = \frac{s'}{2s}; \Gamma^1_{00} = \frac{s'}{2p};$$

$$\Gamma^2_{33} = -\sin(\theta)\cos(\theta); \Gamma^2_{12} = \Gamma^2_{21} = \Gamma^3_{13} = \Gamma^3_{31} = \frac{1}{r}; \Gamma^3_{23} = \Gamma^3_{32} = \frac{\cos(\theta)}{\sin(\theta)}.$$

where the prime means the derivative with respect to r.
Equation for empty space:

$$A_{\mu\nu} = \frac{\partial \Gamma^\alpha_{\mu\nu}}{\partial x_\alpha} - \Gamma^\alpha_{\mu\beta}\Gamma^\beta_{\nu\alpha} = 0. \tag{11}$$

From this equation, we find the components of the tensor Eq. (11), up to a nonzero factor, equal:

$$A_{00} \propto \frac{s'}{p}\left(\frac{s''}{s'} - \frac{p'}{p} - \frac{s'}{s}\right) = 0,$$

$$A_{11} \propto \left(\frac{p'}{p}\right)' - \frac{4}{r^2} - \frac{s's'}{2s^2} = 0.$$

If we equate these components to zero, we get a system of ordinary differential equations. If eliminating from the above the value $\frac{p'}{p}$ we thus obtain a third-order equation:

$$\left(\frac{s''}{s'}\right)' - \left(\frac{s'}{s}\right)' - \frac{s's'}{2s^2} = \frac{4}{r^2}. \tag{12}$$

We have not been able to find an analytical solution to this equation.

When solving the equation numerically, the asymptotic proximity with the Schwarzschild solution was used. This made it possible to use the values of the Schwarzschild solution $1 - \frac{1}{r}$ as boundary conditions close to the expected solution. As expected, in the region of relatively small values, the solution follows the Schwarzschild solution. Then, similarly to the Schwarzschild solution, it rapidly decreases (**Figure 1**). But the difference from the Schwarzschild solution, the solution of Eq. (11) remains positive, although extremely minute. There is a potential well with an almost limiting value of the potential, which is equal to $\varphi = -c^2$. The width of the potential well of the solution is noticeably larger than the Schwarzschild radius $r_g = 1$ (**Figure 1**). No singularities or deviations from the monotonic dependence of the solution are observed in the surrounding neighborhood of the equation. Moreover, there is no indication that the expected singularity is present at the origin, similar to the singularity formed by a point source in known classical problems.

Thus, Einstein's space finally received the status of an independent physical object with parameters specified by the energy-momentum-stress tensor

$$f_{\mu\nu} = -\frac{8\pi G}{c^4}\frac{D\Gamma^\alpha_{\mu\alpha}}{\partial x_\nu}.$$

DOI: http://dx.doi.org/10.5772/intechopen.1004502

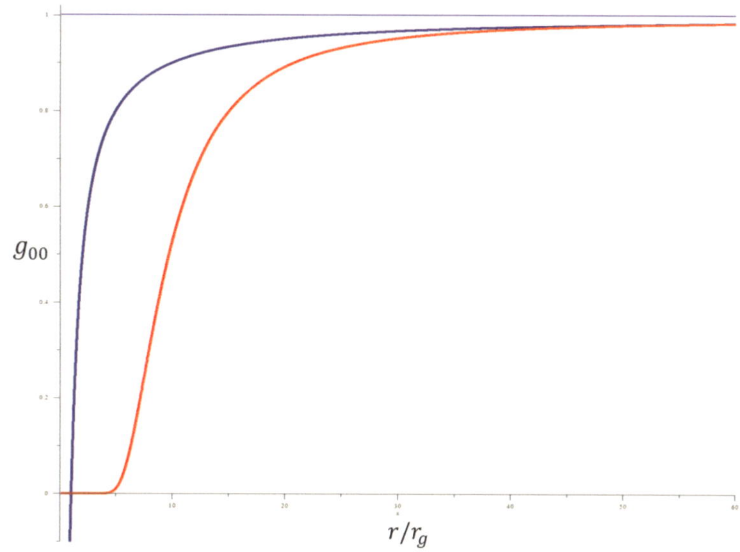

Figure 1.
Dependence of g_{00} on the ratio $\frac{r}{r_g}$. The Schwarzschild solution of the Einstein equation (blue line) and the numerical solution of the exact equation of the gravitational field Eq. (9) (red line).

Einstein's equation is a good approximation but only for gravitational fields that are not too large, therefore it must be replaced by the exact equation of the gravitational field:

$$R_{\mu\nu} - \frac{1}{2}g_{\mu\nu}R = \kappa T_{\mu\nu} \Longrightarrow \frac{\partial \Gamma^{\alpha}_{\mu\nu}}{\partial x_{\alpha}} - \Gamma^{\alpha}_{\mu\beta}\Gamma^{\beta}_{\nu\alpha} - \frac{1}{2}g_{\mu\nu}R = \kappa T_{\mu\nu}.$$

Formally, the elimination of these "small" errors leads to an equation, the solution of which must be a smooth metric tensor. Moreover, the law of conservation of energy-momentum follows directly the equation of the gravitational field.

6. Black holes and neutron stars

There is a potential well with a limiting value of the potential, which is equal to $\varphi = -c^2$. The width of the potential well of the solution is noticeably larger than the radius Schwarzschild. But the difference from the solution Schwarzschild solution of Eq. (11) remains positive, although extremely small. The width of the potential well of the solution is noticeably larger than the radius Schwarzschild $r_g = 1$ (**Figure 2**). But in the vicinity, there are features or deviations from the monotonic dependence of the solution. Moreover, there is no indication that there is a singularity at the origin, similar to the singularity formed by a point source in known classical fields (**Figure 3**).

For applications, the most interesting solutions are for large fields. There was confidence that the fundamental tensor entirely belongs to the original Riemannian space and does not contain invalid values.

The presented results solve the problem of describing super heavy compact objects - black holes and neutron stars. It can be expected that the gravitational field of such objects is a kind of potential wells with extreme but finite depth.

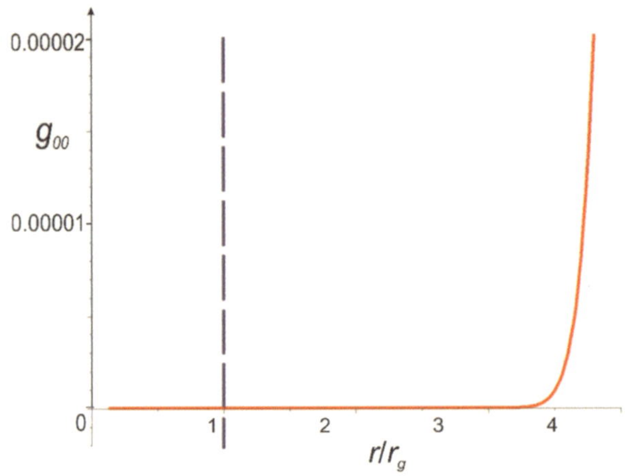

Figure 2.
*Stretched 5000 times **Figure 1**. The horizontal section of the solution of Eq. (9) has a positive value, but it is so small that it graphically merges with the coordinate axis.*

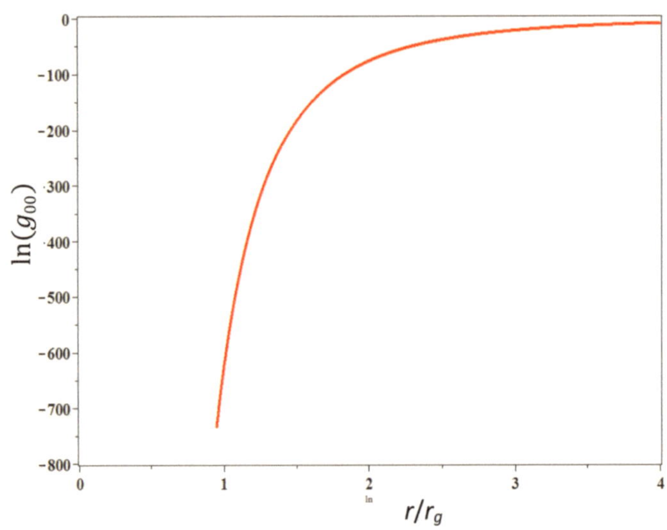

Figure 3.
Dependence of $\ln g_{00}$ *on the ratio* $\frac{r}{r_g}$. *The numerical solution of the exact equation of the gravitational field Eq. (9).*

The ultimate black hole/potential hole test.

Unlike a black hole, thermal radiation can leave the potential well. However, due to the gravitational redshift, the emission maximum should shift toward longer wavelengths. For this reason, it is possible to observe such objects in the radio range.

The filling of the potential well with a substance depends on the equation of the state of the substance and its temperature. This should raise the pit fill level above the extreme value.

Figures 1 and **2** show that the edge of the potential well is slightly blurred. However, this does not prevent us from estimating the radius of the potential well. In **Figure 2**, we see that the radius of the well is slightly larger than $4r_g$. This value is close to the estimated radius of neutron stars, so we can make the assumption that the gravitational component of a neutron star is completely determined by Eq. (8).

Qualitatively, the characteristics of the near gravitational field are similar to those of the field of the Schwarzschild solution. Thus, it will be difficult for the observer to distinguish the details of the field by observing refraction in the electromagnetic bands. However, the observation of gravitational waves makes it possible to observe orbital characteristics up to the moment of collision, which will make it possible to differentiate the various laws of attraction. Although, we repeat, the observation of the low-frequency component of the radiation of heavy objects, in our opinion, remains the most promising way to prove the non-Schwarzschild nature of "black holes."

7. Conclusion

This work was preceded by a series of preprints, starting with preprint [8] and ending with preprints [9] and [10]. The work may claim to change our understanding of both dense compact objects (black holes and neutron stars) and space-time as a physical object.

Author details

Valery Borisovich Morozov
NIITeplopribor, Moscow, Russia

*Address all correspondence to: valerymorozov@gmail.com

References

[1] Alternatives to general relativity. Wikipedia. 2024

[2] Fock VA. The Theory of Space, Time and Gravity. Russian: URSS; 2015

[3] Vizgin VP, Smorodinskii Ya A. From the equivalence principle to the equations of gravitation. Soviet Physics Uspekhi. 1979;**22**:489-513

[4] Einstein A, Grossmann M. Entwurf einer verallgemeinerten Relativitätstheorie und Theorie der Gravitation. Zeitschrift für Angewandte Mathematik und Physik. 1913;**62**:225-261

[5] Einstein A. Die Feldgleichungen der Gravitation. Sitzungsber. preuss. Akademie der Wissenschaften; 1915; **48**(2):844-847

[6] Schwarzschild K. Über das Gravitationsfeld eines Massenpunktes nach der Einstein'schen Theorie. Berlin: Reimer; 1916. p. 189

[7] Landau LD, Lifshitz EM. The Classical Theory of Fields. 4th ed. Vol. 2. Butterworth–Heinemann; 1975

[8] Morozov VB. Einstein's postulate as a correction to Newton's law of gravity. Preprint. ResearchGate; 2017. DOI: 10.13140/RG.2.2.17853.20965

[9] Morozov VB. Energy of space in the gravitational field equation. Preprint. 2024. DOI: 10.13140/RG.2.2.10726.60482

[10] Morozov VB. Energy of space in the gravitational field equation (In Russian). Preprint. 2024. DOI: 10.13140/RG.2.2.22627.94248

Chapter 5

Various Problems of General Relativity with and without a Gravitational Field

Valery Borisovich Morozov

Abstract

A consistent principle of equivalence is formulated. An example of the necessity of strict observance of Einstein's rule for the transformation of tensor coordinates is given. The problem of a homogeneous stationary gravitational field is being solved. All metrics with non-zero curvature. A metric for the space of constant curvature, different from the De Sitter space, is found. A stationary space with a cosmological redshift has been discovered, which creates the effect of accelerated expansion. It is shown that the plane-parallel motion of gas in a force field obeys the law of compression-expansion of the shock adiabat.

Keywords: cosmological redshift, stationary universe, compression-expansion of a gas in a gravitational field, coordinate transformation, gas motion in a potential field, frame of reference, equivalence principle, epsilon-delta method, tangent space

1. Introduction

Einstein's theory of relativity is based on strict definitions and postulates. Neglecting them can lead to misunderstandings or even catastrophic errors.

In Einstein's special theory of relativity, the speed of light is independent of the frame of reference. This makes it impossible to describe the movement of bodies using coordinates and a single time, independent of the speed of these bodies. In this case, the spatial and temporal scales change when moving from one uniformly moving (inertial) frame of reference to another.

Already in the special theory of relativity, we see that this is a law of nature that describes the ratio of the scales of space and time, but not "weighty bodies." Time is on a par with space here. Hermann Minkowski took advantage of this and introduced time t as the Euclidean three-dimensional space with coordinates {x, y, z} as an additional, fourth dimension. In Minkowski's [1, 2] space, the interval between events

$$s^2 = (x_1 - x_0)^2 + (y_1 - y_0)^2 + (z_1 - z_0)^2 - c^2(t_1 - t_0)^2 \tag{1}$$

is a generalization of the Pythagorean theorem. The value of s does not change if the coordinates are changed according to the Lorentz transformation recipe. Here, for the first time, space is considered an independent physical object. Minkowski owns the words:

> *"From now on, time itself and space itself become an empty fiction, and only their unity preserves the chance for reality."*

The next step was taken by Einstein in 1907 [3]. The connection between time and acceleration in accelerated reference systems is found and the principle of equivalence of accelerated reference systems and systems with a gravitational field is established. For several years, Einstein accumulated separate results on the relativistic properties of gravity. Perhaps the most important result was the conclusion that the presence of a gravitational field changes the speed of light, i.e. the speed of light is not a constant depends on the gravitational potential [4]. This makes it impossible to describe gravity in Minkowski's space. The natural way out of this situation is the transition to the curvilinear Riemann space. Fortunately, by that time, the creation of a mathematical apparatus for calculating the parameters of such a space, tensor analysis (absolute differential calculus), had been completed. Einstein resorts to the help of his friend, the mathematician Grossman, and they publish the basic principles for constructing a relativistic theory of gravity (general relativity). It also outlines the basics of tensor analysis.

Here it is already necessary to quote the words of Einstein [5], formulating the subject of the theory of relativity as the study of the physical properties of space:

> *"When it is said here about the ether, then, of course, it does not mean the corporeal ether of the mechanical wave theory, which obeys the laws of Newtonian mechanics and whose individual points are assigned speed. This theoretical notion with the creation of the special theory of relativity, in my opinion, finally disappeared from the scene. On the contrary, we are talking about those physically conceivable real things that, along with weighty matter, consisting of electrical elementary particles, play a role in the structure of the causal relationships of physics. Therefore, instead of the word "ether" one can just as well say "the physical properties of space". In this case, of course, it would be possible to express the opinion that all objects of physics fall under this concept, since, according to consistent field theory, weighty matter or its constituent elementary particles should also be considered a special kind of "field", or special "states of space". However, one has to admit that in the present state of physics, such an idea is premature, since until now all the efforts of theoretical physicists directed towards this goal have failed. Thus, we are now effectively forced to distinguish between "matter" and "fields", although we can hope that future generations will overcome this dualistic view and replace it with a single concept, as the field theory of our days has tried in vain to do."*

Modern evidence confirms that space is indeed not flat. The most compelling evidence for this is the experiment to detect the delay of an electromagnetic signal in a gravitational field, discovered by the group of Irwin Shapiro [6]. Subsequent more accurate observations of nearby relativistic star systems proved that space is not flat - in the presence of a gravitational field, the Pythagorean theorem does not work.

It should be added that the well-known unshieldability of the gravitational field makes the physical space so unusual that attempts to attribute properties of ordinary matter or (quantum) fields to gravity do not look convincing.

General relativity can be compared to a capricious lady trying on outfits and trying to pull on fashionable outfits from someone else's shoulder. I will allow myself to express surprise at the consideration of quantum fields of particles endowed with the ability to interact with matter. This contradicts the observed transparency of matter for the gravitational field. The shading of the gravitational field by the Moon would lead to catastrophes on Earth. In addition, we would observe daily morning and evening tsunamis if such shading took place. In general, attributing the properties of ordinary matter to the gravitational field does not seem reasonable. Speaking of the ether as a degree of the materiality of space, Einstein says [5]:

> *"According to the general theory of relativity, space is inconceivable without ether; indeed, in such a space, not only would the propagation of light be impossible, but scales and clocks could not exist, and there would be no space-time distances in the physical sense of the word. However, this ether cannot be imagined as consisting of parts traceable in time; only weighty matter has this property; in the same way, the concept of motion cannot be applied to it."*

It is noteworthy that almost immediately after Einstein's death, in 1958, David Finkelstein identified the Schwarzschild surface as an event horizon, "an ideal unidirectional membrane: causal influences can cross it in only one direction." At the same time, from the point of view of pure mathematics, the Schwarzschild surface does not belong to the original metric space - the metric on this surface and inside it simply does not exist, i.e. is not a solution to the Einstein equation.

It cannot be said that all problems in Einstein's theory have been solved, and the initial provisions of the theory have been correctly understood and accepted by everyone without exception. The purpose of this chapter is to explain the insufficiently rigorous provisions of the theory and to warn the researcher against errors with the help of non-trivial results [7].

2. Local equivalence principle

In 1907, Einstein considered the equivalence principle [3] as a means to transfer the results obtained for accelerated reference systems to systems with a gravitational field. At the same time, Einstein limited himself to systems with a homogeneous field, which of course is not suitable for arbitrary reference systems. The modern approach to the equivalence principle is distinguished by a variety of interpretations and definitions, for example, [8–12].

We will talk about the local equivalence of the frames of reference of the Riemann space and the Minkowski space. In the cult book [11], Landau and Lifshitz try to prove the local equivalence of a system with a gravitational field to an inertial system:

> *"Formula (85.15) under condition (85.16)[1] allows us to prove the above statement about the possibility of such a choice coordinate system for which all Γ^i_{kl} vanish at any predetermined point (such a system is called locally inertial or locally geodesic, see §87). Indeed, let a given point be chosen as origins and quantities Γ^i_{kl} have in it initially (in x^i coordinates) values $(\Gamma^i_{kl})_0$. We will produce near this point transformation*
>
> $$x'^i = x^i + \frac{1}{2}(\Gamma^i_{kl})_0 x^k x^l \qquad (85.18)$$

[1] The numbering of the quoted passage is preserved

Then

$$\left(\frac{\partial^2 x'^m}{\partial x^k \partial x^l}\frac{\partial x^i}{\partial x'^m}\right)_0 = \left(\Gamma^i_{kl}\right)_0 \tag{85.19}$$

and according to $(85.15)^2$ all Γ^i_{kl} vanish."

Here we see that the authors consider the transition from the first expression to the second by incorrectly replacing small values of the x^i coordinates with differentials, but.

$$x^i = \partial x^i + o(X). \tag{2}$$

The rejection of the finite value $o(X)$, contrary to popular belief, cannot be perceived as a coordinate transformation. Replacing coordinates with their differentials means replacing geodesics with their tangents at the point P.

A more transparent approach is used in Møller's excellent monograph [13] (§ 9.6. Local pseudo-Cartesian coordinates and local inertial systems). Møller comes from the curvilinear coordinates X to the local (near the origin) pseudo-Euclidean system $\check{X}$ using the transformation.

$$\check{x}^i = \frac{\partial \check{x}^i}{\partial x^k} x^k; \tag{3}$$

$$x^i = \frac{\partial x^i}{\partial \check{x}^k} \check{x}^k, \tag{4}$$

here the derivatives are calculated at the origin. We note right away that the inverse transformation (4) is impossible, since transformation (4), as well as transformation (3), is a transformation into the gallium space. After such a transformation, all information about the parameters of the original Riemann space is lost, which cannot be restored.

Let us show that in this case, the equal sign in (3) is inappropriate. Indeed, the equalities must be written in terms of the total differential.

$$\check{x}^i = d\check{x}^i + o(X) \approx d\check{x}^i; \tag{5}$$

Thus, we can only speak of an asymptotic transition (but not vice versa!) from curvilinear coordinates to pseudo-Euclidean ones.

$$\check{x}^i \to d\check{x}^i. \tag{6}$$

As the radius δ of the neighborhood decreases (**Figure 1**), the geodesics begin to look more and more like straight lines. In other words, a small neighborhood of the pseudo-Riemannian space can be arbitrarily close to the inertial frame (Einstein's freely falling elevator) but never coincide with it. In this case, as it should be in a finite region, neither the metric tensor nor, depending on it, the Riemann tensor with its convolutions and Christoffel symbols.

[2] This is the coordinate transformation formula $\Gamma^{\mathrm{i}}_{\mathrm{kl}} = \Gamma'^m_{np} \frac{\partial x^i}{\partial x'^m} \frac{\partial x'^n}{\partial x^k} \frac{\partial x'^p}{\partial x^l} + \frac{\partial^2 x'^m}{\partial x^k \partial x^l} \frac{\partial x^i}{\partial x'^m}$.

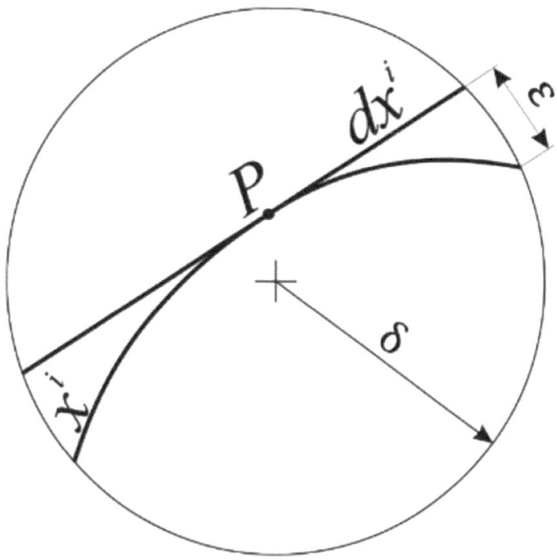

Figure 1.
Transformation (5) is a transition from the original space with coordinates x^i to the inertial tangent space with a common point P.

Let us illustrate this.

In curvilinear coordinates, the geodesic line will never coincide with the tangent in the vicinity of the tangent point, no matter how small this neighborhood is.

Let us formulate the principle of spatial equivalence of a Riemannian space of a neighborhood of a point P to a pseudo-Euclidean space.

If the point P belongs to an open neighborhood Q of a Riemannian space X and has a derivative at the point P, then this space is locally equivalent to the Minkowski tangent space $\check{X}$ at the point P if, for any positive ε, there is a neighborhood radius δ such that for any $x^i \in Q$ will be carried out $\varepsilon > \left|x^i - dx^i\right|$.

3. Coordinate system and reference system

It is necessary to distinguish between the concepts of reference system and coordinate system. The reference system can be specified in various coordinate systems.

When passing from the X coordinates to the X' coordinates [11], the contravariant vector A^μ transforms as a differential:

$$A^i = \frac{\partial x^i}{\partial x'^k} A'^k. \tag{7}$$

The transformation coefficient is called the Jacobi matrix, and the transformation is written

$$A^i = \mathrm{J}^i_k A'^k. \tag{8}$$

where the Jacobi matrix $\mathbf{J}^i_k = \left[\frac{\partial x^i}{\partial x'^k}\right]$.

The transformation of other contra-covariant tensors of arbitrary order is defined similarly to [14] as a sequence of linear transformations of the form (7). For example, the transformation of the covariant tensor of the second-rank

$$A_{ik} = \mathbf{J}^l_i \mathbf{J}^m_k A'_{lm}. \tag{9}$$

It is now clear that such transformations are not arbitrary. They must leave the frame of reference unchanged. General coordinate transformations transform a coordinate system into any other one with the same number of dimensions, including without preserving the reference system and its invariants.

$$x_i = x_i\left(x'_0, x'_1, x'_2, x'_3\right). \tag{10}$$

With such a transformation, the reference system can change, therefore, if we want to explore a certain reference system, we must limit ourselves to linear coordinate transformations (7) or a sequence of such transformations.

However, transformations of the form (7), even if a transformation matrix exists, do not ensure the preservation of the reference system.

Example. A non-linear transformation that cannot be reduced to a linear one.

Converting a Cartesian coordinate system to a spherical one:

$$\begin{cases} r = \sqrt{x^2 + y^2 + z^2}, \\ \theta = arctan\, \dfrac{\sqrt{x^2 + y^2}}{z}, \\ \varphi = arctan\, \dfrac{y}{x}. \end{cases} \tag{11}$$

When trying to obtain an equivalent transformation according to (7), we obtain:

$$\begin{gathered} r = \frac{\partial r}{\partial x}x + \frac{\partial r}{\partial y}y + \frac{\partial r}{\partial z}z = \sqrt{x^2 + y^2 + z^2}; \\ \theta = \frac{\partial \theta}{\partial x}x + \frac{\partial \theta}{\partial y}y + \frac{\partial \theta}{\partial z}z = \frac{x^2 + y^2}{z^2}\left(\frac{x^2 + y^2}{z\sqrt{x^2 + y^2}} - \frac{\sqrt{x^2 + y^2}}{z^2}z\right) = 0; \\ \varphi = \frac{\partial \varphi}{\partial x}x + \frac{\partial \varphi}{\partial y}y + \frac{\partial \varphi}{\partial z}z = \frac{x^2}{x^2 + y^2}\left(-\frac{y}{x^2}x + \frac{1}{x}y\right) = 0. \end{gathered} \tag{12}$$

Such a transformation is, of course, not equivalent to transformation (11). In this example, instead of the expected spherical coordinates, we got a one-dimensional space.

This unexpected result is a good illustration of the danger that lies in waiting for the researcher in case of ill-conceived application of arbitrary transformations of the form (10), which can lead to erroneous results. The phrase "arbitrary transformation of coordinates" should also not be abused, if it does not follow from the context that this is a transformation of coordinates of the reference system (7).

At the same time, one should not forget about the restrictions on relativistic frames of reference. For example, the speed limit should not exceed the speed of light ([13] § 8.7. General accelerated reference systems. The most general allowable coordinate transformations).

4. Some solutions of the Einstein equation

Einstein's equation has proven itself well for solving problems in small and moderate gravitational fields. Let us consider some non-trivial solutions of this equation.

4.1 Uniform fields

The equations of motion in the general theory of relativity [11] allow one to calculate the covariant or contravariant acceleration vector.

$$\frac{d^2x_i}{ds^2} - \Gamma_{i,kl}\frac{dx^k}{ds}\frac{dx^l}{ds} = 0; \frac{d^2x^i}{ds^2} + \Gamma^i_{kl}\frac{dx^k}{ds}\frac{dx^l}{ds} = 0. \tag{13}$$

For a fixed point, the acceleration vectors α_i or α^i are calculated by the formulas.

$$\alpha_i = \frac{d^2x_i}{dt^2} = c^2\Gamma_{i,00}; \alpha^i = \frac{d^2x^i}{dt^2} = -c^2\Gamma^i_{00}. \tag{14}$$

In this work, we will use the scalar acceleration value.

$$\alpha = \sqrt{\alpha_i\alpha^i}. \tag{15}$$

The simplest metric.

$$ds^2 = \left(1 + \frac{\alpha x}{c^2}\right)dt^2 - \left(dx^2 + dy^2 + dz^2\right) \tag{16}$$

has a constant acceleration $\alpha = \alpha_i = \alpha^i$ in the region $\frac{\alpha x}{c^2} > -1$. The rest of the space does not have any physical meaning. Note that the scalar curvature of the metric is negative

$$R = -\frac{2a^2}{(2ax + c^2)^2}. \tag{17}$$

We use a general approach to search for homogeneous accelerated frames of reference. Consider the gravitational field in the metric of the form.

$$ds^2 = s(x)dt^2 - p(x)\left(dx^2 + dy^2 + dz^2\right) \tag{18}$$

and one-dimensional metric.

$$ds^2 = s(x)dt^2 - p(x)dx^2 - dy^2 - dz^2. \tag{19}$$

Calculations of parameters of metrics are carried out. Metric (18) has a curvature tensor and its convolutions are much more complicated than metric (19). We did not consider it necessary to present these data in full. Calculations of vectors (14) give the same result for both metrics, and their scalar acceleration is related to the components of the fundamental tensor:

$$\alpha^2 = \frac{s'(x)^2}{4p(x)} \tag{20}$$

or

$$p(x) = \frac{s'(x)^2}{4\alpha^2}. \tag{21}$$

This makes it possible to write homogeneous metrics.

$$ds^2 = s(x)dt^2 - \frac{s'(x)^2}{4a^2}\left(dx^2 + dy^2 + dz^2\right) \tag{22}$$

and

$$ds^2 = s(x)dt^2 - \frac{s'(x)^2}{4a^2}dx^2 - dy^2 - dz^2 \tag{23}$$

where $s(x) > 0$ is an arbitrary function with a continuous derivative.
We see that the resulting metric has a non-zero scalar curvature.

$$R = -\frac{2a^2}{s(x)^2}. \tag{24}$$

System (22) also has a negative scalar curvature. However, we did not present it because of the cumbersome expression. This suggests that in a flat space, there is no uniformly accelerated metric at all.

A good example of a uniformly accelerated reference system is one with the metric:

$$ds^2 = e^{2ax}dt^2 - e^{4ax}dx^2 - dy^2 - dz^2. \tag{25}$$

With scalar curvature

$$R = -2a^2e^{4ax}.$$

4.2 Universe with constant curvature

At present, the cosmological hypothesis of the expanding Universe is being successfully used. The first attempts to describe a homogeneous stationary Universe on the basis of the general theory of relativity were made by A. Einstein and W. de Sitter. However, soon A.A. Friedman published an article [15], in which he proposed a relativistic model of a homogeneous non-stationary Universe. Somewhat later, E. Hubble discovered that the redshift of the optical lines of galaxies increases with increasing distance to them. This made it possible to interpret the redshift as a Doppler one, related to the speed of the galaxy's receding. Thus, the hypothesis of a homogeneous expanding universe has now become practically indisputable.

We are primarily interested in homogeneous spaces. Not only spaces with a uniform distribution of matter and constant curvature, but also spaces with a constant speed of light.

A metric with a constant speed of light belongs to exceptional metrics. Metrics with a constant speed of light were first considered by Einstein and Fokker [16]. The authors of the article [16] used a similar metric to discover its connection with the Nordström theory, which, as the authors showed, cannot describe the gravitational field, since the speed of light in this theory is constant, which, as the authors of the work showed, is incompatible with the general theory of relativity.

Metrics with a constant speed of light have the form:

$$ds^2 = f^2(c^2 t^2 - dx^2 - dy^2 - dz^2), \tag{26}$$

where f is a function of coordinates. For each metric of this kind, one can construct the Einstein equation by calculating the Einstein tensor for the given function f. Consequently, all metrics of this kind are related to solutions of the Einstein equation.

It is easy to see that in the space given by the metric (26), the speed of light is isotropic and constant. Indeed, the wavefront equation $ds = 0$ for the metric (26) coincides with the wavefront equation in the Minkowski space.

Let us require that the metric of the required space be homogeneous not only in the speed of light but also be homogeneous in curvature, i.e. had a constant scalar curvature R.

The Einstein equation for such metrics has the form.

$$G_b^a = -\frac{1}{4}R\delta_b^a. \tag{27}$$

Einstein's equation is a system of differential equations of ten unknown functions of four space-time coordinates.

It is essential that time is orthogonal to spatial coordinates. This allows you to separate the variables into spatial and temporal components. Then, the complete solution can be represented as two independent solutions:

$$ds_X^2 = A(X)c^2 t^2 - B(X)dx^2 - C(X)dy^2 - D(X)dz^2, \tag{28}$$

and

$$ds_t^2 = A(t)c^2 t^2 - B(t)dx^2 - C(t)dy^2 - D(t)dz^2. \tag{29}$$

The solution of the form (28), which depends on the spatial coordinates, is the well-known metric of the de Sitter space [17]. The metric tensor of this space, expressed in terms of scalar curvature [18], looks like this in spherical coordinates:

$$d(s_X)^2 = \left(1 + \frac{Rr^2}{12}\right)c^2dt^2 - \frac{dr^2}{1 + \frac{Rr^2}{12}} - r^2\left(d\vartheta^2 + sin^2\vartheta d\varphi^2\right). \tag{30}$$

Substituting the metric (29), in spherical coordinates, into Eq. (27), after simple transformations, we obtain the second solution of the equation.

$$d(s_t)^2 = \frac{-12}{Rc^2t^2}\left[c^2dt^2 - dr^2 - r^2\left(d\vartheta^2 + sin^2\vartheta d\varphi^2\right)\right]. \tag{31}$$

Let us immediately note that from one time value to another, a space with a given metric transforms conformally, i.e. all scales of space-time change in the same way. Then, if we use light standards to locally measure linear dimensions and time, we will not be able to detect any space changes. This means the real homogeneity of space-time.

The choice of the multiplier $f(\mathrm{t})$ in the metric (26) sets the proper time scale [14, 19] as a function of time at the origin. In the case of the metric.

$$d(s_t)^2 = \frac{1}{(Ht)^2}\left[c^2dt^2 - dr^2 - r^2\left(d\vartheta^2 + \sin^2\vartheta d\varphi^2\right)\right]. \quad (32)$$

is the dependence of the time τ at the origin of coordinates on the time of the spatial region located at a finite distance from the origin.

$$d\tau = \frac{1}{Ht}dt, \quad (33)$$

here H is a constant, t is the time it takes the light signal to reach the origin. If we have a set of identical generators with a unit period λ_0, then the frequency signal of the generator located at a distance $r = ct$ is a signal from the past and the point of view of the observer at the origin, the period of this generator [20].

$$\lambda - 1 = Ht\lambda_0. \quad (34)$$

Thus, we have a space with a redshift. The offset is given by Eq. (34). However, this law does not coincide with the Hubble law, which states that the dimensionless speed of the removal of galaxies is proportional to the distance to them[3].

$$v = Ht. \quad (35)$$

Usually, deviations from Hubble's law are shown using a graphical comparison with this law. The signal period of the receding signal generator is related to the dimensionless velocity.

$$\lambda = \frac{1+v}{\sqrt{1-v^2}}\lambda_0. \quad (36)$$

This makes it possible to compare the redshift laws (34) and (35). If we substitute the value λ into (34), we get.

$$\frac{1+v}{\sqrt{1-v^2}} - 1 = Ht. \quad (37)$$

Of course, the galaxies in the new space remain motionless, and v is a parameter of this space (**Figure 2**).

The figure shows the dependence of the redshift on the "velocity" of the recession of galaxies as if the redshift were Doppler. Of course, the galaxies in the universe (34) are stationary, and this picture was needed for comparison with the Hubble law.

Thus, in the stationary Universe, a redshift is observed, equivalent to the observed redshift with the effect of "accelerated expansion".

It remains to add that the density of such a universe[4].

$$\varepsilon = \frac{3H^2c^2}{8\pi G} \quad (38)$$

---

[3] In the framework of the Friedman model, the Hubble law is expressed in terms of the dimensional parameter $\dot{a}(t_1) = Ha(t_1)$.

[4] These results are derived from the Einstein Eq. (27).

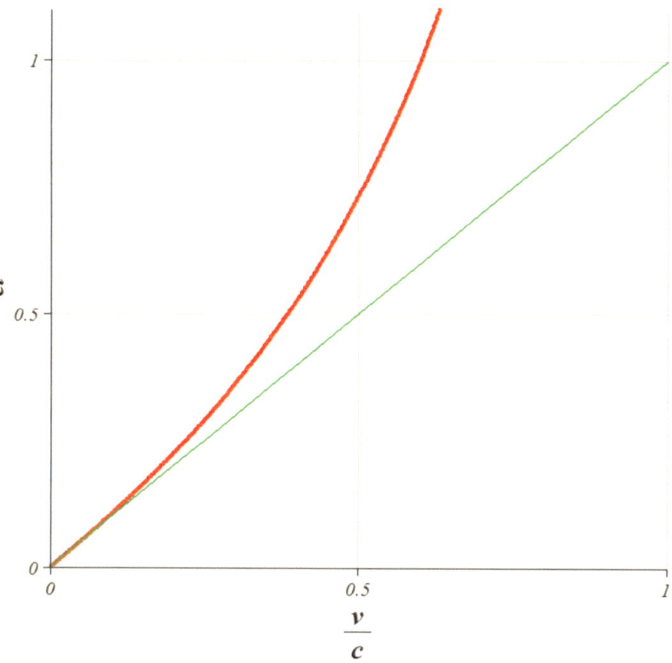

Figure 2.
Redshift in the universe with metric (32). In a stationary universe, stationary galaxies are assigned a speed v according to their redshift for comparison with Hubble's law. The horizontal axis shows the redshift in units of velocity. Vertical distance to the galaxy. The thin green line is Hubble's law. The thick red line is the distance to a stationary galaxy in a stationary universe.

coincides with the Friedmann critical energy density. The motion of test bodies in this Universe is determined by the equation of geodesics with Christoffel symbols.

$$\Gamma^{1}_{10} = \Gamma^{1}_{01} = \Gamma^{2}_{20} = \Gamma^{2}_{02} = \Gamma^{3}_{03} = \Gamma^{3}_{30} = \Gamma^{0}_{00} = -\frac{1}{T};$$
$$\Gamma^{0}_{11} = -\frac{1}{c^2T}; \Gamma^{0}_{22} = -\frac{r^2}{c^2T}; \Gamma^{0}_{33} = -\frac{\sin^2(\theta)r^2}{c^2T}. \tag{39}$$

These 10 quantities describe the behavior of a test particle at a distance cT from the observer. Who knows, perhaps Christoffel's symbols will make it possible to identify the new Universe with ours.

However, the dependence of the Christoffel symbols on time does not allow us to consider such a Universe as ideally homogeneous.

4.3 Homogeneous universe with cosmological redshift

Meanwhile, the solutions of Einstein's equation as applied to the gravitational field show the presence of a redshift, which is possibly a gravitational redshift. However, solutions of the Einstein equation without a gravitational field were demonstrated above, which have a redshift within the parameters according to the Hubble law, but at the same time do not serve to describe the expanding space:

$$ds^2 = \frac{1}{(Ht)^2}\left[dt^2 - dr^2 - r^2\left(d\vartheta^2 + \sin^2\vartheta d\varphi^2\right)\right]. \tag{40}$$

A natural question arises. Are there truly stationary non-trivial homogeneous universes with Christoffel symbols that are independent of time?

Consider a metric of the form:

$$ds^2 = a(t)^2\left[dx_0^2 - dx_1^2 - r^2\left(dx_2^2 + \sin^2\vartheta dx_3^2\right)\right]. \tag{41}$$

The time-dependent Christoffel symbols of this metric are:

$$\Gamma^1_{10} = \Gamma^2_{20} = \Gamma^3_{30} = \Gamma^1_{01} = \Gamma^2_{02} = \Gamma^3_{03} = \Gamma^0_{11} = \Gamma^0_{00} = \frac{\dot{a}(t)}{a(t)};$$

$$\Gamma^0_{22} = \frac{r^2\dot{a}(t)}{a(t)}; \Gamma^0_{33} = \frac{r^2\sin^2\vartheta\dot{a}(t)}{a(t)}. \tag{42}$$

The only solution to the equation.

$$\frac{\dot{a}(t)}{a(t)} = -H = const \tag{43}$$

is an.

$$a(t) = e^{-Ht}. \tag{44}$$

In contrast, the metric with Christoffel symbols independent of time is:

$$ds^2 = e^{-2Ht}\left[dx_0^2 - dx_1^2 - r^2\left(dx_2^2 + \sin^2\vartheta dx_3^2\right)\right]. \tag{45}$$

The equations of motion in the universe do not change with time and do not differ from the equations of the motion of flat space for motionless bodies. Therefore, we can conclude that there is no gravitational field in such a universe, even the Riemann tensor and its convolutions are nonzero.

Metric (2) is described by the Einstein equation:

$$G_{\mu\nu} = diag\left(3H^2, -H^2, -H^2r^2, -H^2r^2\sin^2\vartheta\right) \tag{46}$$

We associate the constant H with the Hubble constant. Then the right side of the equation can be expressed in terms of the (critical) density of the Friedmann universe $\rho_0 = \frac{3H^2}{8\pi G}$:

$$G_{\mu\nu} = 8\pi G\, diag\left(\rho_0, -\frac{1}{3}\rho_0, -\frac{1}{3}\rho_0 r^2, -\frac{1}{3}\rho_0 r^2\sin^2\vartheta\right). \tag{47}$$

The equations of motion in the universe do not change with time and do not differ from the equations of the motion of flat space for motionless bodies. Therefore, we can conclude that there is no gravitational field in such a universe, even though the Riemann tensor and its convolutions are nonzero.

Metric (45) is described by the Einstein equation:

$$G_{\mu\nu} = diag\left(3H^2, -H^2, -H^2r^2, -H^2r^2 \sin^2\vartheta\right) \tag{48}$$

We associate the constant H with the Hubble constant. Then the right side of the equation can be expressed in terms of the (critical) density of the Friedmann universe $\rho_0 = \frac{3H^2}{8\pi G}$:

$$G_{\mu\nu} = 8\pi G\, diag\left(\rho_0, -\frac{1}{3}\rho_0, -\frac{1}{3}\rho_0 r^2, -\frac{1}{3}\rho_0 r^2 \sin^2\vartheta\right). \tag{49}$$

Eq. (48) does not depend on the choice of origin of the spatial coordinates. The proper time at the origin is related to the time t, which is counted from the initial time $t = 0$:

$$d\tau = e^{-Ht}dt \approx \frac{1}{Ht}dt. \tag{50}$$

here t is the time for which the light signal reaches the origin. If we have a set of identical generators with a unit period λ_0, and the frequency signal of the generator located at the distance $r = ct$ is a signal from the past and the point of view of the observer at the origin, it is the period of this generator:

$$\frac{\lambda}{\lambda_0} = e^{Ht} \tag{51}$$

Thus, the space under consideration is redshifted. The offset is given by Eq. (51). However, this law does not coincide with Hubble's law, which states that the speed of the recession of galaxies is proportional to the distance to them.

$$cz = v = Hr. \tag{52}$$

In our case, the galaxies are motionless. However, we can introduce a fictitious speed and, accordingly, a redshift. For this, we use the relativistic Doppler shift formula:

$$\frac{\lambda}{\lambda_0} = e^{Ht} = \sqrt{\frac{1+v}{1-v}}, \tag{53}$$

or

$$r = ct = \frac{c}{2H}\, ln\, \frac{1+v}{1-v}. \tag{54}$$

This allows for the comparison of the redshift laws.

Figure 3 shows the dependence of the redshift on the recession "velocity" of the galaxies as if it is a Doppler redshift. The fact is, of course, that the galaxies in the universe are stationary and this picture (54) was needed to show a comparison with Hubble's law.

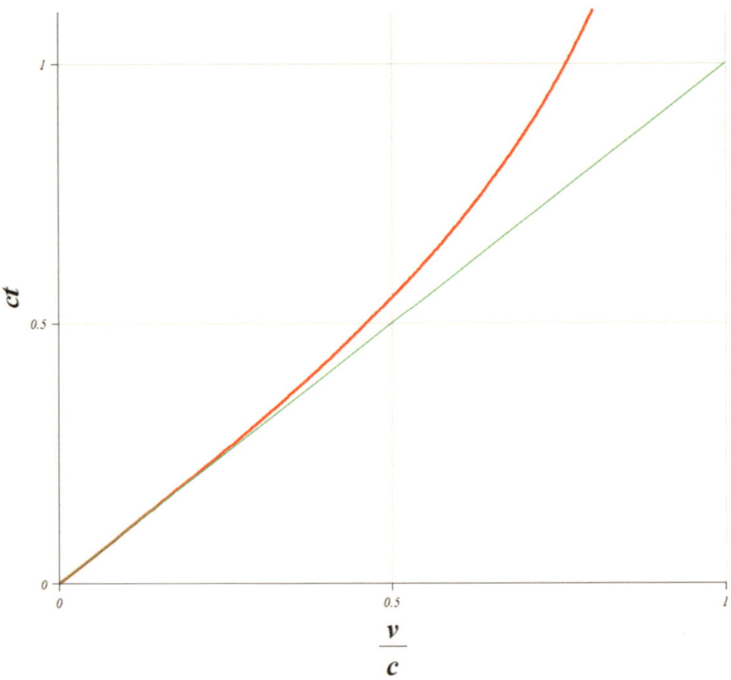

Figure 3.
A redshift in the universe shown through a metric (45). In a stationary universe, stationary galaxies are assigned a speed "v" according to their redshift properties for comparison with Hubble's law. The horizontal axis shows the redshift in units of velocity. The vertical is the distance to the galaxy. The thin green line is Hubble's law. The thick red line is the distance to a stationary galaxy in a stationary universe.

5. Reducing the entropy of gas flows in a gravitational field

A while ago, an amazing result was published [21]. The authors demonstrated a magnet-controlled superconducting diode in a multi-layer structure [Nb/V/Ta]n having no inversion center. During the diode transition from the superconductivity to the normal state, the curve of the current-dependent non-reciprocal resistance was observed when measuring the current direction. Notably, it is thought that the difference in the critical current is due to the magnetic field-induced anisotropy caused by the inversion of space & time. Theoretically, such a superconducting diode rectify some thermal current noises; in so doing it can play a role of the Maxwell's demon, when converting the thermal energy into the current.

Another well–known example of non-reciprocal devices is found in electromagnetic fields. The reciprocity theorem was proved for the electromagnetic radiation in the absence of a magnetic field [22]. In the practice, the non-reciprocal devices are implemented in the form of microwave or optical devices with one-way bandwidth [23, 24]. Those devices are based on the Faraday Effect, i.e. on rotation of a polarization plane of electromagnetic waves in a medium located in a magnetic field.

The Russian physicist Nikolai Nikolayevich Pirogov published a paper back in 1887 in which he considered gas in a gravitational field based on the methods of statistical physics [25]. Here he proves that the temperature of an ideal gas column in a gravitational field and in an equilibrium state does not depend on altitude and derives the now well-known exponential barometric dependence of altitude pressure.

Pirogov did not limit himself to the stationary case and considered gas flows in a potential field. At the same time, he showed that.

> *"the well-known theory of the Carnot cycle is correct only under the assumption that the working substance is not affected by external or long-range forces, or, which is all the same if the action of these forces can be neglected."*

Further, Pirogov gives an example of a heat engine operating due to the expansion of gas in a gravitational field. The operation of the machine was ensured by the supply of heat from the surrounding space. At the same time, the author did not believe that such a machine violates the second law of thermodynamics because this law is applicable only to finite regions of space, while the gravitational field is infinite.

First of all, we are interested in the law of expansion/compression of gas in pipes of constant cross-section. Unlike the well-known (in hydrodynamics) adiabatic gas flow in pipes of variable cross section (for example, in a Laval nozzle), gas expansion under the action of some external forces cannot be described using the usual adiabat (Poisson's adiabat). Fortunately, the problem is solved simply [26] based on the laws of conservation of mass, energy, and momentum. Let us choose two sections in the vertical gas column; now the ratio of pressure and density is described by the relation:

$$\frac{V_2}{V_1} = \frac{(\gamma+1)p_2 + (\gamma-1)p_1}{(\gamma-1)p_1 + (\gamma+1)p_2}, \tag{55}$$

where adiabatic index is $\gamma = C_p/C_v$; C_p is heat capacity at constant pressure; and C_v is constant heat capacity.

A gas flow with a shock wave or in a flow with compression along a potential field is accompanied by an increase in entropy, which can be estimated using the difference in entropy in the final and initial stationary flow [26]:

$$s_2 - s_1 = \frac{1}{12T_1}\left(\frac{\partial^2 V}{\partial p_1^2}\right)_s (p_2 - p_1)^3. \tag{56}$$

It is usually stated that the difference between the entropies of the adiabatic compression of a gas in a plane-parallel flow cannot be negative. This should follow from the second law of thermodynamics. However, N.N. Pirogov drew attention [25] to the fact that the gas flow in a potential field can, depending on the direction of the flow, both compress and expand the gas. But then, according to Eq. (56), we can conclude that the entropy of the expanding gas in the potential field decreases.

The main results of the theory of shock waves [26] allow us to compare two adiabatic processes, the shock adiabat and the isentropic Poisson adiabat.

So the dependence of temperature on pressure in the Poisson adiabat in a polytropic gas.

$$\frac{T_P}{T_0} = \left(\frac{p}{p_0}\right)^{\frac{\gamma-1}{\gamma}}. \tag{57}$$

where γ is the adiabatic exponent. Flow temperature in the shock adiabat.

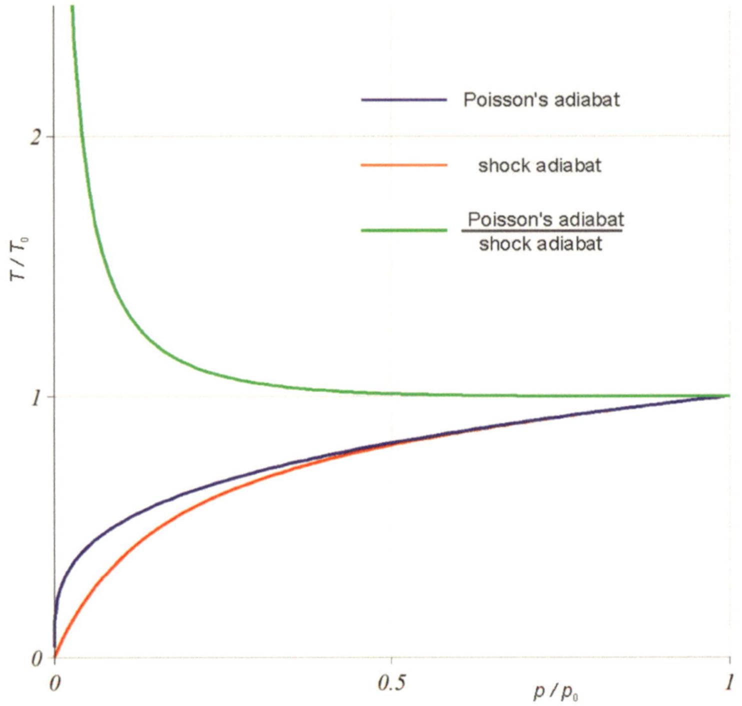

Figure 4.
Expansion of gas with γ = 1.4 in a potential field (red line) compared to the Poisson adiabat (blue line): Dependence of the normalized temperature on the normalized pressure and their ratio (adiabatic efficiency) (green line).

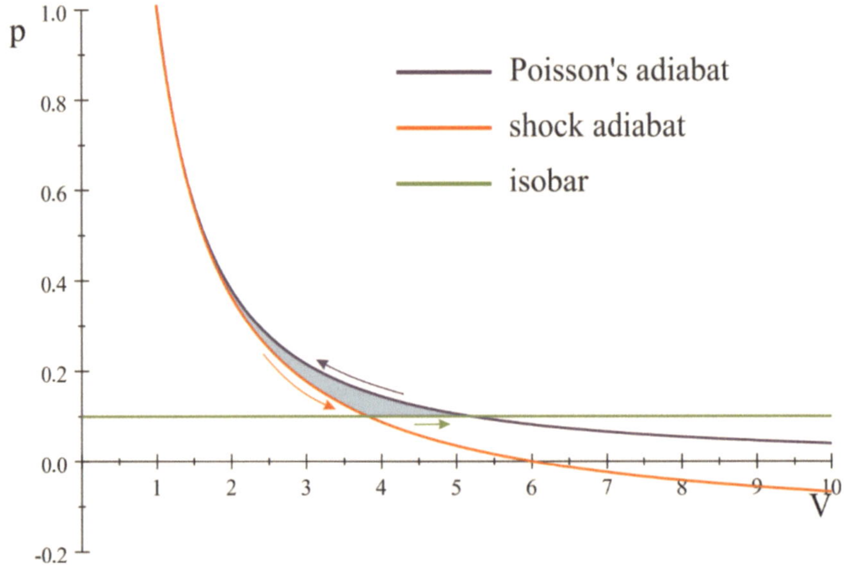

Figure 5.
Hypothetical anti-Carnot cycle.

$$\frac{T_s}{T_0} = \frac{p}{p_0}\left[\frac{(\gamma+1)p_0 + (\gamma-1)p}{(\gamma-1)p_0 + (\gamma+1)p}\right]. \tag{58}$$

In **Figure 4**, we see that when expanding in the shock adiabat, the gas cools faster, and the adiabatic efficiency is always greater than unity. Similar phenomena are observed more often than many people think. Tornadoes, atmospheric cyclones, and almost certainly cosmic jets are stable, thanks to the drop in entropy and the release of the internal energy of the gas/plasma. In this case, the potential force of the vortex motion to the opposing gravitational forces (**Figure 5**).

Author details

Valery Borisovich Morozov
NIITeplopribor, Moscow, Russia

*Address all correspondence to: valerymorozov@gmail.com

References

[1] Einstein A. The Meaning of Relativity. Princeton: Princeton University Press; 1921

[2] Minkowski G. Raum und Zeit. Phis. ZS. 1909;**10**:104

[3] Einstein A. Über das Relativitätsprinzip und die aus demselben gezogenen Folgerungen. Jahrb. d. Radioaktivitat u. Elektronik. 1907;**4**:411-462

[4] Einstein A. Zur Theorie des statischen Gravitationsfeldes. Annals of Physics. Leipzig. 1912;**38**:443-458

[5] Einstein A. About the ether (Russian). Collection of Scientific Papers. T.2. Science. Moscow. 1966. p. 154

[6] Shapiro II. Fourth test of general relativity. Physical Review Letters. 1964; **13**(26):789-791

[7] Morozov VB. Law of conservation of energy of gravitational fields and the gravitational field equation. Parana Journal of Science and Education. Parana. 2022;**8**(8):1-6

[8] Bell JS. Speakable and Unspeakable in Quantum Mechanics. Cambridge: Cambridge University Press; 1993. p. 67

[9] Einstein A. Über die spezielle und die allgemeine Relativitatstheorie (Gemeinverstdndlich). Druck und Verlag von Friedr. Braunschweig: Vieweg & Sohn; 1920

[10] Fock VA. The Theory of Space, Time and Gravitation. 2nd edition. Pergamon Press – Macmillan Company; 1964

[11] Landau LD, Lifshitz EM. Fluid Mechanics. 2nd ed. Vol. 6. Pergamon Press; 1987

[12] Logunov AA, Mestvirishvili MA, Chugreev Yu V. On incorrect formulations of the equivalence principle. Physics-Uspekhi. 1996;**39**: 73-79

[13] Møller C. Theory of Relativity. 2nd ed. Oxford: Oxford University Press; 1972

[14] Cornish N, Blas D, Nardini G. Bounding the speed of gravity with gravitational wave observations. Physical Review Letters. 2017;**119**:161102

[15] Friedmann A. Über die Krümmung des Raumes. Zeitschrift für Physik 10, Nr. 1, S. 377– 386. 1922

[16] Einstein A, Fokker AD. Die Nordströmsche Gravitationstheorie vom Standpunkt des absoluten Differentialkalküls. Annals of Physics. Leipzig. 1914;**44**:321-328

[17] Einstein A. Relativistic theory of asymmetric field (Russian). Collection of Scientific Papers. T.2. Science. Moscow. 1966. p. 849

[18] de Sitter W. On Einstein's theory of gravitation, and its astronomical consequences. Vol. 78. 3rd paper. Oxford: Monthly Notices of the Royal Astronomical Society; 1917. p. 3

[19] Nordstrom G. Träge und schwere Masse in der Relativitätsmechanik. Annals of Physics. Leipzig. 1913;**42**:533

[20] Hawking SW, Ellis GFR. The Large Scale Structure of Space-Time. Cambridge: Cambridge University Press; 1973

[21] Ando F, Miyasaka Y, Li T. Observation of superconducting diode effect. Nature. 2020;**584**:373-376

[22] Vaganov RB, Katsenelenbaum B. Fundamentals of Diffraction Theory (Russian). Moscow: Nauka; 1982. p. 272

[23] Abramov VP, Dmitriev VA, Shelukhin SA. Non-reciprocal devices on ferrite garnets (Russian). Moscow. 1989

[24] Voitovich AP, Severikov VN. Lasers with anisotropic resonators (Russian). Minsk: Institute of Physics of the Academy of Sciences of the BSSR; 1988

[25] Pirogov NN. Applicability of beginnings to systems that are acted upon by external forces (Russian). St. Petersburg: Demakova; 1887. p. 40

[26] Landau LD, Lifshitz EM. The Classical Theory of Fields. 4th ed. Vol. 2. Pergamon Press; 1987

Chapter 6

Modeling and Simulation of Fluid Flow and Energy Systems

Firas Makahleh and Anas Nassar

Abstract

Engineering and science-related problems become more complicated as human knowledge evolves. This complication includes apparatus geometry and operational environment such as extreme variations in pressure and temperature. Analytical solution for such problems needs many assumptions that underestimate the problem under study and could lead to unrealistic results. Moreover, an experimental setup for a certain problem is constrained by the prototype size and each experiment is set up for certain operating conditions. This leads to building up many setups to deal with changes in size and operating conditions and, therefore, the prototype validation becomes very expensive and time-consuming. This calls for modeling and simulation approaches to deal with such engineering problems with the powerful computational capabilities available nowadays. Real-world patterns and processes are roughly modeled by scientific models. They may be refuted because they are representations, which are by definition imperfect. Models, however, are quite helpful for a variety of reasons. They first give us a method to comprehend procedures that would otherwise be incomprehensible to us. They also give scientists a base on which to build new research and theories. Finally, modeling and simulation reduce the time and cost of prototyping.

Keywords: modeling, simulation, energy, fluid, ANSYS-FLUENT, MATLAB

1. Introduction

Modeling is a small-scale representation of a process or a device, using mathematical descriptions implemented by a computer program to predict the behavior of a system at different operating conditions and at any size scale of a process or a system. Simulation is the process of representing the results of the model in a clear and understandable format like figures, tables, and animation. Simulation is used to examine the function of a device that is not subjected to experimentation. A good example of this is modeling and simulation of energetic materials (explosives) where experimentation is almost impossible due to its hazardous effects and very expensive cost [1, 2].

Simulation provides evidence for decision-makers to proceed to industrial and commercial production of the products. Verification and validation of the model are very important and for that sufficient data and information are used to prove the result of any model. Modeling and Simulation have many advantages since they reduce the time and cost of real experimentation, are suitable for severe operation conditions (very low and high temperature, very low and high pressure) in which real experimentation is very expensive, and give information that increases the confidence of the stakeholders in the product.

This chapter discusses some of the engineering applications related to our own experiences. Three subjects are covered; fluid flow and energy systems. Each subject is discussed in three categories. Namely, CFD packages, object-oriented programming, procedural programming, and the most recent usage of Artificial Intelligence (AI) for fluid flow and energy systems.

ANSYS-FLUENT uses the finite volume method to solve the governing equations of fluid flow and heat transfer. This involves dividing the domain of the simulation into a series of small, interconnected control volumes or cells. The values of the variables (such as velocity and temperature) are then calculated at the centroids of these cells. ANSYS-FLUENT uses iterative algorithms to solve these equations and arrive at a converged solution.

Artificial Intelligence (AI) has recently been used in the modeling and simulation of fluid and energy systems [3–8]. The term ML is also used synonymously. Machine learning (ML) arguably has four types: supervised learning, unsupervised learning, reinforcement learning (RL), and deep learning (DL). In supervised learning, the model learns from labeled data. Hence, the input and output are known. Whilst in unsupervised learning, the output is not known and the data are unlabeled. The model learns the similarities and correlations in the unlabeled data. On the other hand, RL, the third type of machine learning [3, 4], attains long-term objectives by direct contact with the environment, without prior knowledge. The RL agent assesses the state of the environment and selects the best course of action. The RL agent is rewarded positively if it does the right action. A negative is created if the agent commits a mistake. In RL, the model learns through trial and error and does not require a big dataset [5].

Finally, DL requires big data and is where the biological brain is mimicked through the concept of neurons. A neuron is the building block of any biological brain. By connecting multiple neurons in a network, an artificial neural network is generated and could learn from data. **Figure 1** shows the basic concept of such a network.

Summary of all four AI (or ML) configurations and methods is shown in **Figure 2**.

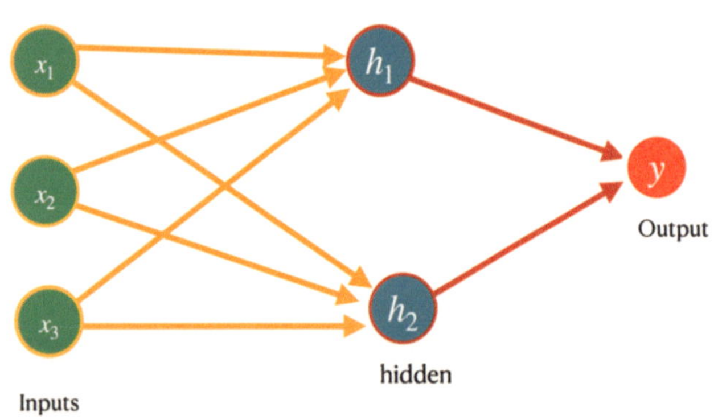

Figure 1.
An artificial neural network configuration showing input, hidden, and output layers [4].

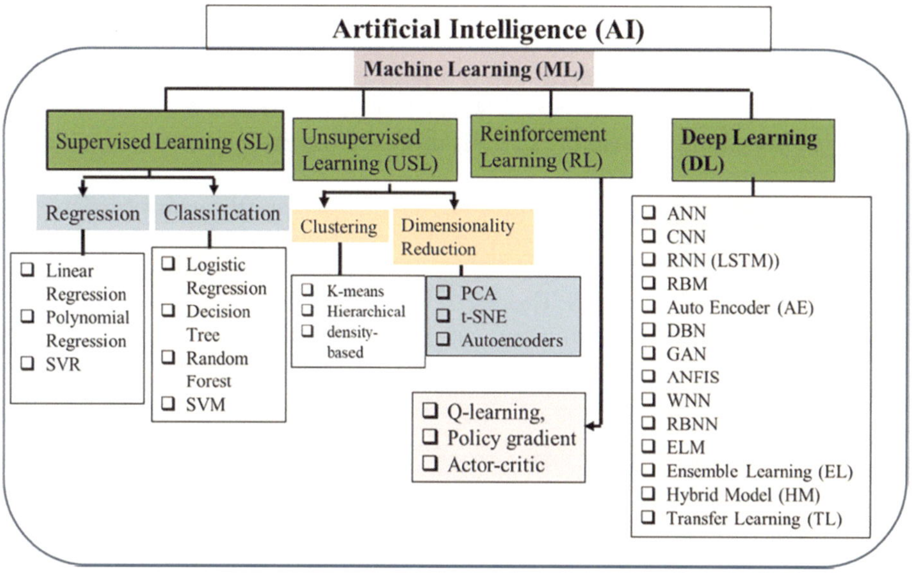

Figure 2.
The four types of ML: Supervised, unsupervised, reinforcement, and deep learning methods [4].

2. Modeling of fluid flow

Fluid flow modeling is very important since it is widely used for many industries that are not limited to aero-planes, vehicles, air conditioning, propulsion, bio-fluids, nano-fluids military projectiles, and missiles. Fluid flow is classified as continuum flow or molecular flow at a low Knudsen number in which molecules are moving freely in its domain. Other five categories of classification include; laminar or turbulent, steady or unsteady, compressible or incompressible, viscous or nonviscous, and rotational or irrotational flow.

2.1 Computational fluid dynamics (CFD)

Computational fluid dynamics (CFD) became the most popular tool used to model fluid flow due to increase in computing capabilities offered by both main frame super computers and even personal computers that could deal with complex geometry and operating conditions. Computational fluid dynamics modeling or CFD is based on the principles of fluid mechanics, utilizing numerical methods and algorithms to solve problems that involve fluid flows.

CFD is classified into difference method (FDM), finite volume method (FVM), finite element method (FEM), lattice Boltzmann method (LBM), and smoothed particle hydrodynamics (SPH). CFD tools include many packages that are commercially distributed like ANSYS-ANSYS-FLUENT, Crew, COMSOL, and PATRAN or Open source CFD tools like Open Foam, Palabos, and MFIX. Another possibility for CFD is the direct numerical solution of fluid flow differential equations by procedural programming or by using object-oriented programs like MATLAB or MODELICA.

CFD tools include many software that are commercially distributed like ANSYS-FLUENT, Crew, and PATRAN. Another possibility for CFD is the direct numerical solution of fluid flow differential equations by procedural programming like C++,

MATLAB, and Fortran or by using object-oriented programs like MATLAB or MODELICA. ANSYS-FLUENT is a general-purpose computational fluid dynamics (CFD) software offered by ANSYS, which is used to model fluid flow, heat, and mass transfer, chemical reactions, and more. ANSYS-FLUENT offers a modern, user-friendly interface that streamlines the CFD process from pre- to post-processing within a single window workflow.

Many examples of using ANSYS-FLUENT will be discussed throughout following paragraphs. ANSYS-FLUENT was used to study the flow of hydrogen in the scramjet engine combustor with different type's fuel injectors at varied velocity, pressure, and temperature as shown in **Figure 3**. The researcher showed that Pylon injector gave better results for stabilized mixing of air-fuel at any Mach number over other types used in the simulation [9]. Mesh and typical results are shown in **Figure 4**. It is very important to

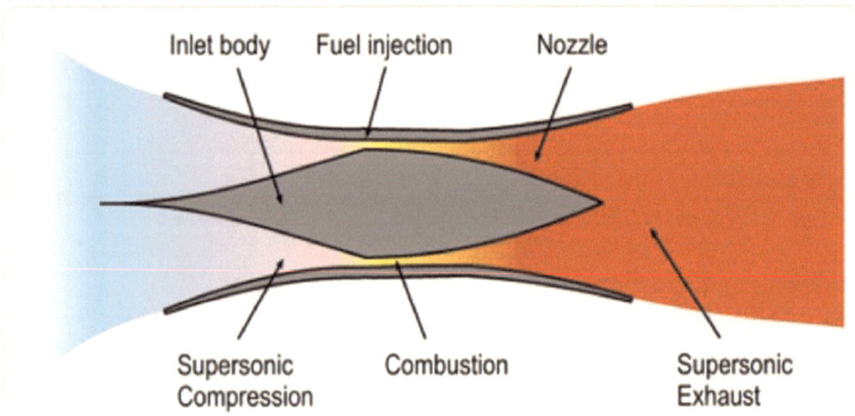

Figure 3.
Scramjet propulsion system [9].

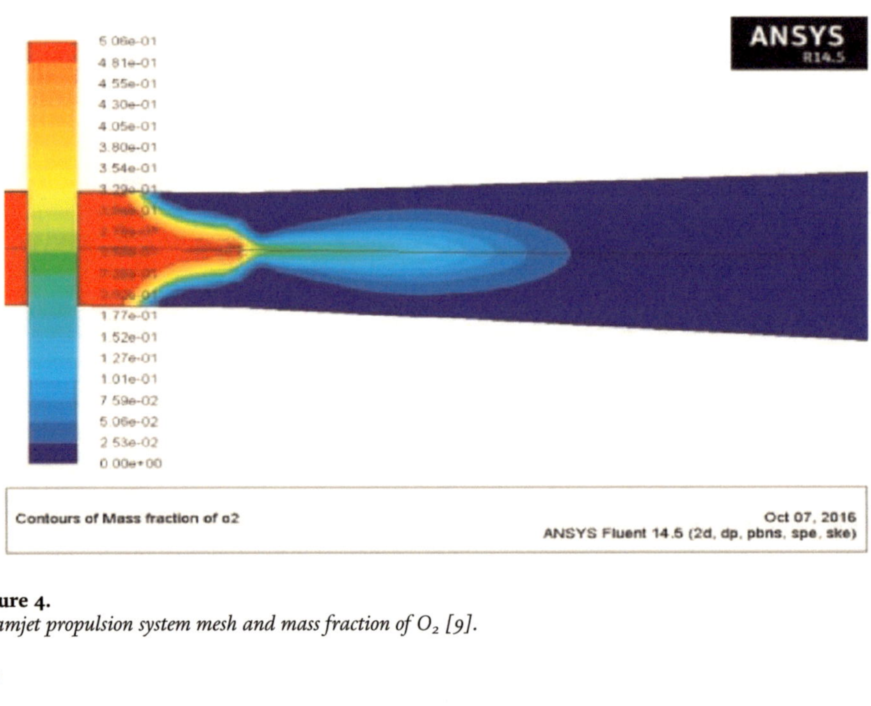

Figure 4.
Scramjet propulsion system mesh and mass fraction of O_2 *[9].*

investigate the mesh sensitivity (mesh independence or mesh convergence) for each modeling problem since this will increase the confidence in the results and so became one of the most important steps in the simulation. Mesh sensitivity provides information on the errors associated with mesh size and type as errors tend to zero as mesh size tends to zero but this will lead to unwanted extra computations that are necessary for model validation. The idea is to start with coarse mesh then this could be refined by the ANSYS-FLUENT program at any time during modeling and simulation.

Another example is where ANSYS-FLUENTCFD package is used to model underwater complex geometry vehicle to figure out the parameters that could be applied to the open-shelf depth setting TUV. The TUV predesign was validated by ANSYS-FLUENT to reduce the cost of repeated prototypes and shorter calculation time when using direct numerical simulation. **Figure 5** showed the model and physical photo of TUV system.

The simulation with ANSYS-FLUENT started with 2D modeling, meshing independence test, specifying boundary conditions on geometry, setting the turbulent model and material, and getting results as shown in **Figure 6**. There are no differences between the fluid dynamic simulation of deep-setting TUV properties and the

Figure 5.
Model and physical photo of TUV system [10].

Figure 6.
Pressure distribution of TUV system [10].

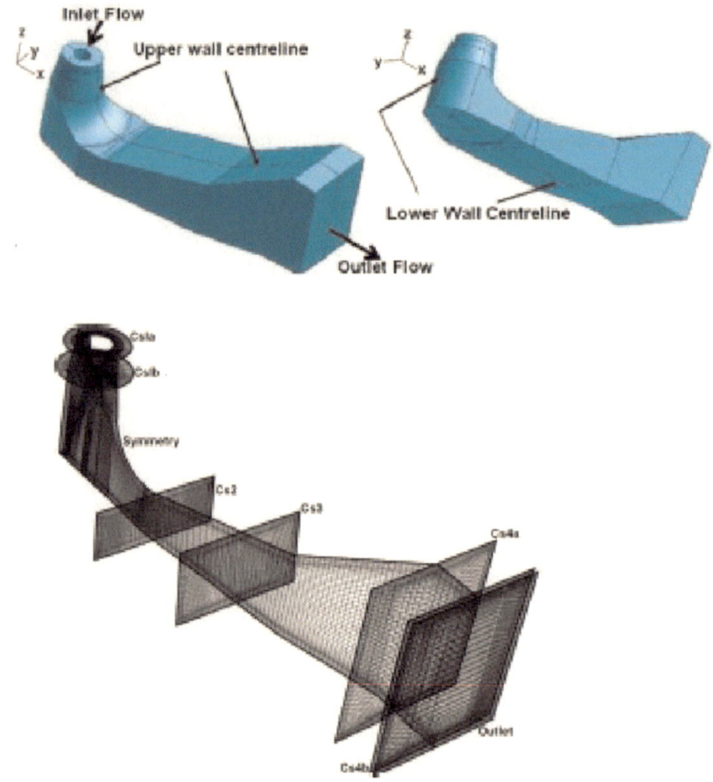

Figure 7.
Geometry and mesh of draft tube [11].

empirical formula. The reliability of simulation results and the simulation results are intuitive within an acceptable range. The deep-setting model and programming can then be created. For pre-post design quantitative analysis, the earlier design stage can be leveraged to shorten design cycles and conduct qualitative research. Consequently, it can lower development costs.

Also, ANSYS-FLUENT could be used to investigate the accuracy of the k-ε models applied to certain applications, for example, the accuracy of k-ε model applied to turbine 99 draft tube as shown in **Figure 7**. Governing equations are given in Eqs. (1) and (2). Two different grid concentrations near the wall y + 1 and y + 50 were used to study the flow behavior. Discussion is based on graphical results and by comparing numerical simulations and experiments in operational mode. It has been investigated how these models relate to the first-order upwind, second-order upwind, third-order upwind (QUICK), and power law schemes.

2.2 Modeling of fluid flow using object-oriented and procedural programming

For simple geometry one and two dimensions modeling, researchers preferred to implement simple programming to save the time during scope stage of the project that imposed many changes during project evolution. Simple programming could be elaborated by using procedural languages like C++ and Fortran or by more representable, easy interactive graphical interface, and object-oriented software like MATLAB and TRNSYS.

2.3 Modeling of fluid flow using object-oriented and procedural programming

For procedural programming, the system is described by a set of differential equations or matrix then solved simultaneously at given initial or boundary conditions. To model the vacuum pressure for the 340 m vacuum chambers at SESAME light source storage ring using molecular pump as shown in **Figure 8** [12], the researcher had two options either to use a 3D modeling that required 3–6 months of modeling the complicated chamber or to use one-dimensional matrix computation with 5 days handwritten program at early stage of the project with daily changes.

To evaluate the pressure value through unit cell, MATLAB program was developed to solve the linear system of balanced equations of three consecutive connected elements. One cell is divided into 152 elements, each one is approximately 11 cm in length except those elements where pump is located 20 cm long, and the total length is approximately 16.3 m.

The governing continuity equations:

$$\mathbf{C_i}\ (\mathbf{P_{i-1}} - \mathbf{P_i}) + \mathbf{C_{i+1}}\ (\mathbf{P_{i+1}} - \mathbf{P_i}) + \mathbf{Q_i} = \mathbf{S_i P_i}\ (i = 1, 2, \dots, n = 152) \tag{1}$$

with the following periodic boundary condition:

$$P_1 = P_{152}$$

where:

$\mathbf{P_{i-1}}$, $\mathbf{P_i}$, $\mathbf{P_{i+1}}$ unknown pressure of the elements **i − 1, i, and i + 1.**

$\mathbf{C_i}$ and $\mathbf{C_{i+1}}$ conductance value **i** and **i − 1** and between element **i** and **i + 1.** evaluated for CO molecules using MOLFLOW.

$\mathbf{S_i}$ is the pumping speed of element **i**

$\mathbf{Q_i}$ is the total gas load of element **i**

The results of MATLAB program shown in **Figure 9** are compared with 3D MOLFLOW software, and experimental data with very good agreement as shown in **Table 1**.

One crucial step toward low-carbon environmental protection, energy conservation, and emission reduction development goals, as well as toward meeting the requirements of sustainable development plans, is clean heating. Another industrial use of orient project MATLAB/SIMULINK for modeling of stratified water tank in order to increase the precision of the stratified hot water storage tank model and simplify the stratified water tank modeling process. The three nodes stratified water tank and SIMULINK model are shown in **Figure 10**.

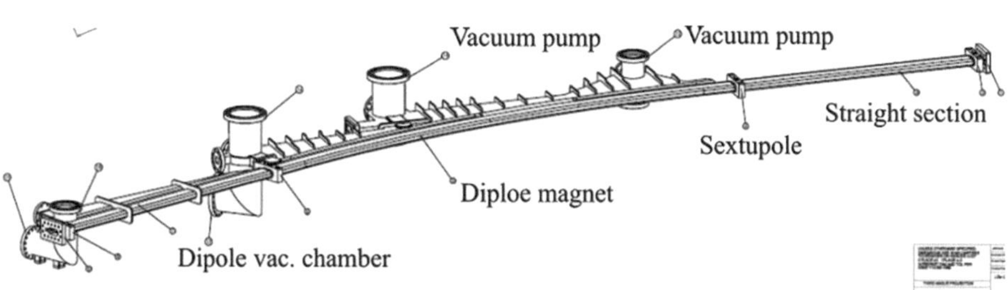

Figure 8.
One cell of SESAME storage ring [12].

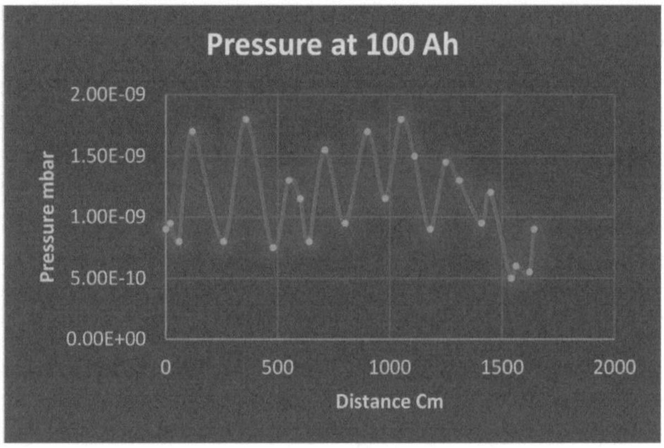

Figure 9.
MATLAB vacuum pressure inside one cell of SESAME storage ring [12].

Dose (Ah)	Current (mA)	Model's results Pressure (mbar)			Measurements Results (mbar)	
		MATLAB	VACCALC	MOLFLOW	With Beam	WO Beam
0.01	10	—	—	1.26×10^{-7}	2×10^{-8}	3×10^{-10}
100	100	1.26×10^{-9} (400 mA)	1.29×10^{-9} (400 mA)	1.4×10^{-9}	3×10^{-9}	1×10^{-10}
500	300	—	—	1.05×10^{-9}	5×10^{-10}	5×10^{-11}
1000	400	—	—	8.75×10^{-10}	2×10^{-10}	1.5×10^{-11}

Table 1.
Model's results VS measurement results [13].

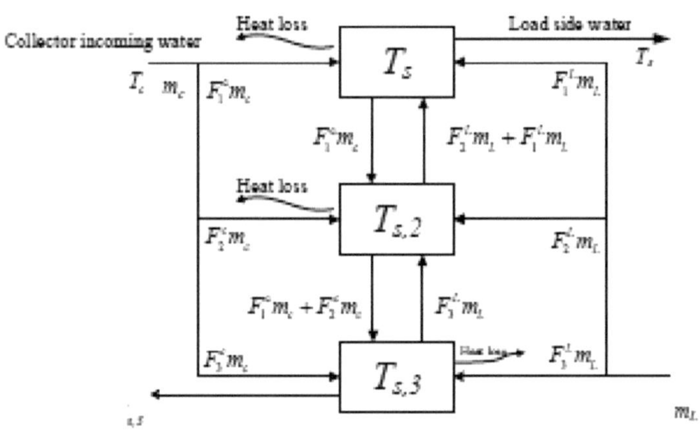

Figure 10.
Three nodes stratified water tank and SIMULINK model [14].

It is confirmed that this procedure is accurate and correct by comparing it to a platform simulation run on the TRNSYS system. A six-node stratified hot water storage tank is also created using this technique. The effect of various inlet and exit

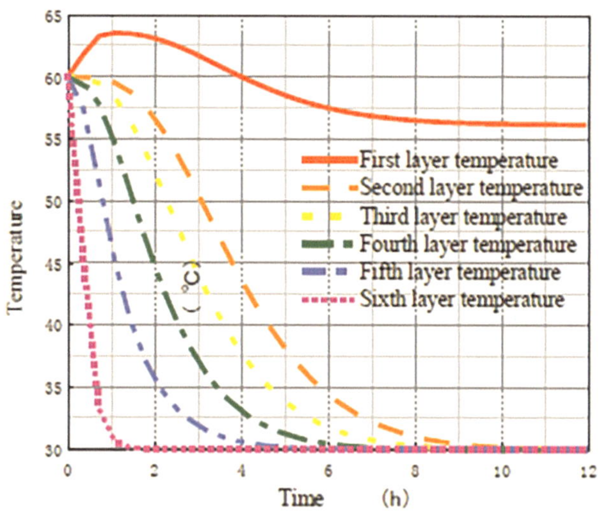

Figure 11.
Results of SIMULINK model at working condition one [14].

water flow rates and temperatures on the stratification effect of the hot water storage tank is examined through the modeling of 8 distinct working situations. Sample of results is shown in **Figure 11**.

The SIMULINK showed that stratification of the hot water storage tank is more likely to occur when the intake water flow on the collector side is higher than the return water flow on the load side. The higher the temperature difference between the inlet water temperature on the collector side and the return water temperature on the load side, the more conducive it is for the stratification of the hot water storage tank when the inlet water flow rate on the collector side and the return water flow rate on the load side.

2.4 Modeling of fluid flow systems by artificial intelligence

Fluid flow has two regimes. Depending on the speed of the flow, which is characterized by the dimensionless Reynolds number, these regimes are called laminar and turbulent. Simulation of fluid flow, in either regime, is called computational fluid dynamics (CFD). Especially in the turbulent regime, CFD has always been an expensive simulation if a high-accuracy solution is required. This is due to the complicated partial differential (PDE) solved in such simulations, Navier-Stokes eqs. A trade-off decision between accuracy and speed is made in such problems. If a full resolution and the highest accuracy are required, the simulation is called direct numerical simulation (DNS). That is, Navier-Stokes equations are completely solved to the finer mesh scale. However, DNS is sometimes impossible and, therefore, other methods are traditionally used. These methods use coarser meshes but reduce the accuracy of the solution. The methods, namely, are Reynolds averaged Navier-Stokes (RANS) [15] and large-eddy simulation (LES) [16].

To overcome CFD simulations, pure ML models have been used [17–22]. However, there could be some caveats that should be looked at when going the route of pure ML models. First off, ML techniques like deep learning frequently cost a lot to train and need a lot of data. Therefore, it is critical to pinpoint instances when ML excels above established, decades-old techniques and may be more precise and effective. The method used to create the training data and whether the cost is included for benchmarking raise further concerns. Transfer learning is a potential field in this context [23]. It is also important to remember that there are deep learning alternatives that could be better suited for specific jobs. Having this said, it is worth noting that a 700× faster than simulation speed could be obtained when such pure ML models are used [17]. **Figure 12** illustrates the general approaches ML used in fluid flow.

The reduced order model (ROM) depends on the fact that even complicated flows frequently display a few prominent coherent structures that might provide basic but useful information about the flow. In order to characterize the fluid in a lower-dimensional, lower-fidelity manner, ROMs explain the development of these coherent structures. By providing optimization and control tasks that need several model iterations or quick model predictions, ROMs act as a quick surrogate model for the more expensive CFD approaches discussed above. This efficiency comes at the expense of generality since ROMs are designed for a particular flow arrangement, offering tremendous acceleration but with a constrained scope of use. In general, the lack of generalization was a major drawback in pure ML models. A hybrid approach that combines machine learning with CFD in a single framework shows more promising results where physics is predicted accurately and the simulation speed is increased [25].

CFD is essential for many science and engineering applications. It is, for example, at the core of climate simulation and modeling and for understanding climate change. However, the available and used paradigm widely used is purely physics-based. That is, the known physical equations (partial differential equations; PDEs) are solved through numerical differentiation and integration. This is very computationally expensive. In the recent literature, data-driven approaches have been used. This includes deep learning. One main drawback of deep learning is that it is physics-illiterate. That is, it is mainly a statistical method that has little or no knowledge of fluid flow physics.

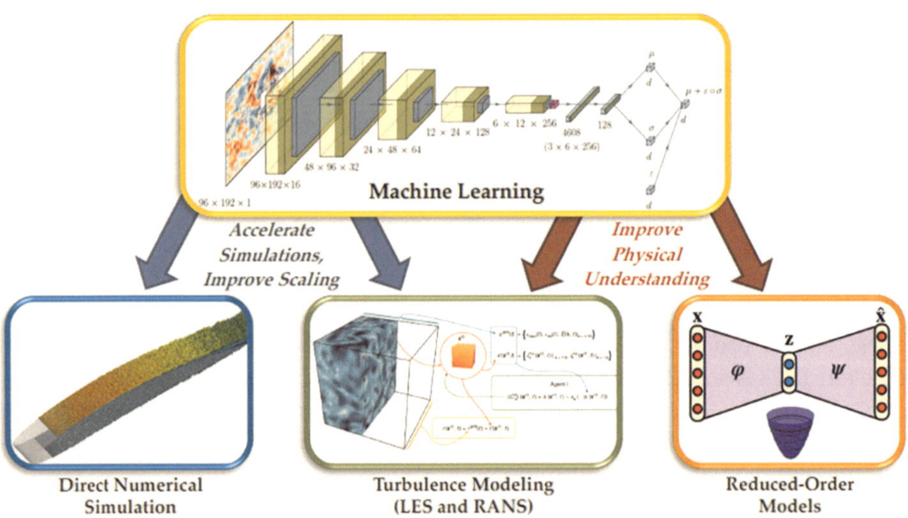

Figure 12.
In the context of direct numerical simulations, turbulence modeling, and reduced-order models, a summary of some of the most pertinent areas where machine learning might improve CFD [24].

Therefore, another paradigm of fluid flow simulation is the physics-informed neural network approach (PINN). A crucial step in improving AI for physical sciences is creating deep learning techniques that can systematically incorporate physical rules. PINNs are the result of intersecting the two approaches; data-driven and physics-based. On top of utilizing the system's physical equations, they use the data-driven supervised neural network to learn the physics. Therefore, they get both advantages. Being consistent with the physics and remain a data-driven approach. Moreover, they could extrapolate beyond the range of available data, as they benefit from being physics-informed. As a result, PINNs are suitable for applications where data are limited. This is particularly appealing for many science and engineering problems. Where data are already scarce.

Such examples of PINN are the modeling of turbulent flow [20] or the prediction of sea surface temperature [26, 27] or the temperature of lakes [28]. Or even in fluid animation applications (such as gaming) where the physical consistency constraint is less stringent [29].

3. Modeling of energy system

Energy is the capacity to carry out work. People have figured out how to transform energy from one form to another and then use it to accomplish tasks, making modern civilization possible. Walking and bicycling, driving cars on roads and boats through water, cooking meals on stoves, making ice in freezers, lighting our homes and workplaces, producing goods, and sending astronauts into space all require the usage of energy. There are many forms of energy: 99, light, motion, electrical, chemical, and gravitational. There are several energy sources, however, they may all be categorized into one of two groups.

Renewable energy and nonrenewable sources of energy. Large energy systems like power plants, wind turbines1, or even nuclear plants are very expensive and so modeling the output of each one is very important to give complete information to decision-makers to avoid misconceptions and low performance of plant before installation. CFD packages, object-oriented software like MATLAB/SIMULINK, TRNSYS, and procedural computing were used to carry out energy performance modeling for many projects to prove the feasibility and technical studies.

3.1 Modeling of energy system by CFD

Solar energy system is widely investigated by researchers with many tools, ANSYS-FLUENT was used to investigate the thermodynamic properties and efficiency of the solar system. For example, the thermal and aerodynamic processes occurring in a solar air heater with L-shaped fins that absorb light are modeled numerically [30]. For the Russian city of Samara, the analysis is conducted using an ANSYS-FLUENT Solver with an inbuilt solar calculator. CFD model and Grid are shown in **Figure 13**.

During the CFD analysis, it was determined how design (the step between the fins) and technological factors (the Reynolds number) affected heat-exchange processes and flow aerodynamics. The thermal and aerodynamic properties of a smooth (un-finned) surface were compared with the findings of a physical experiment in order to confirm the accuracy of the results produced by the CFD model. The dependences determined analytically were also compared with the data derived from the developed model for model verification. A good convergence of the data produced by

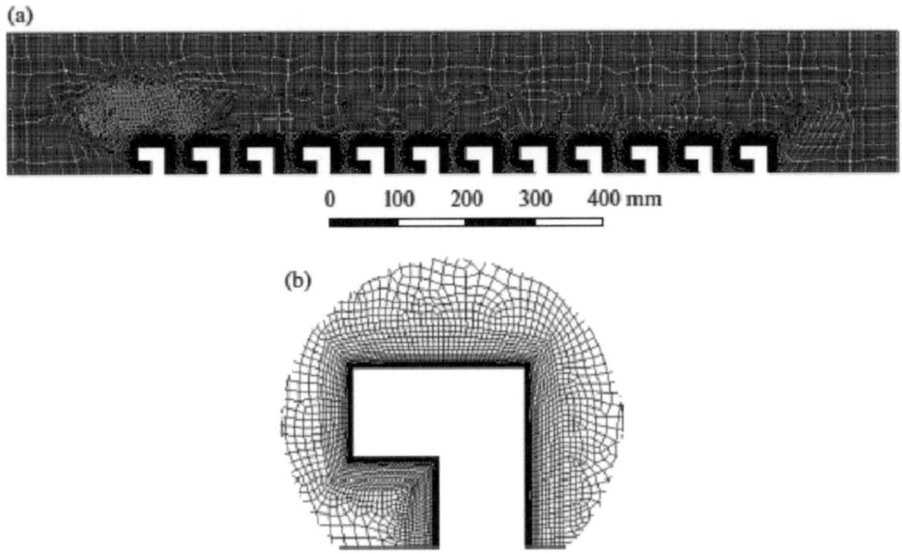

Figure 13.
CFD model and grid [30].

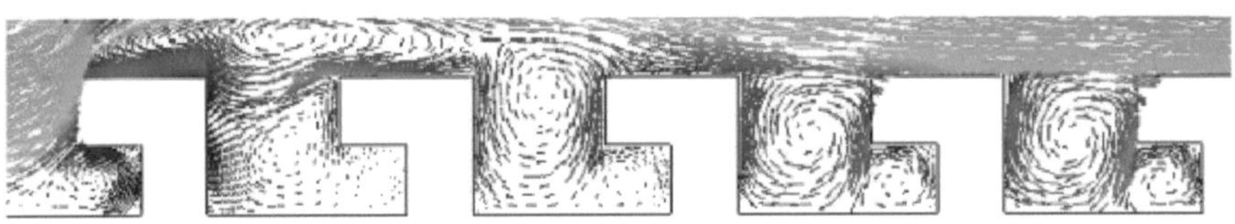

Figure 14.
Velocity contour and velocity vector at Re = 1200 [30].

computer simulation with the outcomes of the physical experiment and the analytical solution is clear from the provided dependencies, which suggest that the ANSYS-FLUENT Solver's parameters were chosen correctly.

The velocity distribution, pressure, and other aspects of the airflow in the solar air-heater box are determined using pictorial outlines in **Figure 14**. In another work, the ANSYS-FLUENT software for operational efficiency of solar air heater model was created whose suitability was confirmed using experimental data. The operational efficiency of the solar air heater has been found to be at its highest for the end of the calendar autumn at an angle of inclination of $\alpha = 60°$ to the earth [31]. Typical results was shown in **Figure 15**.

The researchers have determined how the ribs affect the heat transfer between the air and the light-absorbing surface. The immediate area of the ribs is where vortices and reverse-current zones occur, giving the temperature contour its curvilinear appearance. Research on the SAH's aerodynamic properties has revealed that the ribs have less of an impact than the smooth, light-absorbing surface on the amount of energy used to push air through the air duct [31].

3.2 Modeling of energy system by object-oriented and procedural software

The MATLAB/SIMULINK is widely used to model and validation of Photovoltaic solar cell to produce electricity, pumping system, and many electric devices like

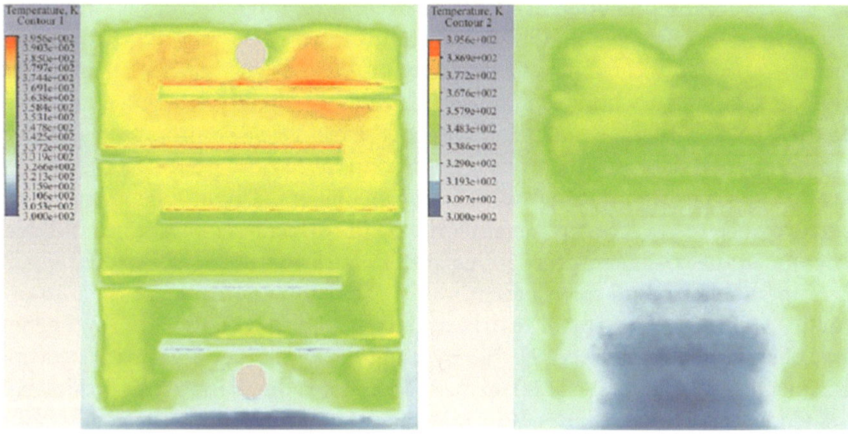

Figure 15.
Temperature contours of the light-absorbing surface (contour 1) and the air at a distance of 14 mm from the surface of the glass (contour 2) at u_{in} *= 1.5 m/s [31].*

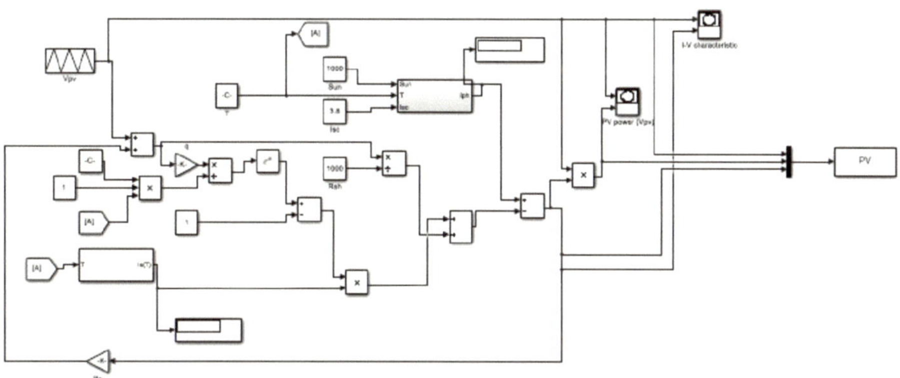

Figure 16.
PV electric circuit and SIMULINK model [32].

electric vehicles. Every important role in using MATLAB/SIMULINK software is to model PV cells (solar cells), PV modules (solar modules), and PV arrays (solar arrays) as well as the analysis of how PV performance changes as a result of changes in various parameters, including temperature, solar radiation, reverse saturation current, series resistance, and shunt resistance [32]. PV electric circuit and model using MATLAB/SIMULINK are shown in **Figure 16**. The PV and IV characteristics curve for the PV array is usually given by this simulation.

Another important object-oriented tool to simulate energy systems is TRNSYS (Transient System Simulation Tool), in which, most simulations are geared toward evaluating the performance of thermal and electrical energy systems.

There are two components to TRNSYS. The first is a kernel-based engine that analyzes the input file, solves the system iteratively, checks for convergence, and graphs system variables. In addition, the kernel offers tools for, among other things, calculating thermophysical characteristics, inverting matrices, running linear regressions, and interpolating external data files. The second component of TRNSYS is a sizable library of parts, each of which simulates the operation of a different system component.

About 150 models are included in the standard library, ranging from wind turbines to electrolyzers, from pumps to multizone structures, from weather data processors to economics routines, and from basic HVAC equipment to cutting-edge developing technologies. Models are built so that users can change already-existing components or create their own features. Excellent example is to use TRNSYS program to analyze both experimental data and theoretical predictions about the performance of a Tri-Generation cooling system using two adsorption chillers. This study aims to analyze the performance of two adsorption chillers and the effects of operational and design parameters on the tri-generation cooling system performance. The system, which has a 240 m^2 parabolic trough solar matrix, was created and confirmed at the Mutah University in Jordan. The trough matrix heated the thermal oil to 260°C and generated 13.7 bar of superheated steam at 210°C. By using the engine's 120°C steam to evaporate brackish water, the power cycle is finished. Each hour, distillation unit produces 150 L of distilled water. The distillation unit rejects heat and is kept in a thermally isolated hydraulic storage tank used to run a special two-stage air-cooled adsorption chiller with a 10 kW cooling capacity. TRNSYS graphical model is shown in **Figure 17**.

The results of TRNSYS include the normalized capacity and COP of adsorption chillers.

One tool preferred by some researchers is to use procedural programming capabilities of MATLAB to model and simulate the performance of energy such astwo-stage adsorption chiller with heat recovery. The chiller schematic used in the simulation is shown in **Figure 18**. The performance of the proposed two-stage adsorption chiller with and without heat recovery utilizing an activated carbon/methanol combination was modeled and simulated using the MATLAB program. The experimental results carried out by Millennium Industries were used to validate the simulation model results. The 10th-order differential equations that made up the model were utilized to

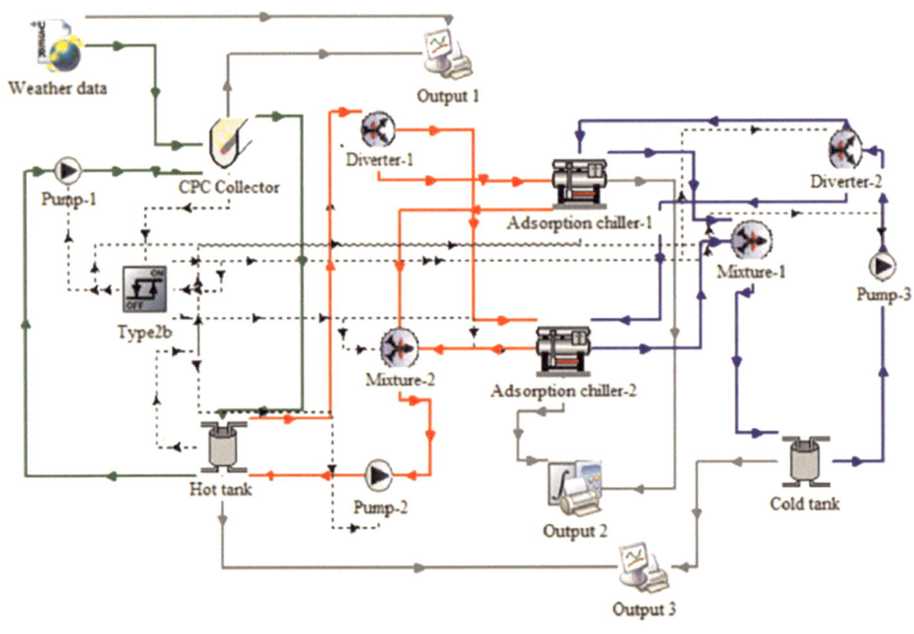

Figure 17.
TRNSYS simulation for solar tri-generation system [33].

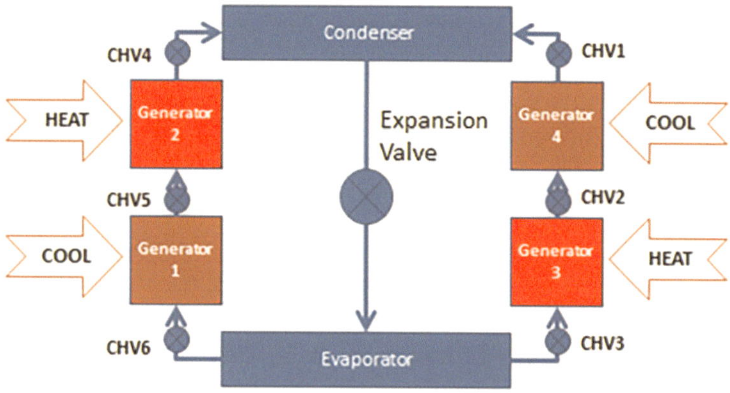

Figure 18.
Schematic of two-stage adsorption chiller [34].

determine the adsorption isotherm and adsorption kinetics, while six of them were used to forecast bed, evaporator, and condenser temperatures.

The energy balance equation used in the simulation:

$$\frac{d}{dt}\left\{\left(M_{ac} * Cp_{ac} + M_{ac} * Cp_m * x_{ads} + M_{Al} * Cp_{Al} + M_{cu} * Cp_{cu}\right)T_{ads}\right\}$$
$$= M_{ac} * Q_{st} * \frac{dx_{ads}}{dt} - M_{ac} * Cp_m * (T_{ads} - T_{eva}) * \frac{dx_{ads}}{dt} + Mc_w * C_{pw} * (T_{ci} - T_{co}) \quad (2)$$

The left side represents the required sensible heat transfer of activated carbon, the metallic part of bed, and the methanol amount absorbed by activated carbon. The first part of right side represents the adsorption heat. The second term represents heat flow from evaporator to adsorbent bed. The third term represents the heat transfer by cooling water. The outlet temperature of the cooling water is calculated by Log Mean Temperature Difference (LMTD) (**Figure 19**):

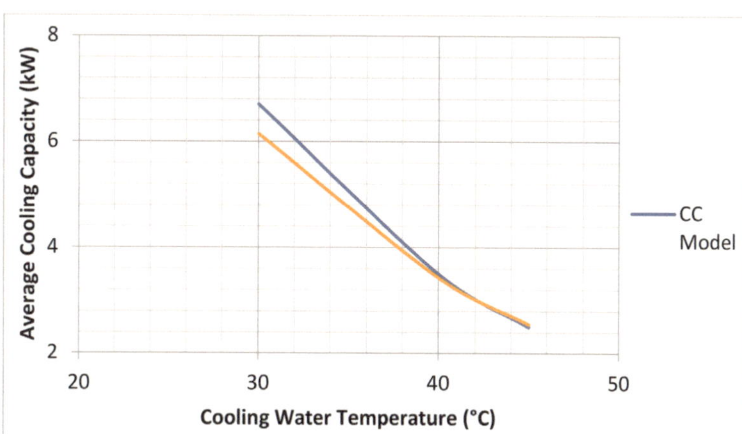

Figure 19.
Chiller experimental versus model cooling capacity [34].

For evaporator:

$$\left(M_{em}(t) * Cp_m + M_{eva} * Cp_{eva}\right)\frac{dTeva}{dt} = -M_{ac} * h_{fg} * \frac{dx_{ads}}{dt} + M_{chw} * C_{pw} * (T_{chi} - T_{cho}) \tag{3}$$

For condenser:

$$\left(M_{cm} * Cp_m + M_{con} * Cp_{con}\right)\frac{dTcon}{dt} = -M_{ac} * h_{fg} * \frac{dx_{des}}{dt} + M_{con} * C_{pw} * (T_{coni} - T_{cono}) \tag{4}$$

it is made clear that the findings of the simulation model for the two-stage air-cooled chiller are well compared with the experimental data in terms of cooling capacity (6.7 kW for the model compared with 6.14 kW from experimental result at same conditions) as shown in **Figure 20**. The coefficient of performance (COP) prediction is quite similar to the value provided by the Carnot cycle operating under the same conditions. To improve cooling capacity and COP, the model optimized the switching, adsorption/desorption, and heat recovery times. The model gives both heat recovery and chiller operation modes, the temporal history of the bed, evaporator, and condenser temperatures as shown in **Figure 21**. The study is significant since it served as the foundation for future built chiller series, which is one of the important needs for simulation in energy industries.

Model could be used also to carryout parametric studies like chiller size, evaporator, and condenser size. Activated carbon mass, bed size, and mass flow rates for cooling, heating, chiller, and condenser are also studied using the model to determine how they affect cooling capacity and COP [35]. Some result of simulation parametric results are shown in **Figure 21**.

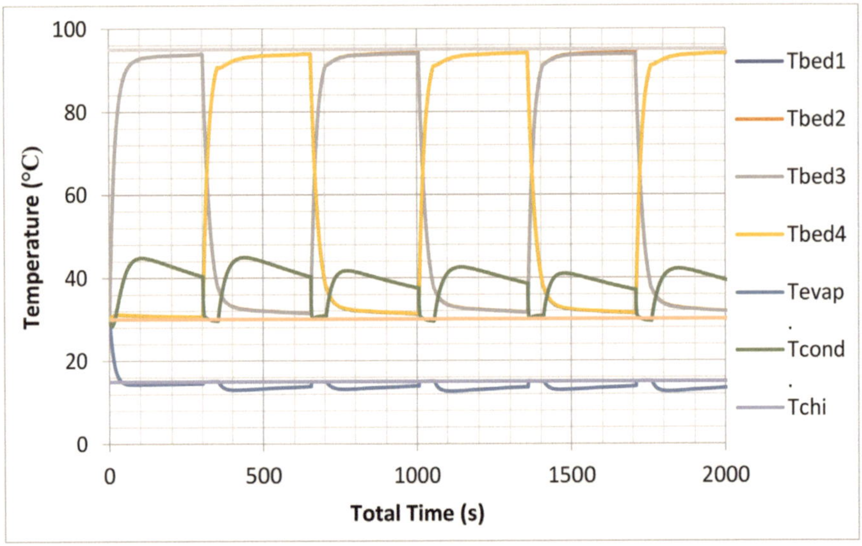

Figure 20.
Temporal history of bed, evaporator, and condenser temperatures [34].

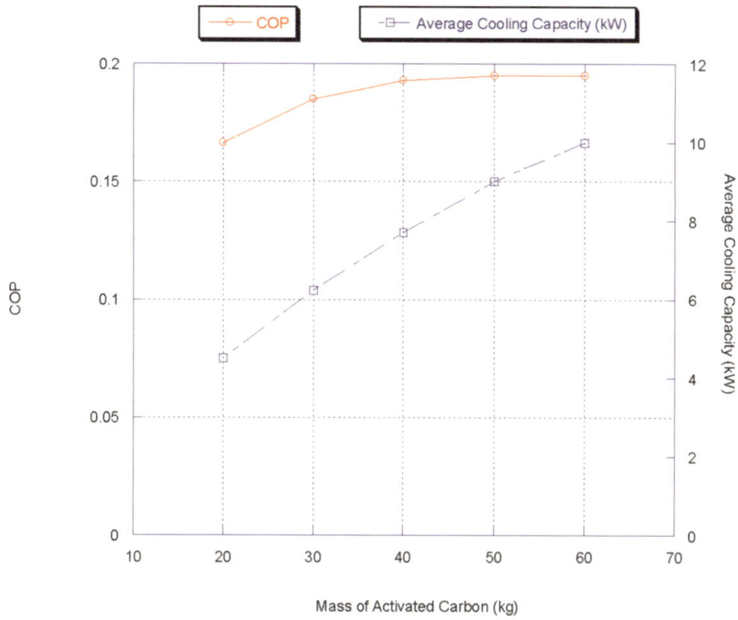

Figure 21.
Effect of activated carbon mass on chiller average cooling capacity and COP [35].

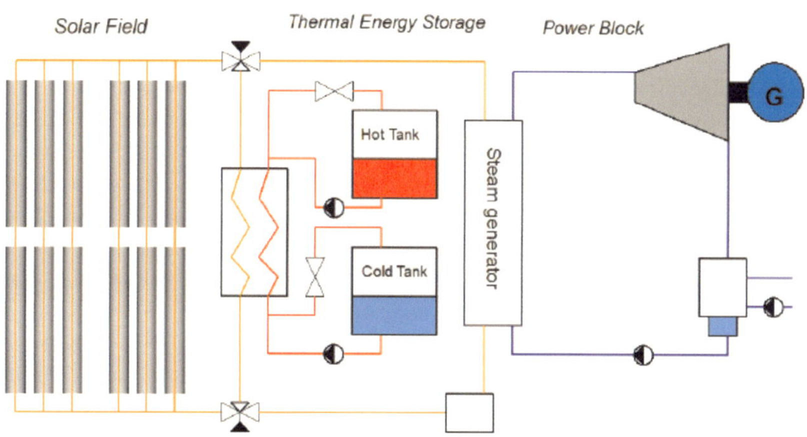

Figure 22.
Schematic of a parabolic trough solar thermal power plant with TES [36].

One of the interesting and user-friendly environments is an object-oriented programming offered by MODELICA, in which a built in graphical library is used to model the energy system operation and efficiency for large-scale plant as shown in **Figure 22**. The model in MODELICA program is shown in **Figure 23**. Results include power generated, solar irradiation, and control strategy for power plant [36].

3.3 Modeling of energy by artificial intelligence

Since 2000, AI has attracted more attention, particularly since 2014, when the number of articles on ML skyrocketed. And in 6 years, ten times greater than they

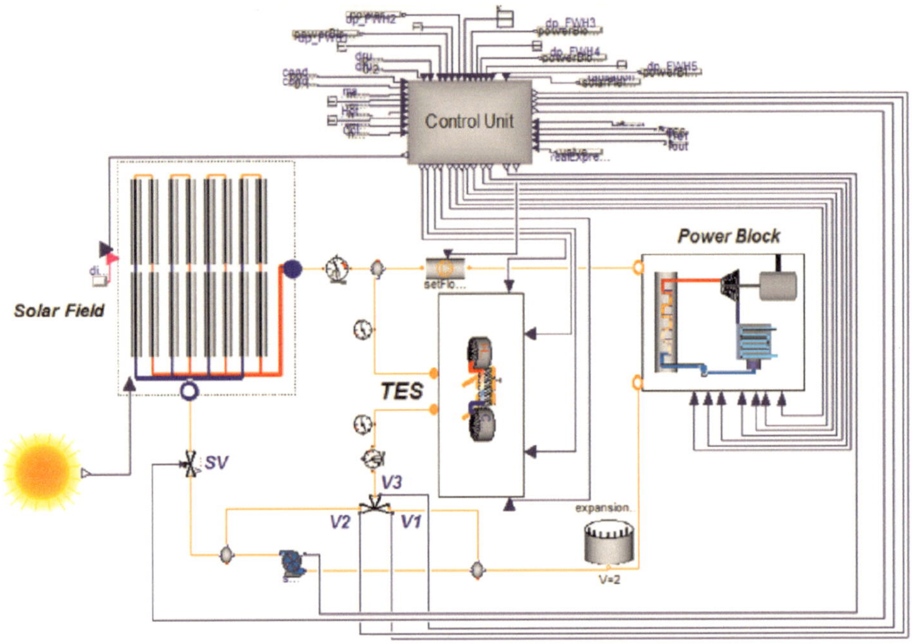

Figure 23.
Model a parabolic trough solar thermal power plant with TES [36].

were in 2014. Despite the fact that several publications in various sectors of energy have offered novel applications of AI and ML in that area, the breadth of energy consumption and its respective domains make it clear that no single book, article, or source could possibly cover all of these topics. However, by looking at the frequency of the AI and ML energy-related keywords in the literature, four areas could be identified: (1) AI-based applications for effectiveness and utilization, (2) machine learning for prediction, (3) algorithm and pattern recognition for learning systems, (4) and administration and transportation of energy sources. Moreover, from 2012 to 2014, the majority of research focused on various fuel technologies, including hydrogen, ethanol, and biofuels. The majority of the literature was dominated by ideas like sustainability and sustainable development during the course of the following 2 years. A lot of attention was also paid to issues, including environmental effects, climate change, greenhouse gas emissions, carbon dioxide, and life cycle analysis. Additionally, attention has been drawn to clean energy sources like solar and wind. AI ideas started to appear around the beginning of 2016. In 2017, forecasting, decision-making, and smart grids were the main topics of study. Later, in 2018, ML approaches for energy use and renewable energy developed. **Table 2** shows the main applications of ML in the energy sector [5].

Another work related to ML in energy systems has compared how well extreme gradient boosting (XGBoost) performs when used in conjunction with artificial neural network (ANN) and degree-day ordinary least square (OLS) regression to create predictive energy models. The effectiveness is tested of each method in terms of the precision of the produced energy models using datasets from cooling electricity and heating gas. Their findings show that XGBoost generates extremely precise predictive energy models. Although XGBoost outperformed ANN and degree-day OLS

Goals	Systems and sections	Techniques and models
Decision making	Solar power	Numerical model
Forecasting	Wind power	Time series analysis
Sustainable development	Stochastic systems	Uncertainty analysis
Energy management	Air conditioning	Fuzzy analysis
Environmental impact	Smart power grid	Genetic algorithm
Performance assessment	Electric vehicle	Support vector
Impact assessment (Social, technical, economic, etc.)	Power transmission network	Multi-objective optimization
	Renewable energy	Random forest
	Energy storage	Decision tree
	Buildings	Artificial neural network
	Internet of things	Pattern recognition
	Energy consumption	Ant colony optimization
		Regression analysis
		Particle swarm optimization
		Data mining
		Deep learning

Table 2.
Main applications of ML in the energy sector.

regression in terms of model accuracy, more research with different energy datasets is necessary for the future to make more definitive claims about their general effectiveness [6].

Special attention should be given to ML applications for renewable energies. The importance of renewable energy sources, such as wind and solar energy, in supplying the world's energy demands is rising. For energy system operators, their fluctuation and unpredictable nature provide serious difficulties. For the system to be stable and reliable, accurate forecasting of renewable energy output is essential. ML and DL algorithms have shown promise as methods for forecasting renewable energy. For a baseline model, traditional ML techniques like linear regression have been successfully applied to the forecasting of renewable energy. It is straightforward, simple to understand, and uses fewer computer resources. However, it could be unable to capture the complex patterns and non-linear correlations present in the data. For forecasting renewable energy, linear regression has been proven to be inferior to the random forest, SVMs, and XGBoost models. Even in the face of noise and outliers, these models are capable of handling non-linear connections, complicated data, and the ability to make reliable predictions. To handle irregularly spaced time series data and identify patterns and trends to produce precise forecasts, specialist time series forecasting techniques like autoregressive models, moving averages, and RNNs are better choices [7].

ANN is also used to predict buildings' energy consumption in terms of heating and cooling electricity consumption. Simulation results were compared of building performance to actual measurements while employing ANN to forecast building energy performance. Energy usage data from the case building for a week was used for training and testing the ANN. With a mean absolute error of 0.9%, the results demonstrate a satisfactory match with the mathematical model [8].

4. Conclusions

Modeling and simulation of systems in mechanical engineering both subdisciplinary of applied mechanics [37–40] or thermal systems "as described in this chapter" became more popular in industry and scientific research due to accelerating in computing power and graphical interface of most of the simulation tools offered commercially or free open access software. CFDs software with all its five techniques are the most interesting for many researchers due to its flexibility and due to the fact that the high cost of experimentation of fluid and energy systems. Modular energy plant system modeling and object-oriented software like MATLAB/SIMULINK, MODELICA or TRNSYS had been widely used last two decades. Libraries are expanded and updated to include more objects representing energy system to fulfill the needs of industry and research for validation of their systems prior to going for prototyping and understanding variant system operating conditions and system efficiency and alternatives for system components. Procedural programming is still in use by some researchers who had the chance to write their own codes for specific system and could be modified to another system with minimum effort by the researcher. Special systems and environments required writing code from scratch since these systems could not be modeled by CFD or object-oriented programming.

Moreover, AI techniques have been used to address the problems of modeling and simulation of fluid and energy systems. In our data-centric era, the application of such techniques will just continue and evolve. Many algorithms have been used across the data-driven modeling and simulation of engineering problems. The availability of data and whether the amount of data is considered a "big data" or not are vital in deciding on the approach taken. Deep learning (DL) neural network (NN) is suitable where data is abundant and considered big data. Whilst ML algorithms are used when the data is comparatively limited. Another interesting topic in the field of modeling and simulation is the physics-informed neural networks (PINNs). PINNs use the data to learn while respecting the physics of the problem at the same time. As physical equations are involved, PINNs usually require less data than what a typical NN would for the same problem. It also serves as a good extrapolator in regimes outside the convex hull of training data. Finally, ML and DL are showing promising results in the modeling and simulation of engineering problems. The investigation of such techniques is still at the early stages and more work needs to be done to fully exploit their capabilities. It is preferable to include a conclusion(s) section, which will summarize the content of the book chapter.

Acknowledgements

I would like to thank my colleagues, at Mechanical Engineering Department of Al-Zaytoonah University, who provided me with their remarkable research in the field of fluid and energy, which highly contributes to the completion of this chapter.

DOI: http://dx.doi.org/10.5772/intechopen.1002220

Author details

Firas Makahleh[1]* and Anas Nassar[2]

1 Al-Zaytoonah University, Amman, Jordan

2 Zarqa University, Amman, Jordan

*Address all correspondence to: f.makahleh@Zuj.edu.jo

References

[1] Nassar A, Rai NK, Sen O, Udaykumar HS. Modeling mesoscale energy localization in shocked HMX, part I: Machine-learned surrogate models for the effects of loading and void sizes. Shock Waves. 2019;**29**:537-558

[2] Nassar A. Physics-Based Machine-Learned Models for Multi-Scale Materials Response to Shock Loads [thesis]. Iowa: The University of Iowa; 2019

[3] Sivamayil K, Rajasekar E, Aljafari B, Nikolovski S, Vairavasundaram S, Vairavasundaram I. A systematic study on reinforcement learning based applications. Energies. 2023;**16**(3):1512

[4] Benti NE, Chaka MD, Semie AG. Forecasting renewable energy generation with machine learning and deep learning: Current advances and future prospects. Sustainability. 2023; **15**(9):7087

[5] Razzaghi P, Tabrizian A, Guo W, Chen S, Taye A, Thompson E, et al. A survey on reinforcement learning in aviation applications. arXiv. 2022:1-14

[6] Entezari A, Aslani A, Zahedi R, Noorollahi Y. Artificial intelligence and machine learning in energy systems: A bibliographic perspective. Energy Strategy Reviews. 2023;**45**:101017

[7] Chakraborty D, Elzarka H. Advanced machine learning techniques for building performance simulation: A comparative analysis. Journal of Building Performance Simulation. 2019;**12**(2): 193-207

[8] Yezioro A, Dong B, Leite F. An applied artificial intelligence approach towards assessing building performance simulation tools. Energy and Buildings. 2008;**40**(4):612-620

[9] Dixit S, Choudhary R, Badholiya S. Mathematical modeling and analysis of different type of fuel injector in scramjet engine using CFD simulation in Fluent. IJRASET. 2017;**5**(2):379-389. DOI: 10.22214/ijraset.2017.2056

[10] Guohua X, Shen X, Kun Y. Modeling and simulation on fluid dynamic depth-setting TUV based on FLUENT. Advanced Materials Research. 2011; **314-316**:1676-1681

[11] Galván S, Reggio M, Guibault F. Assessment study of K-εturbulence models and near wall modeling for steady state swirling flow analysis in draft tube using Fluent. Engineering Applications of Computational Fluid Mechanics. 2011;**4**(4):459-478. DOI: 10.1080/19942060.2011.11015386

[12] Makahleh F, Amro A. Storage Ring Vacuum System Performance Evaluation. Amman: Technical Report; 2005

[13] Al-Najdawi M, Huttel E, Makahleh F, Shehab M, Al-Mohammad H. Vacuum system of Sesame storage ring. In: MEDSI Conference. S Korea. 2016. pp. 71-73

[14] Tan X, Zhu Z, Sun G, Diao L, Cao J. Modeling and simulation of layered water tank based on MATLAB/ SIMULINK. Journal of Physics: Conference Series. 2022;**2173**:012005. DOI: 10.1088/1742-6596/2173/1/012005

[15] Alfonsi G. Reynolds-averaged Navier–stokes equations for turbulence modeling. Applied Mechanics Reviews. 2009;**62**(4):1-20

[16] Lesieur M, Metais O. New trends in large-eddy simulations of turbulence. Annual Review of Fluid Mechanics. 1996;**28**(1):45-82

[17] Kim B, Azevedo VC, Thuerey N, Kim T, Gross M, Solenthaler B. Deep fluids: A generative network for parameterized fluid simulations. Computer Graphics Forum. 2019;**38**(2): 59-70

[18] Li Z, Kovachki N, Azizzadenesheli K, Liu B, Bhattacharya K, Stuart A, et al. Neural operator: Graph kernel network for partial differential equations. arXiv. 2020:1-22

[19] Bhattacharya K, Hosseini B, Kovachki NB, Stuart AM. Model reduction and neural networks for parametric PDEs. The SMAI Journal of Computational Mathematics. 2021;**7**: 121-157

[20] Wang R, Kashinath K, Mustafa M, Albert A, Yu R. Towards physics-informed deep learning for turbulent flow prediction. In: Proceedings of the 6th ACM SIGKDD International Conference on Knowledge Discovery & Data Mining. USA. 2020. pp. 1457-1466

[21] Lusch B, Kutz JN, Brunton SL. Deep learning for universal linear embeddings of nonlinear dynamics. Nature Communications. 2018;**9**(1):4950

[22] Erichson NB, Muehlebach M, Mahoney MW. Physics-informed autoencoders for Lyapunov-stable fluid flow prediction. arXiv. 2019:1-14

[23] Guastoni L, Güemes A, Ianiro A, Discetti S, Schlatter P, Azizpour H, et al. Convolutional-network models to predict wall-bounded turbulence from wall quantities. Journal of Fluid Mechanics. 2021;**28**:A27.99999

[24] Vinuesa R, Brunton SL. The potential of machine learning to enhance computational fluid dynamics. arXiv. 2021:**4**:1-3

[25] Kochkov D, Smith JA, Alieva A, Wang Q, Brenner MP, Hoyer S. Machine learning–accelerated computational fluid dynamics. Proceedings of the National Academy of Sciences. 2021;**118**(21): e2101784118

[26] Solarex. Physics Informed Neural Networks (PINNs): An Intuitive Guide [Internet]. 2022. Available from: https://towardsdatascience.com/physics-informed-neural-networks-pinns-an-intuitive-guide-fff138069563

[27] De Bézenac E, Pajot A, Gallinari P. Deep learning for physical processes: Incorporating prior scientific knowledge. Journal of Statistical Mechanics: Theory and Experiment. 2019;**12**:124009

[28] Jia X, Willard J, Karpatne A, Read J, Zwart J, Steinbach M, et al. Physics guided RNNs for modeling dynamical systems: A case study in simulating lake temperature profiles. In: Proceedings of the 2019 SIAM International Conference on Data Mining. Society for Industrial and Applied Mathematics. USA. 2019. pp. 558-566

[29] Xie Y, Franz E, Chu M, Thuerey N. Tempo GAN: A temporally coherent, volumetric GAN for super-resolution fluid flow. ACM Transactions on Graphics (TOG). 2018;**37**(4):1-5

[30] Pashchenko DI. ANSYS Fluent CFD modeling of solar air-heater Thermo aero-dynamics. Applied Solar Energy. 2018;**54**(1):32-39. DOI: 10.3103/S0003701X18010103

[31] Pashchenko DI. CFD modeling of operating processes of a solar air heater In Ansys-Fluent. Journal of Engineering Physics and Thermo-physics. 2019; **92**(1):73-79. DOI: 10.1007/s10891-019-01908-8

[32] Dasa A, Datta S, Emdadul P, Royb HA. Modeling of solar cell, module,

panel and array using MATLAB/ Simulink. In: International Conference on Mechanical. Industrial and Materials Engineering. USA. 2022

[33] Younes MB, Shaban NA, Altork Y. Modelling and performance assessment of a tri–generation cooling system using two adsorption chillers under Jordanian climate. Case Studies in Thermal Engineering. 2023;**44**:102848. DOI: 10.1016/j.csite.2023.102848

[34] Makahleh F, Badran A, Attar H, Amer A, Al-Almaaitah A. Modeling and simulation of a two stage adsorption chiller with heat recovery, part 1: Physical and mathematical performance model. Journal of Applied Science. 2022;**13**:1-21. DOI: 10.3390/ app12136542

[35] Makahleh F, Badran A, Attar H, Amer A, Al-Almaaitah A. Modeling and simulation of a two stage adsorption chiller with heat recovery, part II: Parametric study. Journal of Applied Science. 2022;**12**:5156. DOI: 10.3390/ app12105156

[36] Montañés RM, Windahl J, Pålsson J, Thern M. Dynamic modeling of a parabolic trough solar thermal power plant with thermal storage using Modelica. Heat Transfer Engineering. 2017;**39**:277-292. DOI: 10.1080/ 01457632.2017.1295742

[37] Alsakarneh A, Tabaza T, Kelly G, Barrett J. Impact dynamics of nonlinear materials: FE analysis. Journal of Applied and Computational Mechanics. 2023; **9**(3):728-738. DOI: 10.22055/ jacm.2022.41487.376

[38] Manna R, Oosthuizen PH. Numerical and experimental study of natural convective heat transfer from two-sided circular and square horizontal plates having a finite thickness. Heat and Mass Transfer. 2020;**56**:2225-2238. DOI: 10.1007/s00231-020-02851-8

[39] Manasrah A. Application and Analysis of Asymmetrical Hot and Cold Stimuli [Thesis]. USA: University of South Florida; 2016

[40] Manasrah A, Crane N, Guldiken R, Reed KB. Asymmetrically-applied hot and cold stimuli gives perception of constant heat. In: 2017 IEEE World Haptics Conference (WHC). USA: IEEE; 2017. pp. 484-489

Section 3

General Simulation Topics

Chapter 7

Analysis of Supersonic Free Jets and Impinging Supersonic Jets on Deflector

Rakhab Mehta

Abstract

The computational fluid dynamics analysis is carried out to analyze the shock and flow characteristics of under- and over expanded supersonic free jets emanating from convergent-divergent nozzles. Influence of exit Mach number on shock cell lengths are analyzed with the help of density contours and schlieren images. A parameter based on an exit Mach number is obtained to characterize the pressure variation along the jet axis which shows them independent of the exit to ambient pressure ratio. Impingement flow fields over axisymmetric and wedge deflector are investigated employing numerically and compared with experimental results. Effects of jet expansion ratio and distance between the nozzle to deflector apex has been studied at various expansion ratios and distances. Impingement load is computed at various conditions. Pressure distributions over a surface of wedge and axisymmetric jet deflector are computed and compared between them. Pressure load on a diaphragm of a solid motor during lift-off of a satellite launch vehicle having four liquid engines is numerically simulated.

Keywords: CFD, compressible flow, convergent-divergent nozzle, expansion, Mach disc, shock, supersonic jets deflector, supersonic free jets

1. Introduction

The phenomenon of under- and overexpanded supersonic free jets emanating from a convergent-divergent nozzle and their impingement over a deflector are employed during the lift-off of a satellite launch vehicle. When a convergent-divergent nozzle deviates from its operating conditions then semi-periodic shock diamonds [1] and Prandtl-Meyer fans appear in the jet field. However, at the design conditions, the traditional smoothly contoured convergent-divergent nozzle with exit flow is a shock-free condition.

Mehta and Prasad [2] studied the effects of various exit Mach numbers and exit-to-ambient pressure ratios on shock-cell lengths with schlieren and density contour plots. The results showed that the Mach disk moves closer to the nozzle exit plane with decreasing pressure ratio, and the length of the first shock cell decreases. Three-dimensional Navier-Stokes equations [3] were solved to estimate the heat flux on a jet deflector caused due to impingement of the rocket exhaust of a canisterized missile. Based on the theory of computational fluid dynamics [4], the flow field of a rocket engine plume is simulated and analyzed by numerical method. Allgood and Ahuja [5] investigated the plume impact characteristics of ARES V propulsion system by numerical method, and the impact shape, surface pressure, temperature and other parameters of a flow deflector were obtained, which can help the practical construction of flow deflector and improve the safety and reliability of space launch.

Zhou et al. [6] have analyzed the four-engine liquid rocket flow field during the launching phase. Jiang et al. [7] presented an overview on progresses and perspectives of the jet impingement research for rocket launching.

Numerical simulations [8] have been performed to investigate the exhaust plume impinging on the wedge-shaped and cone-shaped deflectors. By comparative analysis between the four-engine rockets impinging jet on different deflector. Compared to the wedge-shaped deflector, the cone-shaped deflector could achieve better performance for deflecting with sufficient distance of the sidewalls. The flow fields of the four-engine rocket impinging on the flame deflector under different impingement and uplift angles are simulated. They observed that high temperatures on the deflector surface mainly occur on the impingement point or the cambered surface. A large impingement angle causes the reverse flow intensity to increase whilst a small angle causes the exhaust gas to deflect a little, a suitable uplift angle can smoothly guide the exhaust gas away from the deflector that the best thermal environment of the deflector channel appears at an impingement angle of 25°and an uplift angle of 5°. Numerical solutions of the impingement of an underexpanded axisymmetric supersonic jet on a flat plate at varied angles have been carried out by Wu et al. [9] and Kim et al. [10]. Experimental investigation of underexpanded supersonic jet impingement over inclined plate is carried out by Nakai et al. [11]. Computational analysis of underexpanded jets on inclined plate is carried out by McIlroy and Fujii [12]. Supersonic gas jets in conical convergent-divergent nozzles are studied numerically using the rhoOpenFOAM software solver [13].

The present study attempts to address flow field of supersonic free jets. Phenomena of jet expansion ratio, and distance between nozzle exit plane and deflector apex has been analyzed at various stand-off distance. Overall load coefficient has been obtained and compared with experimental data [14].

2. Supersonic free jets

2.1 Supersonic free jets from Laval nozzle

A conical convergent-divergent nozzle delivers over-expanded, fully expanded and underexpanded jet depending on the nozzle pressure ratio, ratio of stagnation pressure in the settling chamber to the ambient pressure.

2.2 Over expanded jets

If the pressure in the ambient medium to which is discharging is greater than the nozzle exit pressure, the jet is said to be over expanded. Oblique shock waves are formed at the edge of the nozzle exit. These oblique shocks will be reflected as expansion waves from the boundary of the jet. **Figure 1a** schematically diagram delineates the waves in an over expanded jet. Due to this periodic shock cell structure

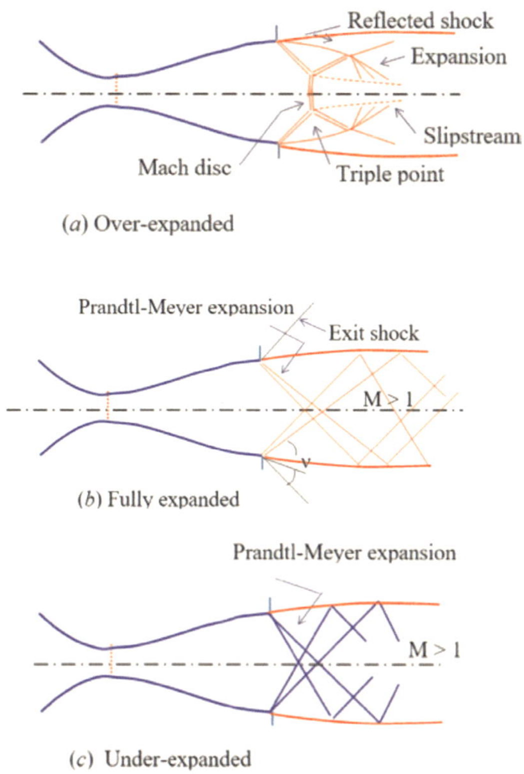

Figure 1.
Schematic sketch of supersonic free jets (a) overexpanded, (b) fully expanded and (c) underexpanded.

is generated in the jet and the wave length of this periodic structure is found to increase with Mach number [15, 16].

Flow has passed through the shock waves will be turned the centre line and the oblique shock wave directed toward the centre line of the nozzle. This process of expansion and compression wave formation continues until the pressure of the jet field becomes the same as the ambient pressure and the flow becomes parallel to the nozzle centre line. These expansion and compression waves which intact with each other lead to the formation of diamond patterns termed shock cells.

2.3 Fully expanded jets

Figure 1b shows schematic sketch of fully expanded jet. The fully expanded flow occurs when ambient pressure is equal to exit pressure then the jet is also alternatively called as correctly expanded [17]. This jet is also wave dominated as an imperfectly expanded jet.

2.4 Under expanded jets

If the pressure in the ambient medium to which is discharging is less than the nozzle exit pressure, the jet is said to be under expanded as shown in **Figure 1c**. Since $p_a < p_e$ a wedge-shaped expansion fans occur at the edge of the nozzle. These waves cross one another and are reflected from the opposite boundaries of the jet as compression waves. The compression waves again cross one another and are reflected on the boundaries of the jet as expansion waves. The gradual compression occurs Mach waves emanating at the wall, and the flow near the wall can be treated as isentropic. The flow properties changes gradually and isentropically and called Prandtl-Meyer angle can be written as

$$\nu = \sqrt{\frac{\gamma+1}{\gamma-1}}\tan^{-1}\sqrt{\frac{\gamma+1}{\gamma-1}(M^2-1)} - \tan^{-1}\sqrt{M^2-1} \tag{1}$$

The Prandtl-Meyer angle is the angle through which sonic flow turns to attain a supersonic local Mach number. When supersonic flow is turned through a convex corner, the flow expands, resulting in an increase in velocity and a drop in pressure. When a M_1 expands to a M_2, according to a Prandtl-Meyer expansion, the turning angle is $\Delta\nu = (\nu_1 - \nu_2) = \pm\delta$.

Supersonic jet emanating from a convergent-divergent nozzle may be overexpanded ($p_e/p_a < 1$), correctly expanded ($p_e/p_a = 1$) or underexpanded ($p_e/p_a > 1$) depending upon stagnation pressure P_0. The basic flow field structures are delineated in **Figure 2**. Generally, it consists of repetitive shock cell, inviscid jet boundary and a shear layer. In the case of $p_e/p_a = 1$, the shock cells consist of weak shock and regular shock reflection. However, for imperfectly expanded jet ($p_e/p_a \neq 1$) cases, formation of Mach disc takes place.

The location of Mach disk [2] moves away from the nozzle exit plane with increase in expansion pressure ratio p_e/p_a. For highly underexpanded case, the flow up to the Mach disc is generally calculated by different methods like method of characteristics. However, the flow downstream of Mach disc cells for viscous analysis. For $p_e/p_a < 1$ cases, the strong jet shocks from the nozzle forms the Mach disc very near to the nozzle exit; hence the estimation of Mach disc location becomes difficult.

The noise generated from the supersonic jet is dominated due to turbulent mixing in the shear layer for $p_e/p_a = 1$. However, for shock associated noise does become appreciable for jet $p_e/p_a \neq 1$. For shock associated noise prediction, it is customary to obtain the shock cell lengths accurately Refs. [2].

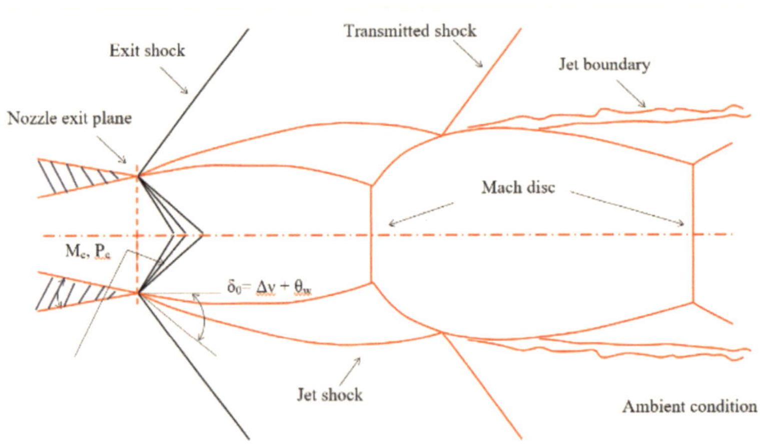

Figure 2.
Schematic sketch of flow field of overexpanded.

During the lift-off condition of the multi-engine rocket, the exhaust jet from the rocket nozzle impinges on the launch pad and is produced complex impingement flow field. The $p_e/p_a < 1$ during the lift-off period. In order to estimate the impingement load on the jet deflector and for prediction of jet noise during the lift-off condition, it is necessary to obtain the free jet properties.

3. Numerical algorithm

A numerical flow simulation is carried out to analysis supersonic free jets and jets deflector using turbulent, compressible Reynolds averaged Navier-Stokes equations. A two-equation turbulence model [18] with compressibility correction [19] is used to solve the governing fluid dynamics equations. A finite volume discretization is carried out in spatial coordinates to compute inviscid and viscous flux vectors. Time evolution is carried out by an explicit multistage Runge-Kutta method to achieve a steady state solution. The numerical algorithm is developed by taking into consideration structured grid arrangement. The numerical results are obtained for nozzle exit Mach number of 2.2, 2.6 and 3.1, exit to ambient pressure ratio of 0.6, 0.8, 1.0 and 1.2 and at different distance from nozzle exit to the apex of the axisymmetric and the double wedge deflector. The numerical scheme is computationally fast, easy to program and efficient. The centre line pressure distributions inside the free jets differs in the presence of the jet deflector. The numerical results are compared with the experimental data and are discussed in the next sections.

4. Experimental facility

All the experimental simulations of supersonic free jet and jet deflector are conducted in an open jet facility as shown in **Figure 3**. High pressure dry air at 4.3×10^6 Pa at ambient temperature is fed through a 15×10^{-3} m diameter pipe line to the settling chamber and nozzle assembly. A pressure regulating valve is used to control the operating pressure. The pressure in the settling chamber is continuously recorded and monitored using a Bourden pressure gauge and a pressure transducer. The experimental set up is coupled with data acquisition. The open jet facility can be operated continuously at the maximum pressure up to about 80 s.

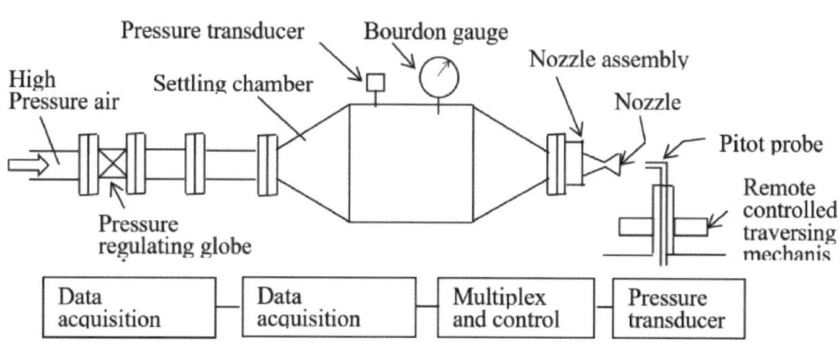

Figure 3.
Schematic sketch of open jet facility.

The conical convergent-divergent nozzles are used for getting supersonic free jets. The nozzles are having semi-divergent angle of 15°. The dimensions of convergent-divergent nozzle and operating conditions are tabulated in **Table 1**. The nozzles *A* and *B* were operated at $p_e/p_a < 1$ and nozzle *C* is operated on $p_e/p_a > 1$. Pitot tube of 1.1 mm outer diameter is used for measuring the centre line pressure. The flow does not separate in the nozzle as per the criterion for 15° semi-conical angle of nozzle [14].

Axisymmetric and wedge deflector model consisted of 70° apex angle having tip bluntness of 0.13 D_e as shown in **Figure 4**, respectively. A curvature of radius 1.2 D_e is provided at a distance of 0.7 D_e from centre. The deflector surface becomes perpendicular to nozzle axis at 1.7 D_e. Both the deflector, that is, axisymmetric and wedge has the same apex angle, radius of curvature and base length. Static pressures on the deflector surface is measured employing scanivalve unit. A computer base data acquisition system is used for data analysis.

5. Numerical results

5.1 Analysis of supersonic free jets

For the sake of brevity, we are displaying only density contour of M_e = 2.6 and p_e/p_a = 0.8 in **Figure 5**. The characteristic features of supersonic free jets such as Mach disk, the intercepting and reflecting oblique shocks are well captured in the density contours. **Figure 6** shows schlieren picture M_e = 2.6 and p_e/p_a = 0.8 obtained from experiment [14].

Nozzle	d^*, mm	d_e, mm	A_e/A^*	M_e	Pressure ratio
A	2.18	30.0	2.005	2.2	$0.36 < p_e/p_a < 1.90$
B	9.2	15.72	2.923	2.6	$0.39 < p_e/p_a < 1.62$
C	10.66	23.0	4.657	3.1	$0.47 < p_e/p_a < 0.88$

Table 1.
Dimensions and operating condition of nozzles.

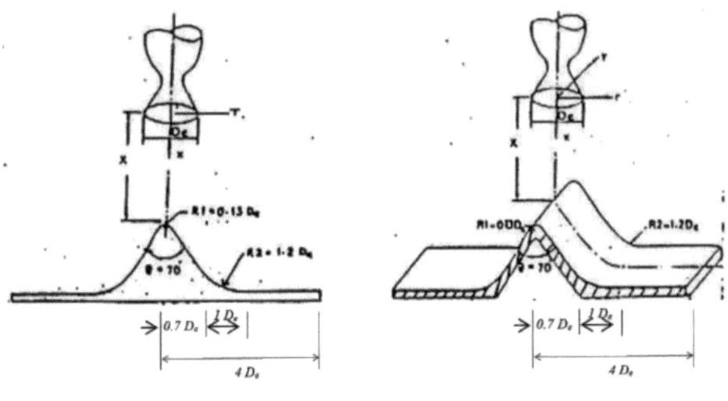

(a) (b)

Figure 4.
Geometrical details of (a) axisymmetric and (b) wedge deflector.

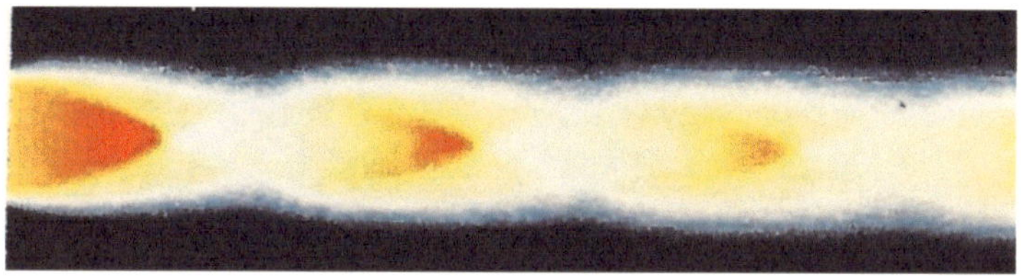

Figure 5.
Density contour $M_e = 2.6$ *and* $p_e/p_a = 0.8$.

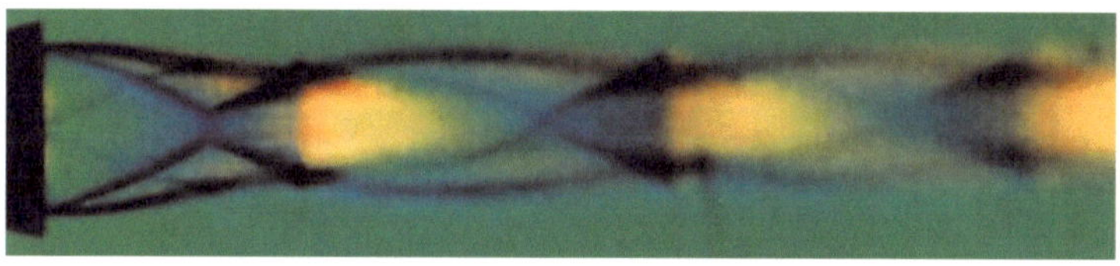

Figure 6.
Schlieren picture $M_e = 2.6$ *and* $p_e/p_a = 0.8$.

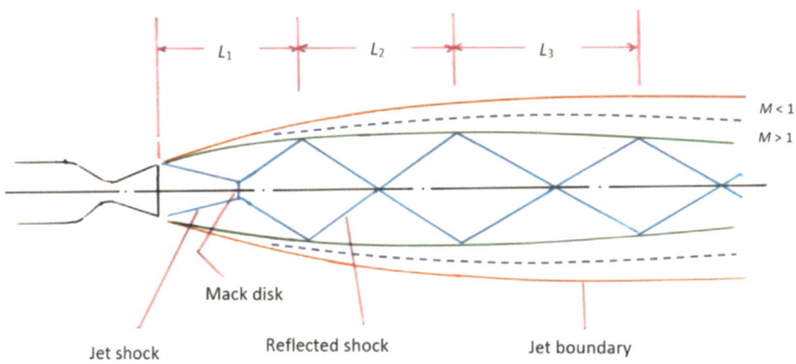

Figure 7.
Schematic sketch of a supersonic free jet.

A schematic sketch of based on the computed density contour and schlieren picture is illustrated in **Figure 7**. Mach disc location L_m, first, second and third shock cell length L_1, L_2, and L_3, respectively are measured from schlieren pictures. All the discontinuities such as jet shock, triple point, slip line; reflected shock and shock cells are clearly visible as described in **Figure 1**. Mach disc location L_m compared with Love et al. [15], first, second and third hock cell length L_1, L_2, and L_3, are compared with Abdel-Fattah [16].

For the sake of brevity, we are presenting density contours in **Figure 8** for $M_e = 2.2$ and $p_e/p_a = 1.2$, $M_e = 3.1$ and $p_e/p_a = 1.2$, 0.6 and 0.4. The characteristic features of supersonic free jets can be easily visualized in the density contours. The first Mach disk has a well defines triple point where the Mach disk, the intercepting and reflecting oblique shocks terminate. The slip line downstream of the Mach disk which separates the subsonic inner core of the jet from the surrounding supersonic flow is also visible.

Figure 9 shows inviscid [2], present viscous and measured static pressure distribution [14] along the centre line of supersonic free jets $M_e = 2.2$ and $p_e/p_a = 1.2$, $M_e = 3.1$ and $p_e/p_a = 0.6$ and 0.8. The viscous results are observed to be in agreement

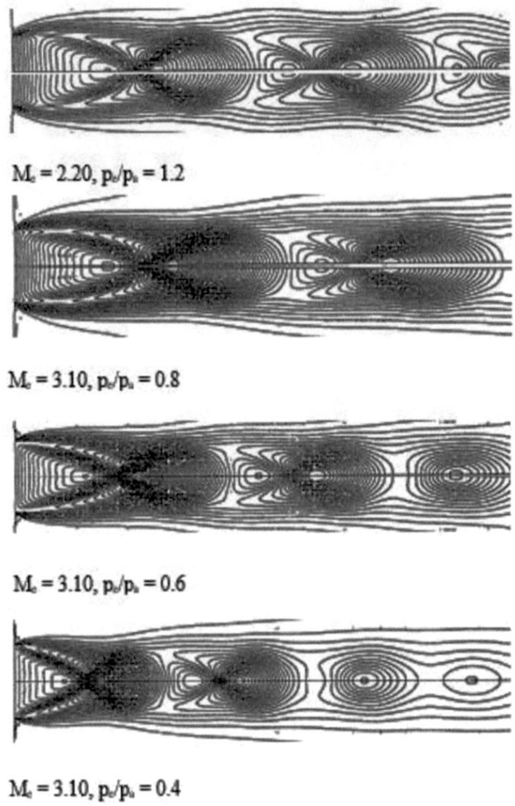

Figure 8.
Density contour of supersonic free jets.

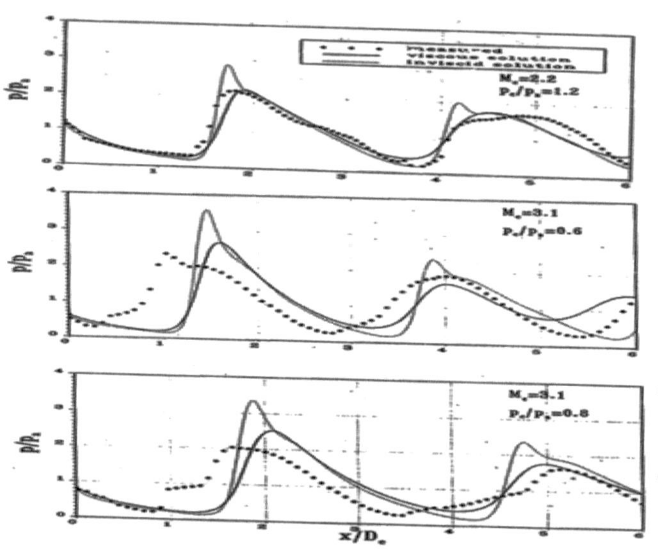

Figure 9.
Pressure distribution along the centre line of supersonic free jets.

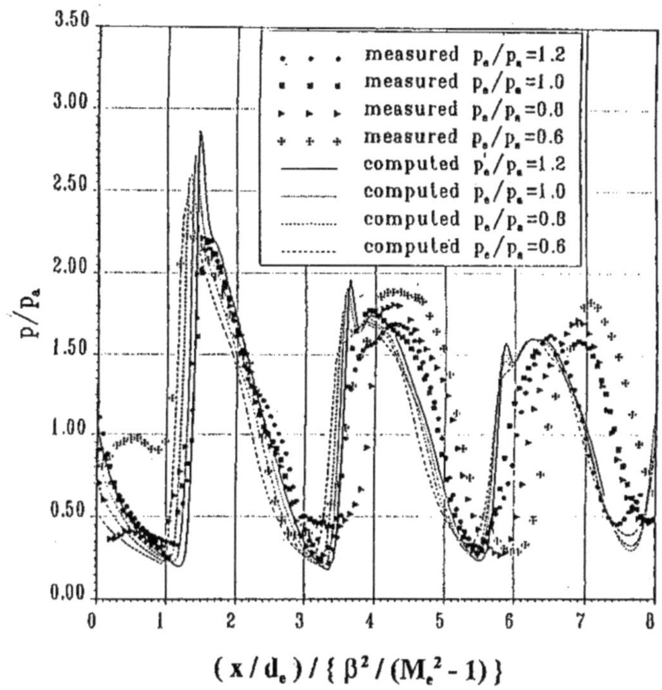

Figure 10.
Pressure variation along the jet axis for M_e = 2.2 *at different* p_e/p_a.

with experimental data [14]. However, there is some discrepancy noticed at far-field which attributed to over prediction of the spreading rate of jets.

Figure 10 shows the variation of measured and computed pressure distribution along the jet axis for M_e = 2.2 at p_e/p_a = 1.2, 1.0, 0.8 and 0.4 and compared with the measured static pressure. A non-dimensional parameter $\left[\beta^2/\left(M_e^2-1\right)\right]$ is obtained to characterize the pressure variation along the jet axis independent of the exit to ambient pressure ratio p_e/p_a. The value of β^2 = $(M_j^2–1)$ for $p_e/p_a \neq 1$ supersonic jet are obtained using Tam and Tanna [20]:

$$\frac{d_j}{d_e} = \sqrt{\frac{M_e}{M_j}\left[\frac{1+\frac{1}{2}(\gamma-1)M_j^2}{1+\frac{1}{2}(\gamma-1)M_e^2}\right]^{\frac{\gamma+1}{4(\gamma-1)}}} \tag{2}$$

$$\frac{p_e}{p_a} = \left[\frac{1+\frac{1}{2}(\gamma-1)M_j^2}{1+\frac{1}{2}(\gamma-1)M_e^2}\right]^{\frac{\gamma}{(\gamma-1)}} \tag{3}$$

L_1/d^* varies linearly with β^2 for $p_e/p_a < 1$ and shows good agreement with the result of Abdel-Fattah [16] and Mehta et al. [21]. A least square straight line fit from these data as.

$$\begin{aligned} L_1/d^* &= 1.04\,\beta^2–1.08 \text{ for } M_e = 2.2 \left(1.03 < \beta^2 < 3.84\right) \\ L_1/d^* &= 1.15\,\beta^2–1.95 \text{ for } M_e = 2.6 \left(1.70 < \beta^2 < 5.81\right) \\ L_1/d^* &= 1.25\,\beta^2–3.85 \text{ for } M_e = 3.1 \left(3.08 < \beta^2 < 8.70\right) \end{aligned} \tag{4}$$

From the density contour measurement of L_2 was carried out in the similar way as.

$$L_2/d^* = 0.87\,\beta^2 \text{ for } M_e = 2.6\ \left(1.03 < \beta^2 < 3.84\right) \tag{5}$$

The L_2/d^* varies linearly with β^2 for all the nozzles irrespective of M_e. in general, it is observed that $L_1 > L_2$ for $p_e/p_a > 1$ and $L_1 = L_2$ for $p_e/p_a = 1$ and $L_1 < L_2$ for $p_e/p_a < 1$. From the density contour is found about $L_3 = 0.94\ L_2$. **Table 2** shows the summary of shock cell length L_1, L_2 and L_3 as a function of β^2.

5.2 Numerical analysis of jet deflector

The impingement flow field in a jet deflector produced due to an over expanded supersonic jet is schematically illustrated in **Figure 11**. It can be observed from the diagram that the flow field consists of a jet shock, Mach disk, reflected shock, wedge shock and compression shock [22, 23]. The impingement flow field data are necessary for the design of the jet deflector.

Mach density contours of the impinging jets have been shown over the axisymmetric deflector in **Figure 12a** and **b** for $M_e = 2.2$ and $X_c/D_e = 2.0$ and $M_e = 2.2$ and $X_c/D_e = 3.0$, respectively. It can be observed from the Mach contour that all the essential flow field features are well captured. Comparison between schlieren picture with Mach contour shows good agreement of flow field.

Figure 13 shows pressure distribution over the axisymmetric and the wedge deflectors for $M_e = 2.2$, $p_e/p_a = 1.2$ and $X/D_e = 2$. The computed pressure distribution for axisymmetric case compares well with the experimental results. The comparisons of pressure variation over axisymmetric and wedge deflector placed at $X_c/D_e = 2.0$,

p_o/p_a	p_e/p_a	β^2	M_j^2	L_1/d^*	L_2/d^*	L_3/d^*
			$M_e = 2.2$			
12.88	1.2	4.36	5.36	2.15	5.5	8.8
10.69	1.0	2.84	4.84	1.75	4.7	7.7
8.55	0.8	3.23	4.23	1.4	3.0	6.5
6.47	0.6	2.50	3.50	1.0	2.9	5.0
			$M_e = 2.6$			
23.95	1.2	6.38	7.38	2.6	6.65	10.95
19.95	1.0	5.76	6.76	2.2	5.8	9.5
15.55	0.8	5.03	6.03	1.76	4.95	8.1
11.97	0.6	4.16	5.16	1.35	3.9	7.1
			$M_e = 3.1$			
19.12	0.8	7.70	8.7	2.2	6.2	10.1
25.59	0.6	6.62	7.62	1.75	5.0	8.3
17.09	0.4	5.70	6.24	1.2	3.6	6.18

Table 2
Summary of shock cells length.

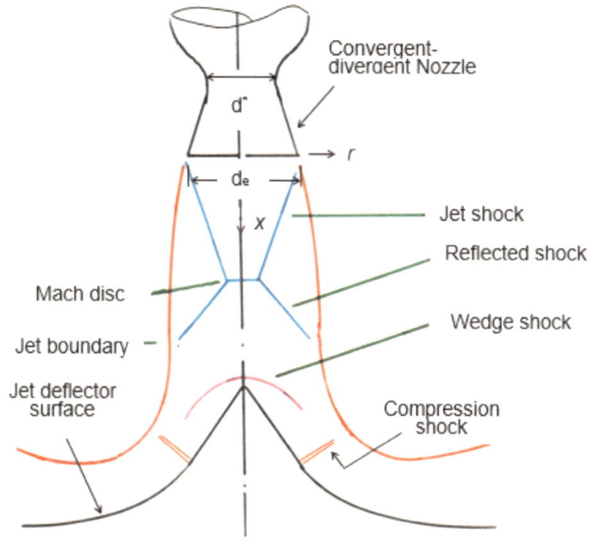

Figure 11.
Schematic of the impingement jets over deflector.

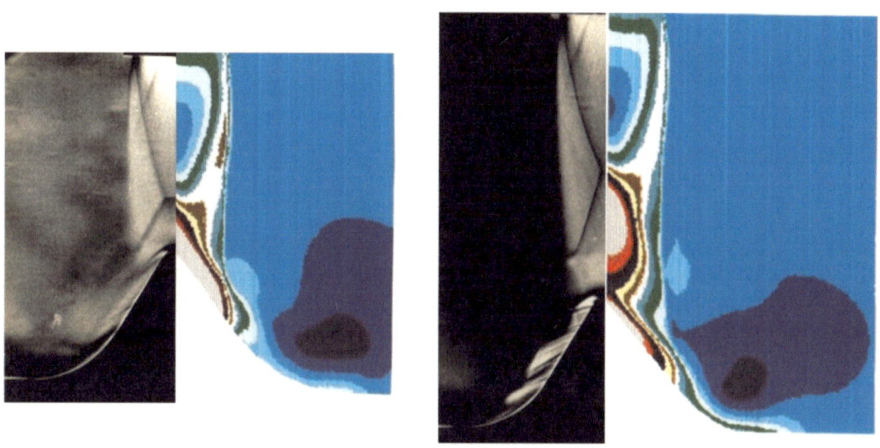

Figure 12.
Comparison of schlieren picture with Mach contour. (a) $M_e = 2.2$ *and* $X_c/D_e = 2.0$ (b) $M_e = 2.2$ *and* $X_c/D_e = 3.0$.

indicate small differences in the apex region till r/D_e of 0.7. the maximum pressure occurs at large distance from the centre for wedge as compared to axisymmetric case.

A quantity, which is of design interest is the overall impingement load, $C_a = L/T$. where T is thrust of the nozzle and L is integrated normal static pressure over the deflector. Load coefficient for various value of p_e/p_a and stand-off distance is shown in **Figure 14**. C_a increases with increase in jet pressure ratio at all distances X/D_e in the range of 2–5. However, at a fixed p_e/p_a the load coefficient decreased with increase in X/D_e. load coefficient for wedge deflector is found to be 40% higher than the

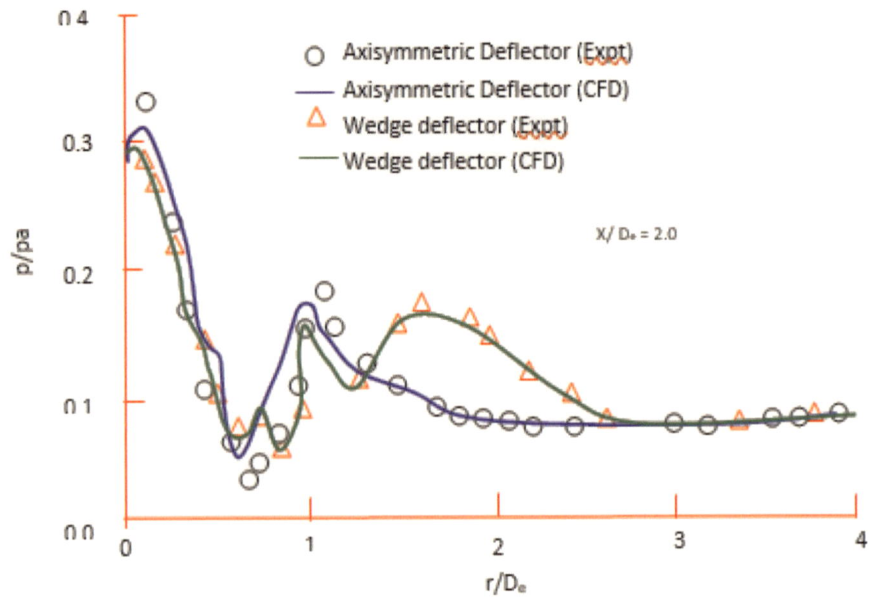

Figure 13.
Pressure distribution over deflector for M_e = 2.2, p_e/p_a = 1.2 *and* X/D_e = 2.

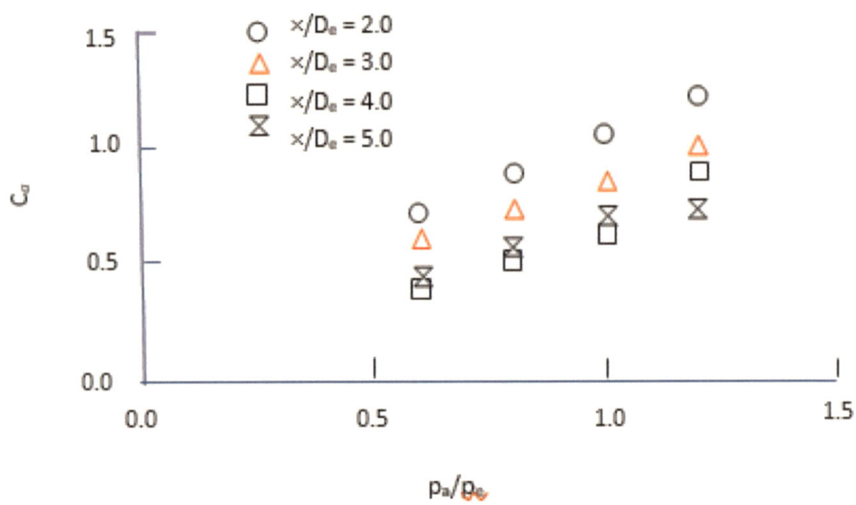

Figure 14.
Variation of load coefficient.

axisymmetric case. The lower load coefficient for the axisymmetric case attributed to the pressure relief in the radial direction.

Pressure distribution over wedge deflector for M_e = 2.2, p_e/p_a = 1.2 and X/D_e = 2 is shown in **Figure 15**. It can be observed from the pressure profile the influence of deflector surface. The pressure distribution is computed using three-dimensional inviscid flow solver.

Figure 16 shows the pressure distribution over different section of the wedge deflector for for M_e = 2.2, p_e/p_a = 1.2 and X/D_e = 2. There kinks are most likely to be

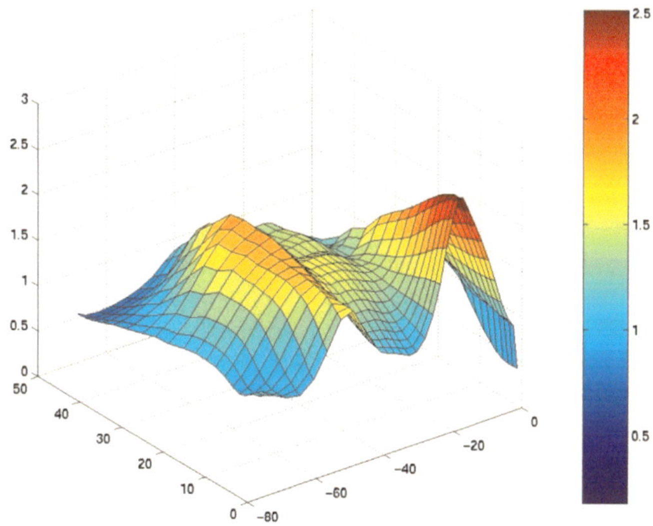

Figure 15.
Pressure variation over wedge deflector for $M_e = 2.2$, $p_e/p_a = 1.2$ *and* $X/D_e = 2$.

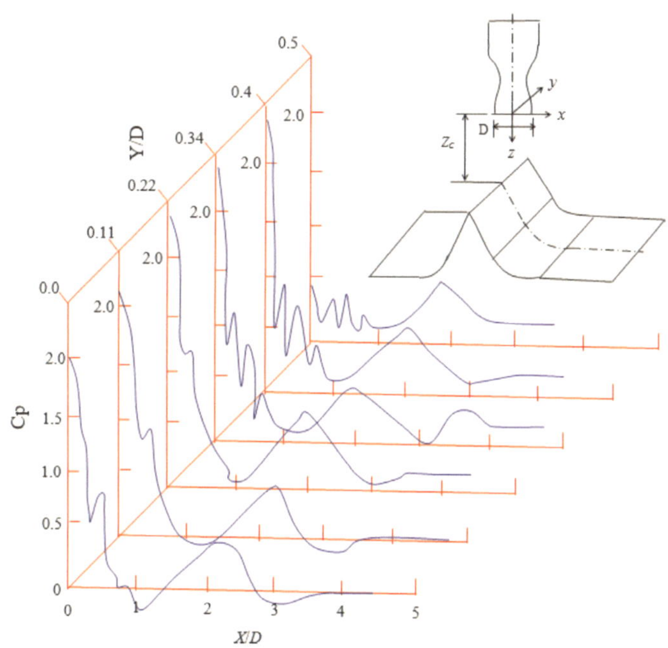

Figure 16.
Surface pressure distribution over wedge deflector.

due to the three dimensionality of the impingement flow field. However, the compression due to the curvature is still present. These surface pressure distributions were used to obtain overall impingement load coefficient. Integration has been done using the trapezoidal rule.

6. Flow field over deflector at lift-off time

During the lift-off condition of the rocket, the exhaust jet from the rocket nozzle impinges on the launch pad and is produced complex impingement flow field. The p_e/p_a is less than 1 during the lift-off period. In order to estimate the impingement load on the jet deflector and for prediction of jet noise during the lift-off condition, it is necessary to obtain the free jet properties.

A three-dimensional numerical simulation is carried using compressible Euler equations. The governing fluid dynamics equations are solved using finite volume method. Time integration is performed using an explicit multistage Runge-Kutta method.

This problem is solved when the L-40 engine is operated during the lift-off time as shown in **Figure 17**. The boundary conditions are enforced by using the idea of image cells on the plane of symmetry. On the nozzle exit plane of L-40 engine, the following conditions are taken for the computational purposes:

$$\rho = 0.081\ \text{kg/m}^3, T = 1144.907\ \text{K}, w = 20.927\ \text{kg/mol}, Cp = 1978.581\ \text{J/kg K}, M_e = 3.689$$

At the nozzle diaphragm of the core motor and solid wall of the jet deflector no normal flow conditions are applied. For quiescent external condition, the ambient pressure is imposed on the remaining sides of the computational domain (**Figure 18**).

Figures 19 and **20** show the velocity field and density contour over the jet deflector. The velocity vector shows complicated flow field and also having recirculating zones. This can also be noticed in the density contour plot. Static pressure distribution along the deflector surface has been shown in **Figure 21**. A higher pressure is observed at a location downstream of the apex of the deflector which is impingement region of

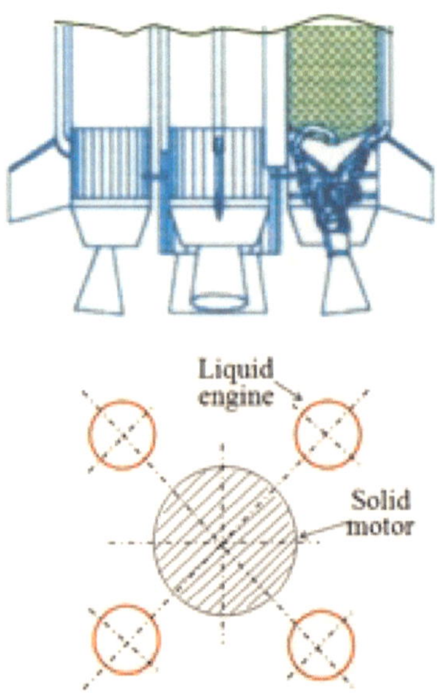

Figure 17.
Launch vehicle configuration at lift-off.

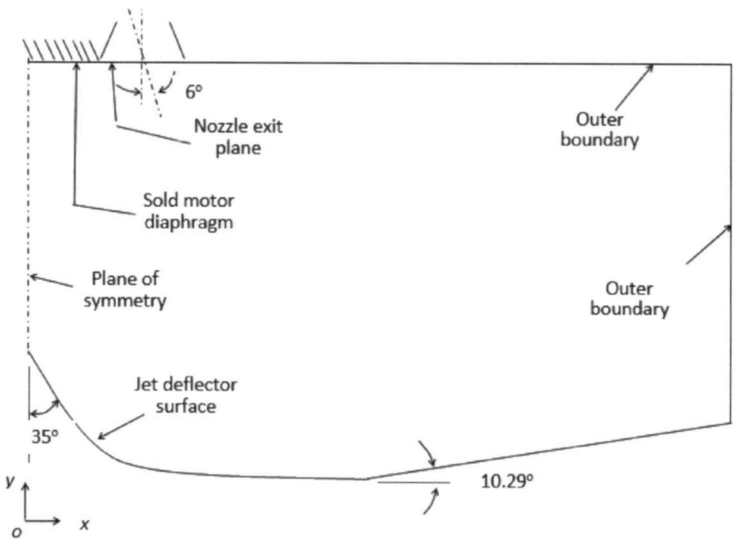

Figure 18.
Computational domain.

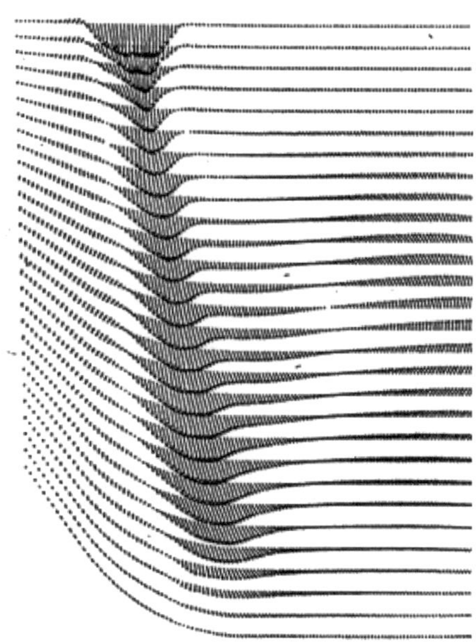

Figure 19.
Velocity field over the jet deflector at the lift-off.

the jets. **Figure 22** shows pressure on the base of the launch vehicle. It can be seen from the pressure distribution that the diaphragm is having pressure above the ambient pressure. It is important to mention that the present numerical analysis may predict higher conservative pressure distribution.

Figure 20.
Density contour over the jet deflector at the lift-off.

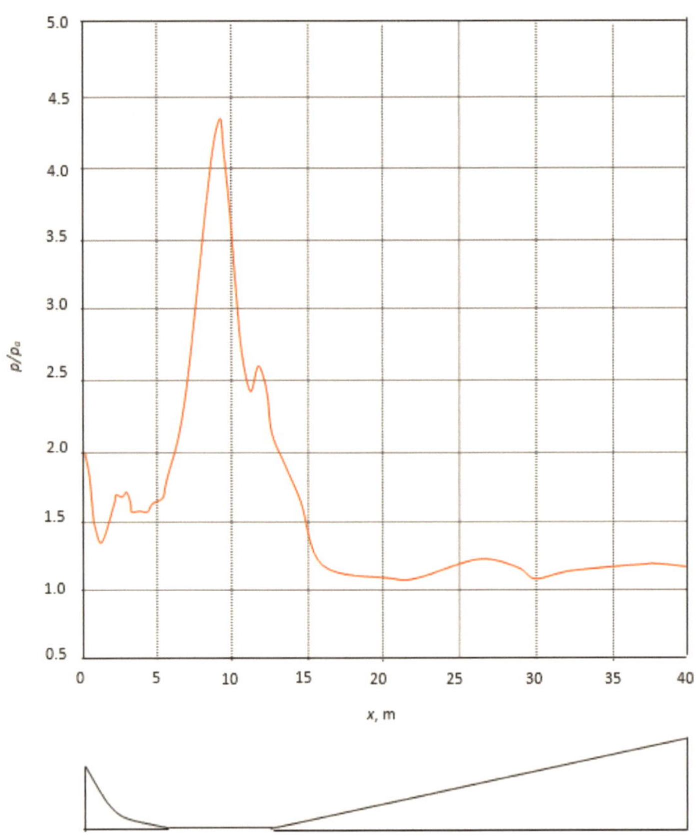

Figure 21.
Surface pressure distribution along the jet deflector.

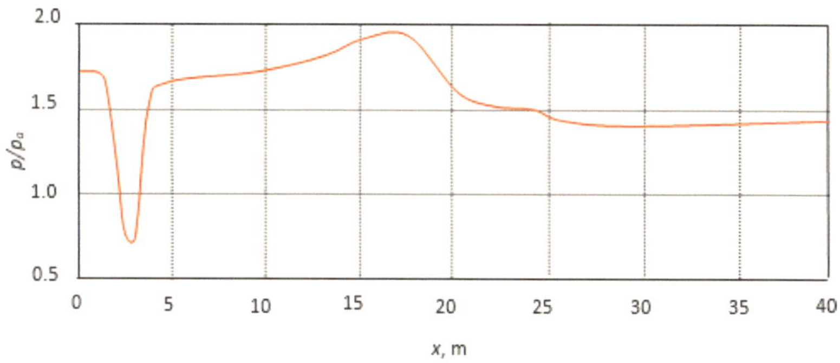

Figure 22.
Pressure distribution on the base of launch vehicle.

7. Conclusion

Numerical simulations are carried out to obtain supersonic free jets emanating from convergent-divergent nozzles at different exit Mach number and operating nozzle pressure ratio. Shock cells are obtained using density contours and a least square straight line is obtained from the numerical and experimental data. Effect of jet expansion ratio and distance between nozzle exit plane to deflector apex has been analyzed for various stand-off distance. Overall load coefficient has been calculated for the axisymmetric and wedge deflector. Pressure distributions on the jet deflector and on the base of rocket have been computed using Euler flow solver.

Nomenclature

d_j	effective jet diameter
M	local Mach number
M_1	Mach number before expansion
M_2	Mach number after expansion
M_e	nozzle design Mach number
M_j	fully expanded jet Mach number corresponding to isentropic expansion of jet from P_0 to p_a
β^2	(M_j^2-1)
P_0	stagnation pressure
p_e	exit pressure
p_a	ambient pressure
pt_2	Pitot pressure along jet
L_m	Mach disc location
L_1	First shock cell length
L_2	Second shock cell length
L_3	Third shock cell length
x	Axial distance from the nozzle exit plane
X_c	distance from nozzle exit plane to deflector apex
δ	deflection angle
γ	ratio of specific heats
ν	Prandtl–Meyer angle

Author details

Rakhab Mehta
Noorul Islam Centre for Higher Education, Kumarcoil, Tamil Nadu, India

*Address all correspondence to: drrakhab.mehta@gmail.com

References

[1] Meyer T. Über zweidimensionale Bewegungsvorgänge in einem Gas, das mit Überschallgeschwindigkeit strömt. Göttingen: Georg-August Universität; 1908

[2] Mehta RC, Prasad JK. Investigation of supersonic free jets emanating from convergent-divergent nozzles. International Journal of Computational Fluid Dynamics. 1998;**10**:81-71. DOI: 10.1080/10618569808961673

[3] Basu S, Saha S, Chakraborty D. Numerical Simulation of Missile Jet Deflector. Journal of Spacecraft and Rockets. 2017;**54**:930-935. DOI: 10.2514/1.A33761

[4] Ding J, Liu Y, He H. Numerical study on the effect of rocket plume to the jet flow deflector in liquid rocket engine test stage. Journal of Physics. 2019;**1300**:012052. DOI: 10.1088/1742-6596/1300/1/012052

[5] Allgood D, Ahuja V. Computational plume modeling of conceptual ARES vehicle stage tests. 43rd AI-AA/ASME/SAE/ASEE Joint Propulsion Conference & Exhibit. Cincinnati, OH, USA: AIAA; 2007. p. 5708. DOI: 10.2514/6.2007-5708

[6] Zhou Z, Zhao C, Lu C, Le G. Numerical studies on four-engine rocket exhaust plume impinging on flame deflectors with afterburning. Defense Technology. 2021;**17**:1207-1216

[7] Jiang C, Han T, Gao Z, Lee C. A review of impinging jets during rocket launching. Progress in Aerospace Science. 2019;**2019**:109. DOI: 10.1016/j.paerosci.2019.05.007

[8] Zhou Z, Zhang L, Le G. Numerical study for the flame deflector design of four-engine liquid rockets. Engineering Applications of Computational Fluid Mechanism. 2020;**14**:726-737. DOI: 10.1080/19942060.2020.1761453

[9] Wu J, Lin T, Luke EA, Tong X, Cinnella P. Comprehensive numerical study of jet-flow impingement over flat plates. Journal of Spacecraft and Rockets. 2002;**39**:357-366. DOI: 10.2514/2.3834

[10] Kim KH, Chang KS. Three-dimensional structure of a supersonic jet impingement on an inclined plate. Journal of Spacecraft and Rockets. 1984;**31**:778-782. DOI: 10.2514/3.26512

[11] Nakai Y, Fujimatsu N, Fujii K. Experimental study of underexpanded supersonic jet impingement on an inclined flat plate. AIAA Journal. 2006;**44**:2691-2699. DOI: 10.2514/1.17514

[12] McIlroy K, Fujii K. Computational analysis of supersonic underexpanded jets impinging on an inclined flat plate. AIAA Paper 2007-3859. 2007

[13] Xiao L, Hao X, Lei D, Tiezhi S. Flow structure and parameter evaluation of conical convergent–divergent nozzle supersonic jet flows. Physics of Fluids. 2023;**35**:066109. DOI: 10.1063/5.0151556

[14] Prasad JK. Experimental and Numerical Results of Free Jets and Impinging Jets. Chennai: IIT Madras; 1993

[15] Love EB, Grigsby CE, Lee LF, Woodling MJ. Experimental and theoretical studies of axisymmetric free jets. Hampton, USA: NASA Langley Research Centre, TRR-6; 1959

[16] Abdel-Fattah AM. Discrete tone emission from high-pressure ratio supersonic jets from convergent-

divergent nozzles. AIAA Journal. 1988; **26**:283-291. DOI: 10.2514/3.9886

[17] Rathakrishnan E. Gas Dynamics. 7th ed. Delhi, India: PHI Learning Pvt. Ltd.; 2020

[18] Jones WF, Launder BE. The prediction of laminarization with a two-equation model of turbulent. International Journal of Heat and Mass Transfer. 1982;**15**:1-26

[19] Sarkar S. The stabilizing effect of compressibility in turbulent shear flow. Journal of Fluid Mechanics. 1995;**282**: 163-186. DOI: 10.1017/S0022112095000085

[20] Tan CKW, Tanna HK. Shock associated noise of supersonic jets from convergent-divergent nozzles. Journal of Sound and Vibration. 1982;**8**:337-358. DOI: 10.1016/0022-460X(82)90244-9

[21] Mehta RC, Prasad JK. Estimation of shock cell structure of axisymmetric supersonic free jets. Indian Journal of Engineering & Materials Sciences. 1996; **3**:141-147

[22] Lamont PJ, Hunt BL. The impingement of underexpanded axisymmetric jets on wedges. Journal of Fluid Mechanics. 1976;**76**:307-336. DOI: 10.1017/S0022112076000657

[23] Prasad JK, Mehta RC, Sreekanth AK. Impingement of supersonic jets on an axisymmetric deflector. AIAA Journal. 1994;**32**:1535-1538. DOI: 10.2514/3.12225

Chapter 8

Application of Simulation Modeling to Assess the Operation of Urban Toll Plazas

Alexander Talavirya, Michael Laskin and Alissa Dubgorn

Abstract

This chapter deals with the simulation of traffic flow through toll plazas (hereinafter referred to as TPs) on an intraurban toll road. Toll plazas (TPs) and their functioning parameters are the sources of traffic congestion risk on toll roads. Three types of TPs are considered: at the exit from the toll road; on the main course of the road; at the exit from the road before the regulated intersection. As a transport micro-simulation methodology, discrete-event simulation of TPs using AnyLogic software is used. Each TP simulation model takes into account the specifics of traffic and user behavior. At the end of the chapter, conclusions are offered on the main factors influencing traffic congestion for each of the TPs. Based on the data obtained from the simulations, recommendations were made to improve the performance of the TP and the traffic light facility, reducing the risks of congestion on the research objects.

Keywords: discrete-event simulation, toll road, toll plaza, toll collection system, traffic congestion

1. Introduction

Over the past two decades, toll roads have played a special role for countries with developing transport infrastructures, including those that are part of domestic and international transport corridors. Despite the widespread adoption of barrier-free toll system technologies, such as multilane free-flow, barrier-type toll roads, where fare collection is done with a stop-over at a TP, are prevalent in a number of countries.

One type of toll road is the intraurban toll road, which connects different parts of a city as well as suburban areas. Unlike high-speed toll roads, they are characterized by high volume of traffic, high volume of labor and recreational correspondence, cramped urban infrastructure of alignment, its transport interchanges, and TPs. Intraurban toll roads are designed to unload the street and road network of the city, accelerate movement between districts and increase the mobility of the urban population using individual cars.

Taking into account territorial restrictions on placement of engineering infrastructure, the task of ensuring maximum capacity of the implemented toll collection system (hereinafter referred to as TCS) with a minimum number of toll lanes and dimensions of the TPs zone is entrusted to the TPs of intraurban toll roads. The TCS must be configured in such a way as to ensure the best possible match with traffic flow, distribute the load on the toll lanes evenly during peak hours, and minimize downtime during the day.

These increased requirements for the configuration of the TP allow for the reduction of congestion risks on the road itself and at its connection points to the street and road network. Congestion on a toll road has a negative impact on a combination of traffic, economic, social, and environmental parameters.

Tasks of maintaining, controlling, and managing traffic flow on a toll road are usually entrusted to the operating organization operator, which manages traffic on a toll road with the help of traffic management system and is designed to ensure the efficiency of using all sections of the toll road and maintaining a high quality of user services provided. Toll roads have a capacity limit. The ability to predict anticipated changes in traffic, risks of traffic congestion, and reduction in the quality of user service is a priority and relevant to the road operator.

Various mathematical and instrumental methods can be used to assess the efficiency of the TCS functioning at the TP, one of which is simulation modeling. Simulation modeling is a highly effective and relatively inexpensive tool for assessing traffic flows, and the quality of functioning of road elements under conditions of predictable road traffic not only in road operation but also in road design.

The Western High-Speed Diameter (hereinafter referred to as WHSD), located in St. Petersburg, Russia, is considered as an example of an intraurban toll road. This project is an excellent example of the simulation method because of the variety of technical solutions that allow different system configurations to be considered. The TCS of this road can be classified as a mixed type, as its sections have features of both open and closed types of TCS, and the TPs combine different types of toll lanes, allowing for both stop and nonstop tolling.

The variety of TCS configurations allows the road operator to control the TP by adapting it to varying traffic flow parameters, the main ones being its intensity, traffic composition, and the proportion of toll collection methods used.

Thus, the need for effective management of TCS configurations at TPs, as well as the use of reliable and flexible performance evaluation tools for system configurations that can be used by the operator to improve the quality of management decision-making, is a relevant issue.

This chapter shows the results of the development and application of a simulation method to evaluate the performance of different types of intraurban TPs:

1. Toll road exit TP;

2. TP on the main course of a toll road;

3. TP at the exit from the toll road before the controlled intersection.

The conducted research allowed us to consider and evaluate the quality of operation of TPs with various configurations of TCS under different traffic flow parameters as well as to assess the risks of traffic congestion on them.

Based on the results of the study, the toll road operator can assess the current and predicted congestion of the TP, as well as plan measures to optimize and upgrade the TCS in advance.

2. Literature review

The methodology of studying TP as a queuing system and applying discrete-time modeling to it was applied by Punitha [1]. In Aksoy et al. [2], using the Fatih Sultan Mehmet Bridge toll simulation model, a number of experiments were conducted using the Vissim software to vary the TP intensity at different numbers of open toll lanes, thereby improving the quality of service. An effective methodology for evaluating the TP operation using the Generic Toll Plaza Simulation (GENTOPS) simulation model has been proposed by Aycin [3]. Similar optimization problems for the operation of toll lanes using simulation methods have been solved by Izuhara et al. [4] and Levinson & Chang [5]. An assessment of the effectiveness of TP modernizations has been investigated by Yosritzal et al. [6]. In this study, it was pointed out that the most feasible capacity increase at TP might be to upgrade the TCS system by switching to Electronic toll collection (hereinafter referred to as ETC) and multilane free-flow rather than increasing the number of existing manual toll lanes.

The use of simulation modeling is common for traffic forecasting problems at TPs. Prediction of average queue length at TP using Long Short-Term Memory model and particle swarm optimization algorithm is used by Peng et al. [7]. Munawar & Andriyanto [8] demonstrated the application of simulation modeling to solve queue waiting time prediction problems using the example of Cililitan Toll Plaza, Jakarta. Simulation modeling has been used to estimate the system delay of vehicles in manual toll lanes and ETC lanes under mixed traffic conditions in Chintaman et al. [9, 10]. However, the results of a study by Yogeshwar et al. [11] show that the level of service during toll collection includes not only the service time at the toll lanes but also many additional factors.

The introduction and application of contactless payment-type transponders at TPs, which significantly increase the toll collection rate, is actively discussed by the scientific community. The effectiveness of this method of tolling has been noted by Huang et al., Qian et al., Hewage et al., Tanino et al., Zhu et al., Ito et al., Roth, Lai et al., Lee, Tseng & Wang, Vats, Vats, Vaish & Kumar, Abuzwidah & Aty, and Holguín-Veras & Wang [12–26] in toll road projects in Taiwan, China, Sri Lanka, and other countries. Electronic toll collection (ETC) devices data are also used to study and analyze traffic flow patterns on toll roads, such studies are described in Fan et al., Weng et al., and Komada et al. [24–26]. Note that simulation modeling techniques can be used to analyze traffic flows using ETC devices data, as shown by Hirai et al., Karsaman et al., and Jehad et al. [27–29] in toll road projects in Japan, Indonesia, and Malaysia.

Zhang et al. [30] discuss the use of VISSIM simulation software to determine the optimal ratio of the manual to ETC lanes. The results of this work allow optimization of ETC operation without increasing the number of lanes. The authors point out that specific traffic conditions of different toll collection stations, such as traffic composition, are not considered in this model. An example of an optimization application for selecting the most appropriate lane allocation at the TP, which relies on queue length measurements, is described in Neuhold et al. [31].

3. Toll roads exit toll plazas

3.1 Simulation model of a roll road exit toll plaza

As a methodology for traffic flow analysis, the use of discrete-event modeling of TP is proposed. In contrast to the generalized methodology of TP modeling which is applied at the toll road design stage, the application of simulation modeling methods at the operational stage allows to take into account and reflect a number of special conditions which could be formed directly at the stage of TP functioning and caused by features of its geographical location, traffic composition, regularity of user correspondence as well as the impact of surrounding transport-logistical and social infrastructure.

As an example, the analysis of the TP Bogatyrskiy Ave./Planernaya Str., located at the exit from the WHSD toll road to the Primorsky District of St. Petersburg (**Figure 1**) simulation model is done.

The toll road uses an open-type barrier tolling system. The Primorsky district has the first largest population in the city and also has active daily labor correspondence. Since the launch of the adjacent toll road section in 2016, there has been systematic congestion at this toll road due to the insufficient capacity of the toll lanes at the TP. The configuration of this TP has six toll lanes, four of which operate in automatic mode and provide electronic toll collection (ETC) using onboard units (hereinafter referred to as OBU). The toll mode provides for nonstop passing through the lanes at a maximum speed of 30 km/h. The remaining two lanes are in manual mode and allow cash or bank card payments to be made to the cashier-operator. The traffic model affects only the TP at the exit to the Primorsky District and does not affect the traffic on the main direction of the toll road and on the return to the road from the district.

The simulation model (hereinafter referred to as SM) has been developed in AnyLogic software using traffic and process modeling libraries. The resulting TP model is presented in **Figure 2**.

Figure 1.
General view of the Bogatyrskiy Ave./Planernaya Str. TP.

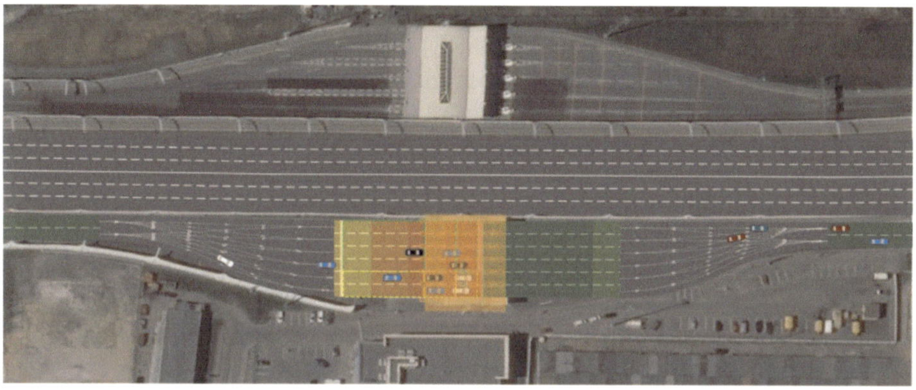

Figure 2.
SM of the Bogatyrskiy Ave./Planernaya Str. TP.

The SM of the TP allows the following traffic flow parameters to be taken into account:

- The intensity of traffic at the TP;
- The composition of traffic;
- Distribution of vehicles by the method of payment;
- Time of service in the automated lane;
- Time of service in the cash payment lane;
- Additional parameters (user behavior and tag failure).

3.2 Queuing analysis

The data for the SM, including data on intensity, traffic composition, distribution of payment methods, and service time, are obtained from video data from the Northern Capital Highway LLC website [32]. The number of vehicles passing through for 10–15 minutes was counted and the results were converted to the number of vehicles per hour. Corresponding calculations were made for all days of the week and time intervals of 24 astronomical hours. The results obtained for the same day of the week and time of day were averaged. There were no significant traffic fluctuations related to major urban cultural events during the time period in question, which could lead to significant changes in the weekly traffic cycle. There are no major logistics centers in the Primorsky district and heavy vehicle traffic is prohibited at the TP, allowing only passenger cars to be taken into account in the SM. The specified length of passenger cars in the model is 5 meters. Distribution of vehicles by payment method for this TP is 80% ETC transactions and 20% manual toll collection. It is assumed that manual toll lanes service time is triangularly distributed with a mean value of 20 (for contactless bank card payments), min of 7 (for ETC), and max value of 45 (for contact bank cards and cash payments) seconds.

An additional parameter of user behavior—tag failure—has been introduced into the model which affects the capacity of the ETC lanes. The "tag failure" parameter takes into account the probability of transponder failure (5%) when passing through ETC lanes in case the vehicle has not been read in the entry antenna, is incorrectly attached, or has a negative balance, that is, is not accepted for payment. It is assumed that tag failure time is also triangularly distributed with a mean value of 7 (in case of wrong OBU fixing), min of 3 (in case of second antenna reading), and max value of 60 (in case of negative balance) seconds.

An additional feature of the TP under study is the bottleneck at the exit of the main road (only two lanes), the six lanes (the TP itself), and the joining of the six lanes back to two lanes after the TP exit. This feature allows one to consider the whole of the TP as a "bottleneck" slowing down the speed of traffic and to estimate the time it takes to pass the section from the toll road exit to the point where the six lanes are connected back to two lanes.

The analysis of the daily traffic volume of the TP allows us to determine the distribution of user service times on the toll lanes as well as to assess the possibility of congestion occurring during the day on a weekly cycle.

3.3 Traffic intensity

Figure 3 shows the graph of observed traffic intensity at Bogatyrsky Ave./ Planernaya Str. TP by days of the week from 0:00 to 24:00 (intensity was calculated in number of vehicles/h).

Figure 3 shows seven schedules (five for weekdays and two for weekends). Weekdays are characterized by higher traffic volumes than weekends. The weekday schedules do not differ much from each other. The peak weekday intensities for the TP fall in the evening rush hour and correspond to work-home correspondences.

3.4 Analysis of toll plaza simulation model

As can be seen from **Figure 3**, the traffic volumes at the TP vary between 60 and 3140 vehicles/h, depending on the day of the week and time of day. We will be interested in the answers to the following questions:

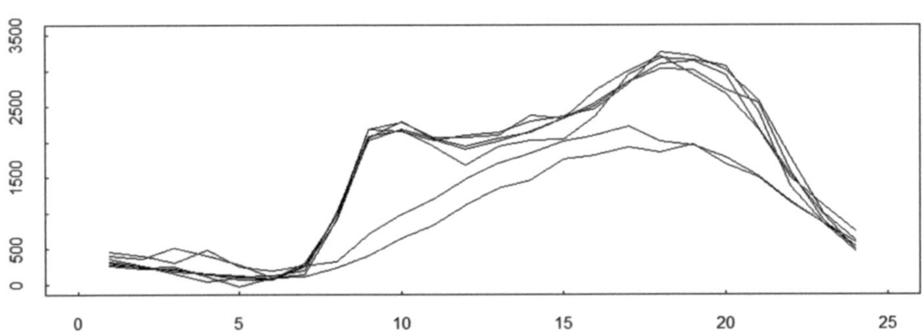

Figure 3.
Traffic intensity at the Bogatyrsky Ave./Planernaya Str. TP by days of the week from 0:00 to 24:00 (intensity was calculated in number of vehicles/h).

- At what traffic intensity will traffic congestion be generated in front of the entrance to the TP?
- How does the intensity of incoming traffic affect the distribution of vehicle service time at the TP?

In contrast to classical mass-transit theory schemes, we define service time as the time it takes for a vehicle to cross a TP from the exit point of the main road (the exit has only two lanes) to the point after the TP, where the six lanes of the TP rejoin the two-lane road (see **Figure 2**).

A number of experiments have been carried out on the SM of the TP. The distribution of service time was studied, with input flow rates from 250 to 3250 vehicles/h, in steps of 250. The occurrence of congestion is considered to be the appearance of a queue of vehicles beyond the entrance area of the TP. The results of the experiments show that congestion starts to form when the traffic intensity is higher than 1010 vehicles/h. Before this intensity is reached, the empirical service time distribution (vehicle passing through the TP zone) is an easily separable mixture of two distributions that approximate the gamma distribution well, each corresponding to vehicle passing time through either the ETC lane or the cash payment lane. Since perturbations in the form of user behavior effects are added to the model, the gamma law of the service time distribution seems to be quite expected, as being more general with respect to the Erlang distribution (see more details in the article by Talavirya A.Yu., Laskin M.B. [33]). **Figure 4** shows the distribution of service times (section traversal) for a traffic volume of 3000 vehicles/h.

Parameters of the gamma distribution shape = 2.1510, rate = 0.0126. The result of the Kolmogorov-Smirnov test for consistency of the empirical distribution with the model distribution with the specified parameters: p-value = 0.7748. At this traffic intensity, the "tag failure" effect appears: there are many vehicles, user behavior and tag failures affect the time for other vehicles to pass the TP, as a consequence, the whole TP starts to operate as a single mass service machine, the mixture of service time distributions at different payment methods is indivisible. At such high intensity, the service time distribution obeys the gamma distribution law with parameters where the distribution has significant left-hand asymmetry. **Table 1** presents the parameters of the gamma law distribution of the service time at high input flow intensities.

At high input flow rates, in an evolving queue environment, the mathematical expectation of the service time is significantly different from the most probable service time.

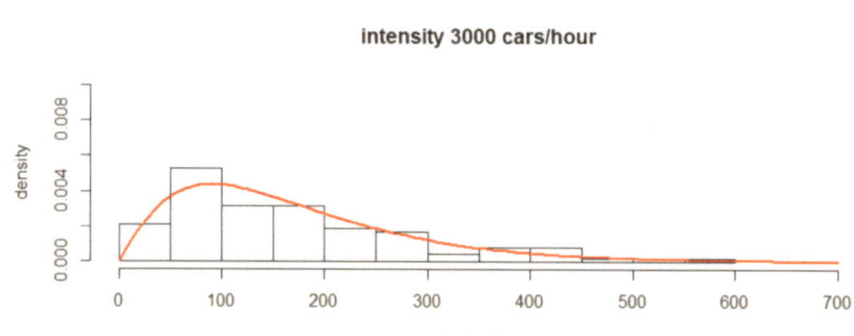

Figure 4.
Empirical distribution of service times at an input flow rate of 3000 vehicles/h.

Intensity	Gamma distribution parameters			Expectation	Mode
	shape	rate	p-value		
1250 vehicles/h	1.521	0.0132	0.0617	115	39
1500 vehicles/h	2.0381	0.0127	0.3758	161	82
1750 vehicles/h	1.9516	0.0137	0.3035	142	69
2000 vehicles/h	2.533	0.0149	0.1739	170	103
2250 vehicles/h	2.5919	0.0135	0.8186	192	118
2500 vehicles/h	2.2586	0.0109	0.6657	207	115
2750 vehicles/h	2.2658	0.0147	0.7156	154	86
3000 vehicles/h	2.1511	0.0126	0.7748	170	91
3250 vehicles/h	2.2402	0.0133	0.2608	169	93

Table 1.
Parameters of the gamma law distribution of the service time at high input flow intensities.

3.5 Determining the queue length when passing through a TP

When managing traffic on a toll road, the most practical relevance for the road operator is when traffic congestion occurs when entering the TP zone. In the context of congestion, the TP's indicators considered are:

- Number of vehicles in queue and length of queue;
- The waiting time for a vehicle to cross the line of entry into the TP area.

The following assumptions were used in the consideration of these indicators:

1. The TP area is considered to be the stretch of road that includes the TP, 135 meters of roadway before entering the TP and 150 meters after exiting the TP, that is, the entire stretch of road near the TP that is widened compared to the main two-lane road;
2. The queue of vehicles is considered to be a hoarding located in front of the line of approach to the TP, the volume of the hoarding is unlimited;
3. Queuing is defined as vehicles in front of the line of approach to the TP zone moving at a speed close to zero with a minimum distance between vehicles in heavy traffic. Provided the intensity of input and output flows is approximately equal, queue formation is of random nature, the queue can form and disappear. As the intensity of incoming traffic increases and the capacity of the TP is reached, the queue begins to increase, so we consider the queue formed if the last 135 meters before the line of entrance to the TP zone is occupied by vehicles;
4. Vehicles can only leave the queue by moving forward through the TP, there is no other way to leave the queue;
5. The queue starts to form when the intensity of incoming traffic reaches a certain threshold (for the given TP and its lanes configuration) value λ^*;

6. The exit flow is assumed to be stationary Poisson (with constant intensity λ^*) as long as there is a queue in front of the TP zone, the intensity of the exit flow is assumed to be equal to the intensity of vehicles crossing the end line of the TP zone;

7. The inlet flow is stationary Poisson (with constant intensity λ) or non-stationary (with variable intensity $\lambda(t)$).

3.6 The length of the queue

Let X be a random variable, the number of vehicles leaving the TP zone per unit time, distributed according to the Poisson distribution with parameter λ^*, that is

$$P(X = m) = \frac{\lambda^{*m}}{m!} e^{-\lambda^*} \tag{1}$$

while Y is a random variable, the number of vehicles arriving at the start of the queue, per time unit, distributed according to the Poisson distribution with parameter λ, that is

$$P(Y = n) = \frac{\lambda^n}{n!} e^{-\lambda} \tag{2}$$

then $V = Y - X$ is also a discrete random variable and its distribution law is given by the discrete function

$$f(k, \lambda, \lambda^*) = P(V = k) = e^{-(\lambda+\lambda^*)} \left(\frac{\lambda}{\lambda^*}\right)^{\frac{k}{2}} I_{|k|}\left(2\sqrt{\lambda \times \lambda^*}\right). \tag{3}$$

where $I_{|k|}(x)$ is a modified Bessel function of the first kind, at $k = 0, \pm 1, \pm 2, \dots, \pm \infty$. Such a distribution is obtained by John Gordon Skellam [34]. Then (a detailed description is given in paper [33]):

$$f(k, \lambda, \lambda^*) = P(V = k) = e^{-(\lambda+\lambda^*)} \times \sum_{m=0}^{\infty} \frac{1}{\mathrm{m!\Gamma(m+|k|+1)}} \lambda^{m+|k|} \times \lambda^{*m} \tag{4}$$

Main characteristics of the random variable V: $E(V) = \lambda - \lambda^*$, the variance of V: $D(V) = \lambda + \lambda^*$, standard deviation $\sigma(V) = \sqrt{\lambda + \lambda^*}$, coefficient of asymmetry:

$$\gamma_1 = \frac{\lambda - \lambda^*}{(\lambda + \lambda^*)^{\frac{3}{2}}} \tag{5}$$

Obviously, this distribution is symmetric only in the case of $\lambda = \lambda^*$.

Figure 5 shows the Skellam distribution probability functions (the probabilities are defined for integer values of k) for different values of the parameters λ, λ^*.

$\lambda^* = 29.17$,

$\lambda = 14.17; 17.17; 20.17; 23.17; 26.17; 29.17; 35.17; 41.17; 47.17; 53.17; 59.17$,

The intensities are given as vehicles/minute (14.17 corresponds to an intensity of 850 vehicles/h, 59.17 corresponds to an intensity of 3550 vehicles/h). It is obvious from **Figure 5** that:

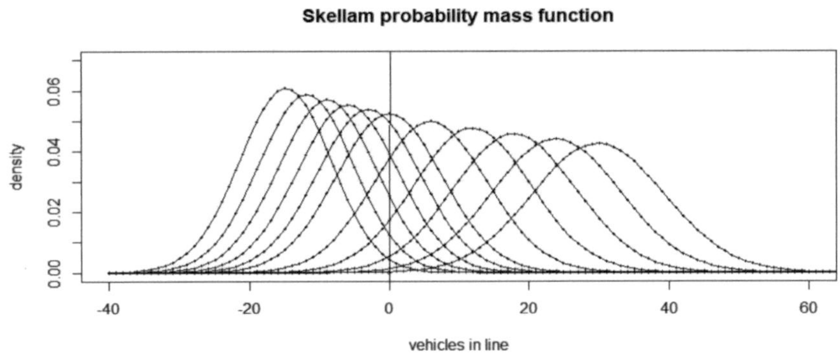

Figure 5.
Examples of Skellam distribution probability mass functions for different input-output ratios. When $\lambda \leq \lambda^*$ the maximum (mode) is less than zero, when $\lambda \geq \lambda^*$ the maximums are greater than zero.

- when $\lambda=\lambda^*$, the probability of queue existence is 0.5, the probability of its absence is also 0.5;
- when $\lambda\ \lambda^*$ the probability of queue existence is less than 0.5, it is positive and asymptotically tends to 0 when $\lambda \to 0$;
- when $\lambda>\lambda^*$ the probability of queueing is greater than 0.5 and asymptotically tends to 1 when $\lambda \to +\infty$.

The length of the queue $V^*(t_0, \tau)$, formed during the time interval $[t_0, t_0 + \tau]$ at variable input intensity $\lambda(t)$ will be a discrete random variable with a distribution law (see [33] for details):

$$\begin{cases} P(V^* = 0) = e^{-(\Lambda(t_0,\tau)+\lambda^*\tau)} \sum\limits_{k=0}^{-\infty} \left(\dfrac{\Lambda(t_0,\tau)}{\lambda^*\tau}\right)^{\frac{k}{2}} I_{|k|}\left(2\sqrt{\Lambda(t_0,\tau)\times\lambda^*\tau}\right), k \leq 0, k \in Z \\ P(V^* = k) = e^{-(\Lambda(t_0,\tau)+\lambda^*\tau)} \left(\dfrac{\Lambda(t_0,\tau)}{\lambda^*\tau}\right)^{\frac{k}{2}} I_{|k|}\left(2\sqrt{\Lambda(t_0,\tau)\times\lambda^*\tau}\right), k > 0, k \in Z \end{cases} \tag{6}$$

where $\Lambda(t_0, \tau) = \int\limits_{t_0}^{t_0+\tau} \lambda(t)dt$, and t_0 means the moment the queue starts to form.

The standard library function of the statistical package R can be used to calculate the values of the Bessel functions.

Table 2 shows the calculated values of the mathematical expectation and the most probable value of the queue length occurring per unit time for the TP under consideration. The calculated values for output flow intensity $\lambda^* = 29.17$ highlighted in bold in **Tables 2–4**.

3.7 Waiting time in the queue

As before, $X \in Pois(\lambda^*)$ is a random variable, the number of vehicles leaving the TP zone per unit time, and $Y \in Pois(\lambda)$ is a random variable, the number of vehicles arriving at the start of the queue per unit time. Then, X_t is the number of vehicles

Input intensity vehicles/min.	Queue existence probability	Mean	Mode	Skewness
14.17	0.01	0.022	0.000	−0.05
17.17	0.04	0.100	0.000	−0.04
20.17	0.11	0.324	0.000	−0.03
23.17	0.22	0.817	0.000	−0.02
26.17	0.37	1.699	0.000	−0.01
29.17	**0.50**	**3.040**	**0.000**	**0.00**
35.17	0.79	7.048	6.000	0.01
41.17	0.93	12.281	12.000	0.02
47.17	0.98	18.056	18.000	0.03
53.17	0.99	24.010	24.000	0.03
59.17	1.00	30.002	30.000	0.04

Table 2.
Average queue length, most probable queue length, probability of queue occurrence, coefficient of asymmetry of queue length distribution formed per unit time at output flow intensity $\lambda^* = 29.17$, *and input flow intensities* $\lambda = 14.17; 17.17; 20.17; 23.17; 26.17; 29.17; 35.17; 41.17; 47.17; 53.17; 59.17$.

leaving the TP zone in time t is also a random variable and $X_t \in Pois(\lambda^* t)$. It can be shown (see details in [33]) that the distribution density of the random variable T, the time in which all the vehicles arrive at the beginning of the queue per unit time (with intensity λ), with the intensity of outgoing flow λ^* will leave the queue, can be written as

$$f^*(0, \lambda, \lambda^* t) = \lambda^* f(0, \lambda, \lambda^* t) = \lambda^* \times e^{-(\lambda + \lambda^* t)} \times I_0\left(2\sqrt{\lambda \lambda^* t}\right) \quad (7)$$

Mathematical expectation, variance, and standard deviation of the value T (waiting time in the queue of vehicles arriving at the beginning of a unit time interval):

$$E(T) = \frac{\lambda + 1}{\lambda^*}, D = \frac{\lambda + 1}{\lambda^{*2}}, \sigma = \frac{\sqrt{\lambda + 1}}{\lambda^*} \quad (8)$$

Figure 6 shows the distribution functions of the random variable T, with different ratios of λ and $\lambda^* : \lambda^* = 29.17$,
$\lambda = 14.17; 17.17; 20.17; 23.17; 26.17; 29.17; 35.17; 41.17; 47.17; 53.17; 59.17$.

Numerical characteristics of time distribution T can be obtained either by calculation (e.g., expected value, variance, standard deviation) or by simulation in R statistical package. **Table 3** shows the values of the mathematical expectation and the most probable value of T obtained by modeling in R and the mathematical expectation, variance, and standard deviation obtained by Eq. (8).

3.8 Changing the dimensionality of intensities and time scales

When changing the time scale, it should be taken into account, that in Eqs. (6) and (7) at an increase of values of the exponent argument and the modified Bessel

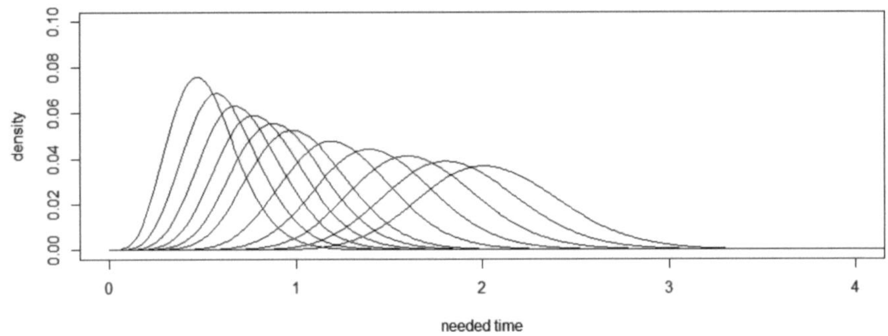

Figure 6.
Distribution density of the random variable T, *with* $\lambda^* = 29.17$ *and* λ *ranging from 14.17 to 59.17 = vehicles/min (850 to 3555 vehicles/h).*

Input intensity vehicles/min.	R simulated values		Calculated values		
	Mean time	Modal time	Mean time	Variance	Standard deviation
14.17	0.520	0.480	0.520	0.02	0.13
17.17	0.623	0.580	0.623	0.02	0.15
20.17	0.726	0.680	0.726	0.02	0.16
23.17	0.826	0.790	0.829	0.03	0.17
26.17	0.931	0.890	0.931	0.03	0.18
29.17	**1.034**	**0.990**	**1.034**	**0.04**	**0.19**
35.17	1.240	1.200	1.240	0.04	0.21
41.17	1.446	1.400	1.446	0.05	0.22
47.17	1.651	1.610	1.651	0.06	0.24
53.17	1.857	1.820	1.857	0.06	0.25
59.17	2.063	2.063	2.063	0.07	0.27

Table 3.
Mathematical expectation, the modal value of T *obtained by modeling in R, and mathematical expectation, variance, and standard deviation obtained by Eq. (8).*

function (first kind, order 0) at numerical calculations even within really existing values of intensities and the time intervals of interest, the machine zero is quickly formed. Therefore, in the calculations, it is better to change the ratios of intensities and time scales so that the function arguments in Eqs. (6) and (7) allow for obtaining numerical values. Suppose that we want to obtain an estimate of queue waiting time in hours, knowing the intensity expressed in TC per hour. The intensities $\lambda^* = 29.17$, $\lambda = 14.17; 17.17; 20.17; 23.17; 26.17; 29.17; 35.17; 41.17; 47.17; 53.17; 59.17$, previously dimensioned in TC per minute are converted to TC per hour: $\lambda^* = 1750$, $\lambda = 850; 1030; 1210; 1390; 1570; 1750; 2110; 2470; 2830; 3190; 3550$. Such dimensionality is also not yet suitable for calculating exponents and values of Bessel functions. On the basis of actual intensities on the considered road, it seems reasonable to consider hourly intensities in hundreds of vehicles per hour ($\lambda^* = 17.50$, $\lambda = 8.50; 10.30; 12.10; 13.90; 15.70; 17.50; 21.10; 24.70; 28.30; 31.90; 35.50$), the time unit is 1 hour. We obtain a similar figure and parameters of time distribution laws, but already in hours (**Table 4**).

Figure 7 shows the distribution functions of the random variable T, with different ratios of λ and λ^* : $\lambda^* = 17.5$, $\lambda = 10.3; 12.1; 13.9; 15.7; 17.5; 21.1; 24.7; 28.3; 31.9; 35.5$.

The asymmetry of the obtained distribution laws is insignificant (the modal value differs from the mean between 3.6% and 13.4%), but the standard deviation is rather large.

When the input intensity is variable ($\lambda = \lambda(h)$), h —time of day), we consider the input intensity parameter as $\Lambda(h_0, \tau) = \int_{h_0}^{h_0+\tau} \lambda(h)dh$, ... h_0 — the time (time of day) of the onset of road congestion.

In Laskin et al. [35], it is shown that the daily intensities given in **Figure 3** can be well approximated by the first n terms of the trigonometric series. This model has been applied to the study of queue lengths over the course of a day.

Input intensity 100 vehicles/h	R simulated values		Calculated values		
	Mean time (hour)	Modal time (hour)	Mean time (hour)	Variance	Standard deviation (hour)
8.5	0.54	0.47	0.54	0.03	0.18
10.3	0.65	0.57	0.65	0.04	0.19
12.1	0.75	0.67	0.75	0.04	0.21
13.9	0.85	0.78	0.85	0.05	0.22
15.7	0.95	0.88	0.95	0.05	0.23
17.5	**1.06**	**0.98**	**1.06**	**0.06**	**0.25**
21.1	1.26	1.19	1.26	0.07	0.27
24.7	1.47	1.39	1.47	0.08	0.29
28.3	1.67	1.60	1.67	0.10	0.31
31.9	1.88	1.80	1.88	0.11	0.33
35.5	2.09	2.01	2.09	0.12	0.35

Table 4.
Mathematical expectation, modal value of T *obtained by modeling in R, and mathematical expectation, variance, and standard deviation obtained by Eq. (8) in hours.*

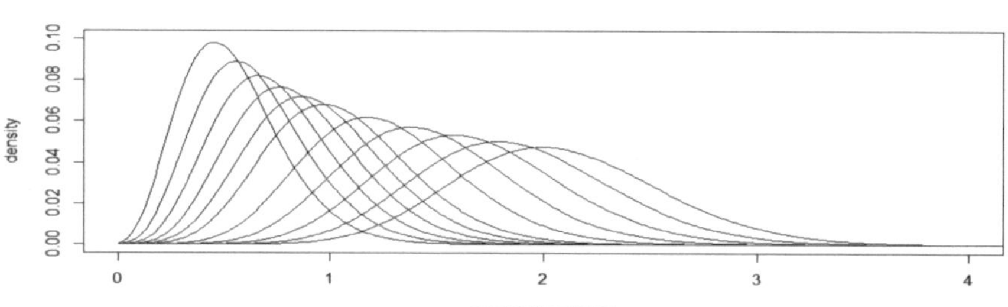

Figure 7.
Distribution density of the random variable T, *with* $\lambda^* = 17.5$ *(100 vehicles/h) and* λ *ranging from 8.50 to 35.55 (100 vehicles/h).*

When examining the queue length and waiting time for vehicles in the queue before entering the TP zone, we consider the accumulated intensity as

$$\Lambda(h_0, \tau) = \int_{h_0}^{h_0+\tau} \lambda(h)dh \tag{9}$$

where h_0 is the moment at which a period of time τ of length begins to be counted.

We estimate the queue length by the mean value $E(V) = \Lambda(h_0, \tau) - \lambda^* \tau$ with standard deviation $\sigma(V) = \sqrt{\Lambda(h_0, \tau) + \lambda^* \tau}$. We estimate the waiting time in queue θ by the mean $E(\theta) = E(T) * \tau - \tau = \frac{\Lambda(h_0, \tau)+1}{\lambda^* \tau} \times \tau - \tau = \frac{\Lambda(h_0, \tau) - \lambda^* + 1}{\lambda^*}$ with standard deviation $\sigma(\theta) = \sigma(T) = \frac{\sqrt{\Lambda(h_0, \tau)+1}}{\lambda^* \tau}$.

The corresponding calculations were performed in the R statistical package environment, for the following input and output flows and TP parameters:

- An example of changing input intensity observed during October 2019 was chosen for modeling (shown in **Figure 2**); a trigonometric series was used for approximation, similar to the article by Laskin et al. [35];
- The outbound flow intensity is considered to be equal to the minimum inbound flow intensity at which stable traffic congestion is formed when entering the TP zone; it was determined by simulation methods under the following parameters of six toll lanes operation mode: four lanes with automatic transponder signal reading, two lanes providing cash payment (the second lane is mandatory as backup), readout failures (user behavior)-5% of total number of vehicles in the flow, percentage of vehicles equipped with a transponder. The simulations with different entry flow intensities show that with these TP parameters a stable inlet congestion is formed at an entry flow intensity of 2550 vehicles/h;
- The estimation of the queue length with the distribution law $P(V^*(t_0, \tau) = k)$ (6), as seen in **Figure 5**, depends on the ratio $\Lambda(h_0, \tau)$ and $\lambda^* \tau$, if these values are close, the queue can appear and disappear, but as the value $\Lambda(h_0, \tau)$ increases, the probability that congestion exists increases. Therefore, under the conditions of the persistent congestion observed in the simulation, we assume the difference between $\Lambda(h_0, \tau)$ and $\lambda^* \tau$ to be large enough to use Eqs. (7) and (8) to estimate the average queue length and its standard deviation for the estimation judgments:

$$E(V^*) \approx \Lambda(h_0, \tau) - \lambda^* \tau, \tag{10}$$

$$\sigma(V^*) \approx \sqrt{\Lambda(h_0, \tau) + \lambda^* \tau} \tag{11}$$

- The estimation of the queueing time also depends on the parameters $\Lambda(h_0, \tau)$ and $\lambda^* \tau$. We use the following representations for the estimated queueing time judgments:

$$E(T) \approx \frac{\Lambda(h_0, \tau) + 1}{\lambda^* \tau}, \tag{12}$$

$$\sigma(T) \approx \frac{\sqrt{\Lambda(h_0, \tau) + 1}}{\lambda^* \tau} \tag{13}$$

The results are shown in **Figures 8** and **9** and **Table 5**.

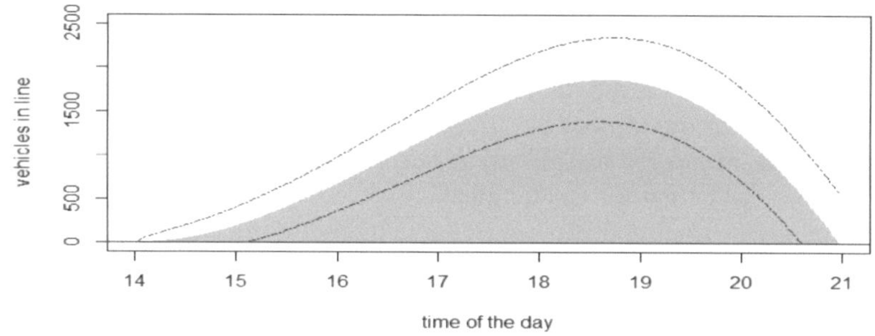

Figure 8.
Estimation of average number of vehicles in queue during the working day. The blue line is a corridor of +/− 3σ.

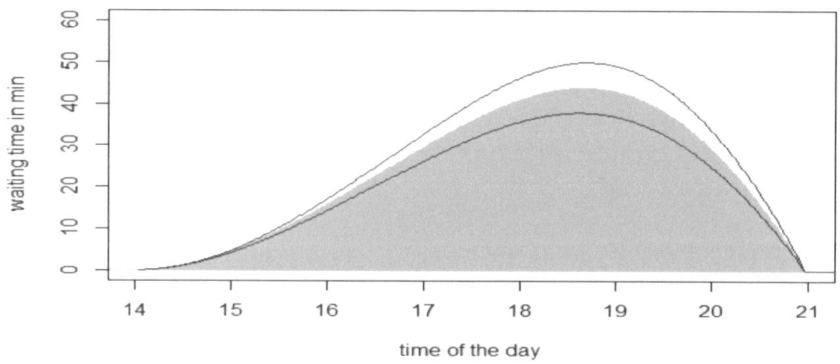

Figure 9.
Estimation of waiting time (in minutes) for vehicles in the queue from the moment of approach to the beginning of the queue to the entrance to the TP zone, depending on the time of day when the vehicle arrived at the beginning of the queue. The blue lines show a corridor of possible deviations of +/− 3σ to the average waiting time.

Time of the day	Average queue length in vehicle	Average queue length in vehicle-3σ	Average queue length in vehicle +3σ	3-lane queue in km	4-lane queue in km	Average waiting time (min.)	Average waiting time (min.)-3σ	Average waiting time (min.) + 3σ
14:00	—	—	—	—	—	—	—	—
15:00	189	—	403	0.5	0.38	4.5	4.2	4.7
16:00	667	356	977	1.8	1.33	16.0	14.0	17.0
17:00	1250	865	1634	3.3	2.50	30.0	26.0	33.0
18:00	1739	1295	2184	4.6	3.48	41.0	36.0	46.0
18:40	1862	1385	2340	5.0	3.72	44.0	38.0	50.0
19:00	1825	1331	2320	4.9	3.65	43.0	37.0	49.0
20:00	1238	704	1772	3.3	2.48	29.0	25.0	34.0
20:30	676	126	1227	1.8	1.35	16.0	13.5	18.4
21:00	—	—	—	—	—	—	—	—

Table 5.
Time of day, average queue length in vehicles, corridor of possible deviations to average queue length +/− 3σ, estimate of average queue length in km, average waiting time in queue before entering the TP zone, and corridor of possible deviations to average waiting time +/− 3σ.

3.9 Time of existence of the queue

The queue begins to form at 843 minutes from the beginning of the day (00:00), which corresponds to the time of day 14:03.

The queue ceases to exist at 1259 minutes from the beginning of the day (00:00), which corresponds to the time of day 20:59.

Thus, a queue consisting of slow-moving vehicles on the Bogatyrsky St./ Planernaya St. TP exists from 14:03 to 20:59 (2:03 PM to 8:59 PM). The queue length is permanently changing, the estimate of its average value is shown in **Figure 8**. The blue dotted lines show the corridor +/− 3σ for estimating the queue length.

3.10 Maximum queue length, waiting time in queue

The maximum number of vehicles in the queue is estimated at 1863 vehicles, which with a queue of four lanes (4 lanes on the main toll road to the exit to the TP), would be 3.72 km in length. The longest queue length corresponds to 1120 minutes, which corresponds to a time of day of 18:40. The range of possible changes in queue length at the time of greatest length is between 1386 and 2343 vehicles.

Table 5 shows:

1. The average queue length in vehicles;

2. Corridor of possible (+/−3σ) changes to queue length;

3. Estimation of average queue length in kilometers;

4. Average waiting time in the queue before entering the TP zone;

5. Corridor of possible changes.

There is a short (160-meter) stretch of road that has two lanes before exiting to the Bogatyrsky TP. When congestion occurs in front of the toll road, the queue exits beyond the TP zone onto the main course of the toll road, which has four lanes, so the queue length has been estimated for a queue formed of four lanes. The distance between two vehicles standing one behind the other in dense traffic is 8 meters.

The waiting time in the queue before exiting the main turn and entering the TP zone is an estimation and reflects only the case when vehicles in the flow are moving without rearrangements and are lined up, for example, in 4 lanes, when approaching the beginning of the queue. In reality, this condition is not systematically fulfilled, so the dispersion of waiting time of vehicles in the queue can increase significantly due to the actions of unruly drivers. This circumstance does not affect the waiting time of the queue.

To confirm that the calculated values of the queue length and queueing time before entering the TP zone correspond to the actual road traffic situation, we use data from the search and information mapping service Yandex.Maps for St. Petersburg. This service displays both the current traffic situation in real time and forecasted congestion length (in km) on the WHSD road section before the Bogatyrsky St./Planernaya St. TP based on statistical data. **Table 6** shows the lengths of the forecasted congestion (in km) on weekdays in February 2020 due to the low traffic intensity on weekends, there is no congestion then.

Weekday/Time	Monday	Tuesday	Wednesday	Thursday	Friday
16:00	0.9	2.36	2.16	1.74	2.36
16:30	1.7	3.02	2.94	3.39	3.65
17:00	3.22	4.9	3.87	3.68	4.89
17:30	4.9	5.05	4.75	3.86	4.89
18:00	5.29	5.09	4.75	4.65	4.91
18:30	5.29	5.13	4.75	4.65	4.91
19:00	5.29	4.91	4.75	4.65	4.67
19:30	5.29	4.75	4.64	4.65	3.69
20:00	4.93	3.87	3.75	3.03	2.36
20:30	2.25	0.91	1.64	2.46	0.1
21:00	0	0.93	0	0	0
21:30	0	0	0	0	0

Table 6.
The lengths of the forecasted congestions (in km) on weekdays in February 2020, based on data from the Yandex. Map map service.

As shown in **Table 6**, the queue time and length during peak hours at the Bogatyrsky St./Planernaya St. TP, which were obtained for the developed simulation model, are close to the forecasted values obtained from empirical data collected by the Yandex.Maps map service.

In addition, the real-time congestion information for the WHSD road section before Bogatyrsky St./Planernaya St., based on Yandex.Maps service information, was monitored during peak hours. As an example, the results from the monitoring on Thursday, January 21, 2021, at 18:30, are shown in **Figure 10**.

As shown in **Figure 10**, the length of the congestion, formed during rush hour on weekdays, reaches 4.92 km. The observed value is close to the obtained results of average maximum number of vehicles in the queue, indicated in **Table 5**.

4. Toll plaza on the main course of a toll road

4.1 Toll plaza on the main course of a toll road

In order to assess the effectiveness of the TCS for the operation of the TP on the main course of a toll road, the SM of the TP "Main course behind the Ring Road (North) towards Primorsky Ave," located on the northern section of the WHSD, was developed. A general view of the developed SM of the TP is shown in **Figure 11**.

The existing configuration of the TP "Main course behind the Ring Road (North) towards Primorsky Ave" has nine toll lanes, five of which operate in automatic mode, and the remaining four lanes operate in manual mode.

To assess the current performance of the named TP a SM was implemented taking into account the current configuration mode of the TCS (five automatic and four manual lanes) as well as the current share of ETC devices used by users is 91%.

Figure 10.
Congestion (4.92 km length) on the WHSD formed during peak hours at the Bogatyrsky St./Planernaya St. TP at 18:30, 21.01.2021, according to the Yandex.Maps map service.

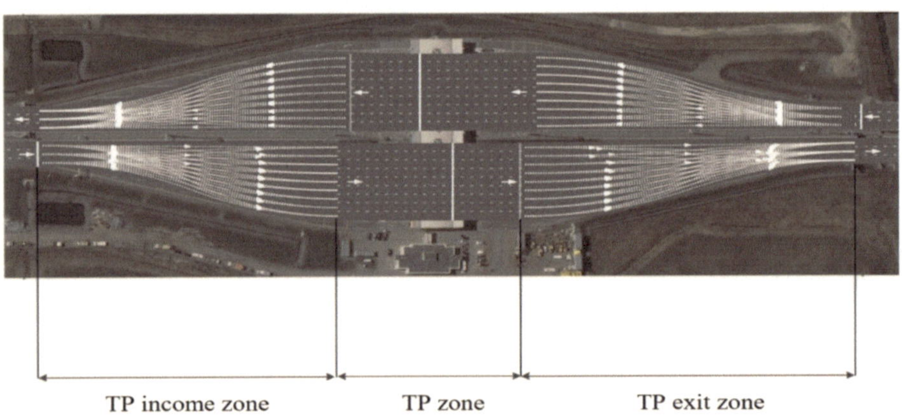

Figure 11.
The SM of the TP "Main course behind the Ring Road (North) towards Primorsky Ave." TP zone.

In order to solve the problem of forecasting the TP efficiency in the current configuration of the TCS, three additional SMs were also implemented with projected values of ETC devices used by 2022—93, 95, and 97%.

In order to solve the problem of optimizing TP efficiency, SMs with the following lane configurations has been implemented:

- A configuration with six automatic and three manual toll lanes;
- A configuration with seven automatic and two manual toll lanes.

The above models have been considered in two different modes of user behavior: 10 and 5% of the errors committed by ETC devices users when passing the automatic toll lanes.

It should be noted that a configuration with seven automatic toll lanes and two manual toll lanes is the maximum permissible for this TP. Reason: the TP must be able to provide cash and bank card payment in the TCS configuration, and a minimum of two manual lanes must be available, subject to redundancy, in case one of the lanes fails.

For each of the configurations, SMs were implemented with existing and projected ETC devices usage values. Thus, for each of the three TCS configurations, four SMs were implemented to assess changes in system performance when the ETC devices pass rates increased over a range of values of 91, 93, 95, and 97%, and user behavior error levels of 10 and 5%.

4.2 Observed traffic volume at the TP

Data on intensity, traffic composition, and time of service were obtained by a visual count of passing vehicles through the TP from an online camera located on it from the road operator's website [32] during the period from October to November 2019. On the traffic direction under study, during weekdays from Monday to Friday, the traffic intensity was observed to not exceed 3500 vehicles per hour. At weekends, the traffic flow schedule varies but also does not exceed 3500 vehicles per hour. The intensity data for the five working days of the observations is shown in **Figure 12**.

Similar to 3.8, the daily intensities shown in **Figure 12** can be well approximated by the first n terms of the trigonometric series. Such a model is constructed with the first nine terms (19 coefficients) of a trigonometric series of the form. A similar pattern is observed for holidays and weekends, for which approximating functions can be constructed. As part of the study, we are interested in two questions:

1. At what intensities and parameters (fraction of ETC devices users, fraction of user behavior errors) of the incoming flow of incoming traffic will congestion be generated at the TP?

2. What measures can the operator take to reduce the risk of traffic congestion within the existing configuration of a TP consisting of nine toll lanes?

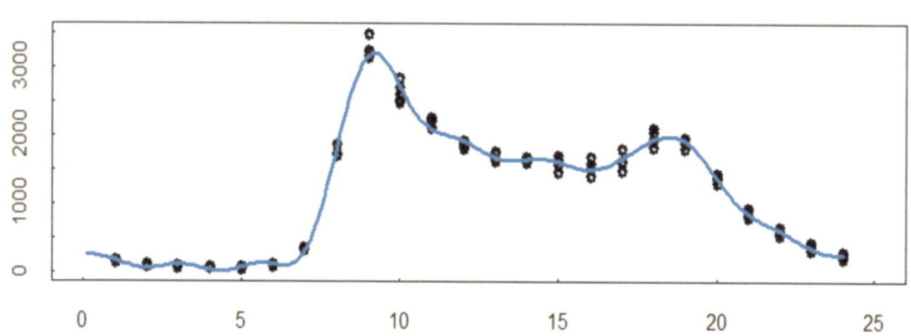

Figure 12.
Observed traffic volumes on weekdays and the approximating variable intensity curve at ПВП the TP "Main course behind the ring road (north) towards Primorsky Ave." traffic—From 0:00 to 24:00 (intensity—in number of vehicles/h).

4.3 Time allocation for passing the TP zone

The TP zone, comprising nine toll lanes separated by safety islands, is defined as a section of toll road beginning at the approach area of the toll road when the 3-lane main course is extended and ending at the exit area of the toll road where the nine toll lanes are converted back to three lanes. We separate the concepts of the service time directly at the toll lane (which would be the baseline when applying classical mass service theory) and the driving time to the TP zone. In this case, we are interested in the time it takes for a vehicle to travel the entire stretch of road from the TP entrance area to the TP exit area. We assume that congestion occurs in situations where vehicles cannot cross the line of division of three lanes into nine and have to queue up to this line.

The distribution of service times for manual and automatic toll lanes is specified in the SM parameters. The study of traffic at different intensities of the incoming traffic has shown that at low intensities the distribution of service times is a mixture of split distribution laws, obviously corresponding to the passage of vehicles through manual and automatic toll lanes. From the practical point of view, these cases are not of interest, as the TP can handle the load.

At high intensity, the distribution of TP driving times at manual and automatic toll lanes is an indistinguishable mix of distributions. At high intensities, the role of user behavior errors increases, leading to significant flow changes at the TP zone and causing interference to other road users. The distribution of TP driving The TP Zone, comprising nine toll lanes separated by safety islands, is defined as a section of toll road beginning at the approach area of the toll road when the 3-lane main course is extended and ending at the exit area of the toll road where the nine toll lanes are converted back to three lanes. We separate the concepts of the service time directly at the toll lane (which would be the baseline when applying classical mass service theory) and the travel time to the TP zone. In this case, we are interested in the time it takes for a vehicle to travel the entire stretch of road from the TP entrance area to the TP exit area. We assume that congestion occurs in situations where vehicles cannot cross the line of division of three lanes into nine and have to queue up to this line.

The distribution of service times for manual and automatic toll lanes is specified in the SM parameters. The study of traffic at different intensities of the incoming traffic has shown that at low intensities the distribution of service times is a mixture of split distribution laws, obviously corresponding to the passage of vehicles through manual and automatic toll lanes. From the practical point of view, these cases are not of interest, as the TP can handle the load.

At high intensity, the distribution of TP driving times at manual and automatic toll lanes is an indistinguishable mix of distributions. At high intensities, the role of user behavior errors increases, leading to significant flow changes at the TP zone and causing interference to other road users. The distribution of TP driving times in these cases is well approximated by a gamma distribution law (shown in **Figure 13**).

The mathematical expectation of a random variable distributed in accordance to Gamma distribution *Gamma* (k, θ) is $k \times \theta$, and the modal (most likely) value is $(k - 1) \times \theta$, where k is a shape parameter, and $\theta = \frac{1}{rate}$.

When congestion occurs and the income flow intensity continues to increase, the parameters of the approximating gamma laws can vary, but the estimates of the mean TP zone driving time do not change significantly when the inlet intensity increases when the congestion is already in place.

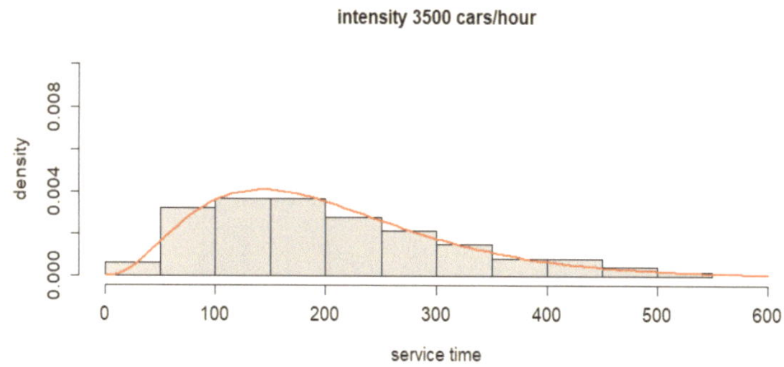

Figure 13.
Service time distribution with an input flow of 3500 vehicles/h, TCS configuration: Four manual and five automatic lanes, 91% transponders in the flow, and 10% user error rate. The red line is gamma density law distribution with parameters shape = 3.2821, rate = 0.016, p-value of Kolmogorov-Smirnov test –0.99.

Before the congestion occurred, the average time to cross the TP zone was calculated as a weighted average of the average TP zone crossing times for automatic (0.91) and manual lanes (0.09). After the occurrence of congestion, the average TP zone driving time was calculated as the mathematical expectation of the approximating density of the gamma law distribution. In the case under consideration, congestion occurs at an input flow rate of 1650 vehicles per hour if the user error rate is assumed to be 10% and at a rate of 3200 vehicles per hour if the user error rate is assumed to be 5%, that is, halving the user error rate significantly increases the capacity of the entire TP. If congestion has already occurred, the average driving time through the TP stabilizes and ceases to increase (this does not include waiting in line before entering the TP), and user error rates are no longer as important.

If congestion has already occurred and the income traffic intensity continues to increase, the congestion cannot be relieved until the inbound traffic intensity decreases. The maximum traffic intensity observed at this TP did not exceed the value of 3500 vehicles per hour. Already in the current situation, congestion at the TP is generated with 1500–1600 vehicles per hour (see **Figure 12**), that is, during peak hours.

With increasing traffic flows, the frequency of traffic congestion and queue lengths will only increase. The natural way to prevent congestion seems to be:

- Increasing the share of ETC devices in traffic flow;
- Reconfiguring TPs by increasing the number of automatic toll lanes;
- Road operator's implementation of a set of information measures aimed at reducing user behavior when using ETC devices to pass through the automatic toll lanes.

Tables 7 and **8** show the simulation results for different intensities, number of automatic lanes, shares of ETC devices users in the flow, and error levels of user behavior.

Tables 7 and **8** show that.

- Increasing the proportion of ETC devices in the TP flow, at a user behavior error level of 10% has almost no effect on the intensity threshold at which congestion begins to form (thresholds 1650, 1700, 1750, 1550), the time to pass the TP zone in this case does not change significantly;

Traffic intensity	Uses behavior (%)	TP configuration	ETC proportion (%)	Gamma distribution parameters			Expectation	Mode	Congestion at the entrance
				Shape	rate	p-value			
1600	10	4 manual + 5 ETC	91	2.9375	0.0277	0.086	106	70	No
1650	10	4 manual + 5 ETC	91	1.8637	0.0146	0.538	128	59	Yes
3500	10	4 manual + 5 ETC	91	3.2064	0.0158	0.738	203	139	Yes
1650	10	4 manual + 5 ETC	93	3.2898	0.0380	0.233	87	60	No
1700	10	4 manual + 5 ETC	93	3.0552	0.0215	0.735	142	95	No
3500	10	4 manual + 5 ETC	93	3.3242	0.0168	0.322	198	139	No
1700	10	4 manual + 5 ETC	95	2.5553	0.0248	0.077	103	63	Yes
1750	10	4 manual + 5 ETC	95	2.3994	0.0221	0.067	109	63	Yes
3500	10	4 manual + 5 ETC	95	3.5340	0.0187	0.749	189	136	Yes
1500	10	4 manual + 5 ETC	97	2.4283	0.0219	0.017	111	65	No
1550	10	4 manual + 5 ETC	97	2.0937	0.0111	0.465	189	99	No
3500	10	4 manual + 5 ETC	97	4.3142	0.0237	0.680	182	140	No
1600	10	3 manual + 6 ETC	91	2.2374	0.0212	0.630	106	58	No
1650	10	3 manual + 6 ETC	91	2.4938	0.0237	0.069	105	63	No
3500	10	3 manual + 6 ETC	91	2.9655	0.0157	0.781	188	125	Yes
1550	10	3 manual + 6 ETC	93	2.9357	0.0425	0.160	69	46	Yes
1600	10	3 manual + 6 ETC	93	2.0936	0.0178	0.391	118	62	Yes
3500	10	3 manual + 6 ETC	93	3.0855	0.0177	0.507	174	118	Yes
1500	10	3 manual + 6 ETC	95	2.5761	0.0350	0.035	74	45	No
1550	10	3 manual + 6 ETC	95	2.6170	0.0267	0.155	98	61	Yes
3500	10	3 manual + 6 ETC	95	3.2146	0.0200	0.960	161	111	Yes
1450	10	3 manual + 6 ETC	97	3.5774	0.0502	0.037	71	51	No
1500	10	3 manual + 6 ETC	97	2.6137	0.0152	0.694	172	106	Yes

DOI: http://dx.doi.org/10.5772/intechopen.1002003

Traffic intensity	Uses behavior (%)	TP configuration	ETC proportion (%)	Gamma distribution parameters			Expectation	Mode	Congestion at the entrance
				Shape	rate	p-value			
3500	10	3 manual + 6 ETC	97	2.7377	0.0159	0.623	172	109	Yes
1800	10	2 manual + 7 ETC	91	2.4565	0.0225	0.204	109	65	No
1900	10	2 manual + 7 ETC	91	1.8873	0.0131	0.082	144	68	Yes
2000	10	2 manual + 7 ETC	91	2.3521	0.0199	0.109	118	68	Yes
2100	10	2 manual + 7 ETC	91	2.5431	0.0155	0.591	164	99	Yes
2500	10	2 manual + 7 ETC	91	3.0474	0.0182	0.705	168	113	Yes
3000	10	2 manual + 7 ETC	91	3.0389	0.0154	0.759	198	133	Yes
3500	10	2 manual + 7 ETC	91	2.5843	0.0151	0.538	171	105	Yes
1650	10	2 manual + 7 ETC	93	4.7607	0.0881	0.155	54	43	No
1700	10	2 manual + 7 ETC	93	1.7889	0.0168	0.061	106	47	Yes
3500	10	2 manual + 7 ETC	93	3.3748	0.0167	0.810	202	142	Yes
1650	10	2 manual + 7 ETC	95	2.7946	0.0439	0.126	64	41	No
1700	10	2 manual + 7 ETC	95	2.6577	0.0278	0.085	96	60	Yes
3500	10	2 manual + 7 ETC	95	3.6861	0.0259	0.583	142	104	Yes
1450	10	2 manual + 7 ETC	97	4.6575	0.0795	0.003	59	46	No
1500	10	2 manual + 7 ETC	97	1.5720	0.0113	0.099	139	51	Yes
3500	10	2 manual + 7 ETC	97	3.1908	0.0196	1.000	163	112	Yes

Table 7.
Simulation results for control flow parameters and lane configurations at a user behavior error rate of 10%.

Traffic intensity	Uses behavior (%)	TP configuration	ETC proportion (%)	Gamma distribution parameters			Expectation	Mode	Congestion at the entrance
				Shape	rate	p-value			
3150	5	4 manual + 5 ETC	91	3.4468	0.0212	0.956	163	116	No
3200	5	4 manual + 5 ETC	91	3.6123	0.0222	0.833	163	118	Yes
5000	5	4 manual + 5 ETC	91	3.5510	0.0215	0.929	165	119	Yes
3000	5	4 manual + 5 ETC	93	3.5311	0.0217	0.616	162	116	No
5000	5	4 manual + 5 ETC	93	3.4889	0.0216	0.968	162	115	No
3000	5	4 manual + 5 ETC	95	3.2896	0.0202	0.990	163	113	No
5000	5	4 manual + 5 ETC	95	3.6504	0.0220	0.738	166	121	No
3000	5	4 manual + 5 ETC	97	7.3778	0.0852	0.893	87	75	No
5000	5	4 manual + 5 ETC	97	6.0563	0.0458	0.983	132	110	No
3000	5	3 manual + 6 ETC	91	2.6839	0.0260	0.057	103	65	No
5000	5	3 manual + 6 ETC	91	3.0190	0.0225	0.376	134	90	No
3000	5	3 manual + 6 ETC	93	4.5039	0.0524	0.176	86	67	No
5000	5	3 manual + 6 ETC	93	3.2217	0.0305	0.185	106	73	No
3000	5	3 manual + 6 ETC	95	3.6269	0.0365	0.188	99	72	No
5000	5	3 manual + 6 ETC	95	6.4878	0.0705	0.706	92	78	No
3000	5	3 manual + 6 ETC	97	10.3129	0.1499	0.319	69	62	No
5000	5	3 manual + 6 ETC	97	4.3676	0.0379	0.896	115	89	No
3000	5	2 manual + 7 ETC	91	3.3526	0.0397	0.061	84	59	No
5000	5	2 manual + 7 ETC	91	2.5611	0.0189	0.096	135	82	No
3000	5	2 manual + 7 ETC	93	5.6887	0.0710	0.051	80	66	No
5000	5	2 manual + 7 ETC	93	3.4629	0.0320	0.143	108	77	No
3000	5	2 manual + 7 ETC	95	9.4472	0.1353	0.121	70	62	No
5000	5	2 manual + 7 ETC	95	5.1386	0.0555	0.411	93	75	No
3000	5	2 manual + 7 ETC	97	5.2843	0.0660	0.400	80	65	No
5000	5	2 manual + 7 ETC	97	4.5880	0.0435	0.309	105	82	No

Table 8.
Simulation results for control flow parameters and toll lanes configurations at a user behavior error rate of 5%.

- Configuring TP to increase the number of lanes and increase the proportion of ETC devices users with a 10% level of user error marginally affects the threshold intensity at which congestion begins to build and the time to cross the TP, congestions still begin to build at vehicle throughput rates of 1500–1750 vehicles per hour;

- Significant increase of threshold intensity at which congestion is formed is to reduce user behavior error rate to 5% with some reduction in tailing time, but the risk of short-term congestion at the exit of the TP zone. When there is a large number of vehicles in the queue, high speed of automatic lanes and reduction of user errors, the payment process in the lanes is relatively fast, but when vehicles leave the lanes at the same time, they begin to interfere with each other in the TP exit area, which leads to traffic slowdown and the possibility of jamming at the exit of the TP. Note that it is not possible for congestion to occur simultaneously at the TP entrance and exit points. If congestion occurs at the entrance point, the efficiency of the lanes is reduced, thus eliminating the possibility of congestion at the exit point.

4.4 Estimation of traffic density based on a toll plaza simulation modeling

There is a well-developed traffic flow theory based on models borrowed from hydrodynamics and described in [36]. One of the main flow indicators in such models is the flow density, expressed as the number of vehicles per unit length (e.g., kilometer) and dependent on the speed of traffic. The existence of a TP on a toll road allows for the estimation of this flow density. This can be done using either data from the IT system of the road operator or the results of simulations. Simulation results are even more preferable as the road operator's data is only the result of observations under current conditions, while the simulation allows for viewing yet non-existing modes and assessing risks of traffic congestions, possible financial losses for variable TP and flow parameters, driver errors, etc.

A flow density estimation method based on observations of the number of vehicles passing through the TP zone in random time t, distributed according to the gamma law with parameters k, μ can be proposed.

Following [37], we find the distribution of the random variable X, that is, the number of vehicles arriving at or leaving the TP at random time t. Let the incoming vehicles flow obey Poisson distribution, with parameter λ. We define λ^* as the intensity of the Poisson flow of vehicles leaving the TP area under the existing congestion (i.e., $\lambda > \lambda^*$). Observations of traffic flow on the toll road show that the assumption of Poisson flow is quite reasonable, despite the existence of several lanes on the main roadway: the event is considered to be the crossing by the front bumper of a vehicle of the line of the beginning of the approach zone to the TP or the end of the exit zone from the TP (**Figure 11**).

The intervals between such events are well approximated by an exponential distribution law. Consider the random variable X, which is the number of vehicles arriving at the TP in random time t distributed according to the gamma law with parameters k, μ. The probability p_m that in random time t distributed according to the gamma law with parameters k, μ, exactly m vehicles will arrive at the TP (a detailed description is given in [38]).

$$p_m = \frac{\lambda^m}{(\lambda+\mu)^m} \times \frac{1}{m!} \times \frac{(m+k-1)!}{(k-1)!} \times p_0, p_0 = \frac{\mu^k}{(\lambda+\mu)^k} \tag{14}$$

The mathematical expectation:

$$E(X) = \frac{\lambda k}{\mu} \tag{15}$$

Standard deviation:

$$\sigma(X) = \frac{\lambda}{\mu}\sqrt{k\left(1+\frac{\mu}{\lambda}\right)} \tag{16}$$

If the transit time t of the vehicle through the TP zone is random and distributed by gamma law with parameters k, μ, then, the inverse quantity $v = \frac{1}{t}$ is distributed according to the inverse-gamma distribution with a density:

$$g(v) = \frac{\mu^k e^{-\frac{\mu}{v}}}{\Gamma(k)} v^{-k-1}. \tag{17}$$

The mathematical expectation of a random variable v :

$$E(v) = \frac{\mu}{k-1} \tag{18}$$

Mode:

$$Mode(v) = \frac{\mu}{(k+1)} \tag{19}$$

Dispersion:

$$D(v) = \frac{\mu^2}{(k-1)^2(k-2)} \tag{20}$$

Standard deviation (for $k > 2$):

$$\sigma(v) = \frac{\mu}{(k-1)\sqrt{k-2}} \tag{21}$$

The physical meaning of the random variable $v = \frac{1}{t}$ is the speed at which the vehicle passes the TP zone. In this Equation, the length of the TP zone looks like a unit of length. Let L be the length of the TP zone in meters (or kilometers as a decimal). Then, for the random variable $v = \frac{L}{t}$ Eqs. (14)–(16) will take the form:

$$E(v) = \frac{L \times \mu}{k-1} \tag{22}$$

$$Mode(v) = \frac{L \times \mu}{(k+1)} \tag{23}$$

$$(\text{for } k > 2)\ D(v) = \frac{L^2 \times \mu^2}{(k-1)^2(k-2)} \tag{24}$$

$$(\text{for } k > 2)\ \sigma(v) = \frac{L \times \mu}{(k-1)\sqrt{k-2}} \tag{25}$$

Thus, an analysis of the TP SM makes it possible to determine the number of vehicles passing through the TP area in a random, gamma-distributed time and to determine the speed in the TP of length L in the usual dimension "km/h." In fact, this makes it possible to estimate the number of vehicles on a section of length L at a speed distributed according to the inverse-gamma distribution. The flow density parameter plays an important role in traffic flow models that rely on models borrowed from hydrodynamic problems.

For example, **Figure 13** shows the random time distribution of vehicles passing through the TP zone in the existing queue mode, obtained from the simulation of the TP at the toll road exit described in Section 3.1 of this chapter. The value of λ^*, at which the steady queueing begins, was obtained as 1250 vehicles/h. The values of the parameters of the gamma distribution law (see **Figure 13**) for the passage time of the TP zone shape = 2.1510 *(k)*; rate = 0.0126 *(μ)*. During a random time t equal to the time of TP passage by one vehicle, $E(X) = \frac{\lambda^{exit} k}{\mu} = \frac{1250\ vehicles/h \times 2.1510}{0.0126} = \frac{0.35\ vehicles/sec. \times 2.1510}{0.0126} = 60$ vehicles. The average time of passing through the TP zone (mathematical expectation of gamma law of distribution t) is $\frac{k}{\mu} = \frac{2.1510}{0.0126} = 170$ seconds. The length of the TP zone is 285 meters. It may seem that the average speed is $\frac{285\ meters}{170\ seconds} = \frac{285 \times 60 \times 60}{1000 \times 170} = 6.03\ km/h$. However, this is not the case. It should be noted that there is a pronounced asymmetry to the gamma distribution law in this case, for example, in **Figure 13** we see that the mode of the distribution law t (no analytical expression is defined) in this example is less than 100 seconds and much less than the mathematical expectation. Therefore, the average speed of movement should be determined based on the expectation of the inverse-gamma of the distribution law of the value v. The average speed will be $E(v) = \frac{\mu}{k-1} = \frac{0.0126 \times 285\ meters}{2.1510 - 1} = 3.2\ m/sec$, which is $\frac{3.2 \times 60 \times 60}{1000} = 11.52\ km/h$. The density of outgoing traffic from this TP, under the conditions of existing traffic congestion in front of it, when driving at 11.52 *km/h*, will be 60 vehicles per 285 *meters* or $\frac{60}{0.285} = 211$ vehicles per 1 km of the road. Given a bumper-to-bumper spacing of 10 meters, this equates to 2110 meters, that is, 1 km of dense traffic on a dual carriageway.

5. Toll plaza at the exit from the toll road before the controlled intersection

5.1 Simulation model of the toll plaza at the exit from the toll road before the controlled intersection

Consider the WHSD interchange at the intersection with Shuvalovsky Ave. This interchange is also located in the Primorsky District of St. Petersburg, a large part of which is included in the interchange gravity zone.

The Shuvalovsky interchange allows to relieve the traffic load both from the street and road network of the district and from the TP at the exit from the WHSD to Bogatyrsky Ave., which was assessed by the authors earlier in Section 3 as well as in the articles [33, 39].

The exit from the toll road is a TP consisting of six toll lanes, with four lanes operating in automatic mode and two lanes in manual mode combined with ETC payment mode. A general view of the TP at the Shuvalovsky Ave. exit is shown in **Figure 14**.

The connection of the toll road exit to the street and road network of the district is *via* a controlled intersection located 475 meters behind the TP and providing exits for traffic on Shuvalovsky Ave. and Planernaya St. The connection to the controlled intersection is in four traffic lanes, three of which prescribe forward traffic on Shuvalovsky Ave. and one providing a right turn onto Planernaya St.

The general view of the toll road exit connection to the regulated intersection is shown in **Figure 15**.

A special feature of this exit is the so-called bottlenecks before the TP and the controlled intersection—two-lane sections of road at the exit from the main course of the road before the entrance area to the TP and a two-lane section of road at the exit from the TP before the approach area to the traffic light facility. This feature allows the road section to be considered as a sequence of "obstacles" slowing down traffic speeds and affecting its capacity.

The main traffic correspondences of the area are home-work and work-home. The congestion on the outbound TP located on Shuvalovsky Ave. usually occurs during evening rush hours. A SM has been developed to assess the traffic conditions in the interchange area. The model study area starts from the WHSD exit and ends with the exit to Shuvalovsky Ave. and Planernaya St. behind the regulated intersection and includes the TP and the traffic light facility, the current operation modes of the TP and traffic lights are reproduced. The general view of the SM of the interchange is shown in **Figure 16**.

Figure 14.
General view of the TP at the Shuvalovsky Ave.

Figure 15.
General view of the toll road exit connection to the regulated intersection.

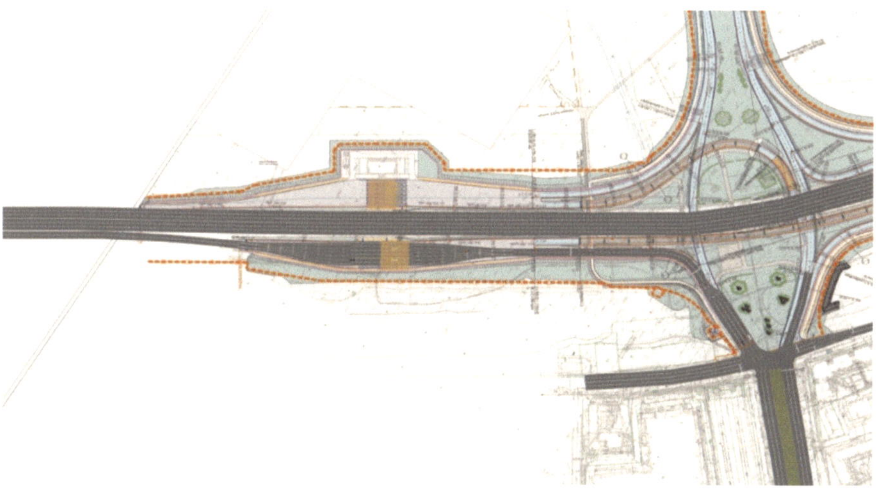

Figure 16.
The general view of the SM of the interchange at the intersection with Shuvalovsky Ave.

The distribution of manual and automatic toll lane service times is defined in the SM parameters.

The general view of the developed SM of the TP at Shuvalovsky Ave. is shown in **Figure 17**.

The SM of the controlled intersection is aimed at evaluating the efficiency of the traffic light phasing that ensures the exit of the vehicles from the interchange area. Thus, this traffic light facility reproduces the regulation of correspondence in the directions of traffic "TP-Shuvalovsky" and "TP-Planernaya St.". It should be noted that these directions are regulated separately as the right turn from the right lane is regulated by an additional traffic light section. The actual duration of the traffic light phases is shown in **Table 9**.

Figure 17.
The general view of SM of the TP at Shuvalovsky Ave.

Traffic light phase	Straight ahead (Shuvalovsky Ave.), sec.	Turn right (Planernaya Str.), sec.
Red	80	56
Yellow	3	—
Green	32	58
Yellow	4	—

Table 9.
Actual duration of traffic light phases after exiting the toll road.

A general view of the SM of the toll road exit connection to a controlled intersection is shown in **Figure 18**.

The sequential location of the TP and the traffic light facility can lead to the following traffic situations:

Figure 18.
General view of the SM of the toll road exit connection to a controlled intersection.

1. In case of insufficient capacity of the TP and a high volume of traffic passing through the crossing point, traffic congestion can occur in front of the TP;

2. If there is insufficient operating time for traffic lights providing an exit from the TP, there can be a risk of congestion in front of the intersection. Since the regulated intersection is 475 meters from the toll road, if the length of congestion exceeds this distance, there may be an additional risk of increased service time for users at the exit of the toll road.

With the help of the traffic interchange developed by the SM, the following were carried out:

1. Analysis of the capacity of the TP at the exit of the toll road;

2. Analysis of traffic capacity of traffic interchange, including functioning TP and controlled traffic light object;

3. Evaluation of the possibility of optimizing the operation of the traffic light facility to increase the capacity of the interchange.

5.2 Assessing the capacity of the toll plaza

The capacity limit is the capacity at which, when passing through the toll lanes, the vehicles queued beyond the approach area of the TP—the line beyond which the widening of the road section from two to six lanes begins. The operation of the traffic light behind the checkpoint was not considered to assess the capacity of the TP.

In this study, traffic flow parameters were selected, advising data for the year 2021. For example, the proportion of traffic flow users using ETC devices to pay at an automated lane is 93%. In this location, the assumption of homogeneity of the traffic flow consisting of passenger cars is acceptable, as there are practically no heavy vehicles at the studied interchange.

A number of simulations were carried out with traffic intensity ranging from 100 to 4000 vehicles/h in steps of 50 vehicles/h in order to estimate the maximum load on the TP. A study of the operation of the TP at different traffic intensities revealed that at low intensities the service time distribution was a mixture of split distribution laws, corresponding to the passage of vehicles through manual and automatic toll lanes.

At high intensities with vehicle queues forming outside the approach area to the TP, the distribution of time spent in automatic and manual toll lanes formed an undivided mixture of distributions that approximates well to the gamma distribution law. The distribution of service times at the TP with 4000 vehicles/h is shown in **Figure 19**. This distribution has a strong right-hand asymmetry. An indirect characteristic of this asymmetry is the significant difference between the mean and the most likely passing time at the TP.

As a result of the simulation experiments, it was found that the throughput capacity of the developed SM of the TP at the set traffic flow parameters corresponds to a capacity of 1700 vehicles/h. The distribution of the service time at the TP at 1700 vehicles/h is shown in **Figure 20**.

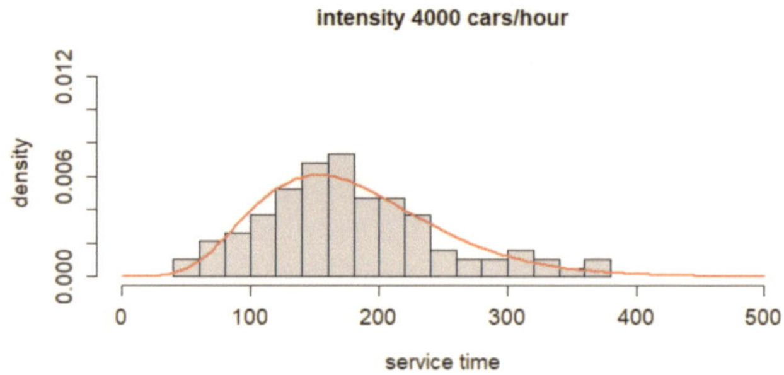

Figure 19.
The distribution of service times at the TP with 4000 vehicles/h.

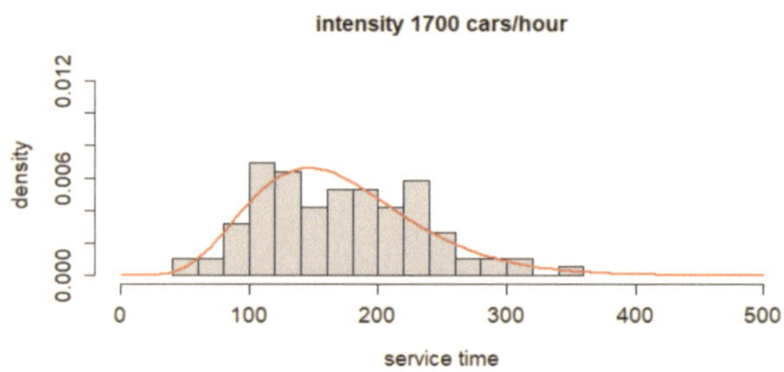

Figure 20.
The distribution of the service time at the TP at 1700 vehicles/h.

5.3 Interchange capacity analysis

In order to analyze the traffic capacity of the interchange, simulation experiments were carried out to investigate the traffic flow time through the TP and the controlled intersection under actual parameters of the phase duration of the traffic light providing the exit from the TP.

A number of simulation experiments were conducted with traffic volumes ranging from 100 to 4000 vehicles/h in increments of 50 vehicles/h.

As a result of these simulations, it was found that with the existing traffic light phasing parameters, traffic congestion begins to form in front of the intersection when the intensity exceeds a value of more than 900 vehicles/h. When the queue of vehicles exceeds a distance of more than 475 meters, congestion occurs in the exit area of the TP, resulting in reduced service times in the toll lanes and the appearance of a second congestion before entering the TP area. An example of traffic congestion in the TP exit area is shown in **Figure 21**. The distribution of service times at the TP with a traffic volume of 900 vehicles/h is shown in **Figure 21**.

At an inbound traffic intensity of 900 vehicles/h, there is no longer a separable mixture of vehicle travel time distributions through the TP zone, but the observed empirical distribution (**Figure 22**) is well approximated by an exponential law (a particular case of a gamma distribution).

5.4 Interchange capacity optimization

Let us consider the process of accumulation of vehicles on the road section from the exit from the TP zone to the traffic light facility (its length is 475 meters, and the

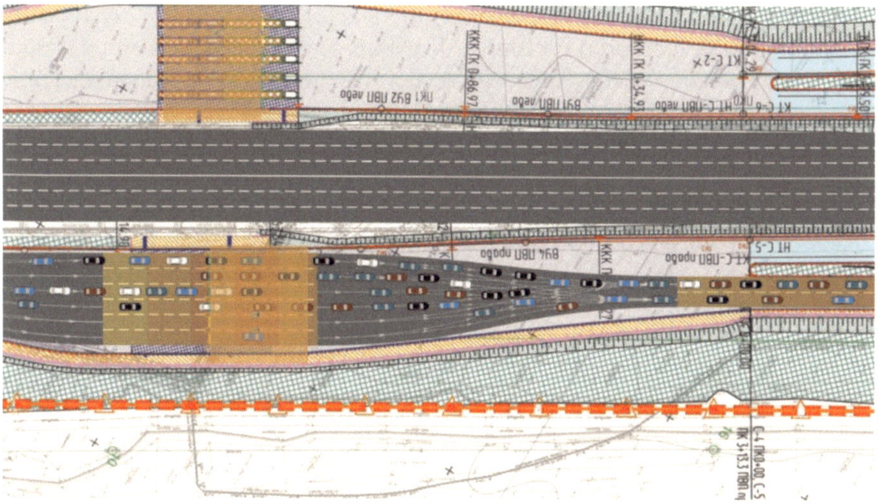

Figure 21.
Traffic congestion in the TP exit area at the intraurban toll road exit to the urban street and road network.

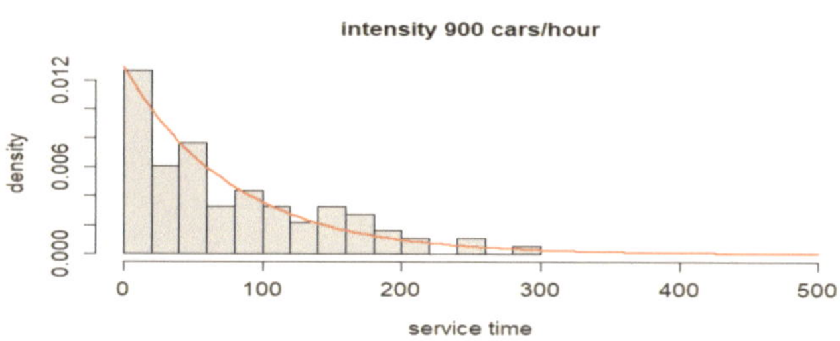

Figure 22.
Distribution of service times at the TP with an input flow rate of 900 vehicles/h.

maximum capacity is about 160 vehicles, at the rate of 8 meters per 1 vehicle, two-thirds of the section are two lanes, one-third of the section is four lanes). Let the inbound flow of vehicles be modeled by the Poisson flow of vehicles with intensity λ (the moment of vehicle arrival is considered to be the moment when the front bumper of the vehicle crosses the end line of the TP zone). Within one cycle of the traffic light object, we will consider this intensity as constant. The moment when the vehicle leaves the whole junction zone, we will consider crossing by the rear bumper of the vehicle of the line on which the traffic light is installed. At opening of a green light, the first vehicles in lanes (if they are there) start moving increasing speed from zero value. The following vehicles cross the end line of the junction area at a higher speed. Consequently, when crossing the traffic light line, there is a flow with variable intensity $\mu(t)$ within the green (and yellow) phase of the traffic light from 0 to τ. Let the full cycle of the traffic light object is T, the duration of the green and yellow phases is τ, and the red phase is $T - \tau$. Let X be a random variable, let the full cycle of the traffic light object is T, and the duration of the green and yellow phases is λT, that is

$$P(X = m) = \frac{\lambda T^m}{m!} e^{-\lambda T} \quad (26)$$

Let Y be a random variable, the number of vehicles leaving the interchange area (crossing the traffic light line) for time τ, distributed according to Poisson distribution with changing parameter $\mu(t)$. Then, the accumulated intensity of the vehicles leaving the interchange area $M(0,\tau) = \int_0^{\tau} \mu(t)dt$. In the remaining time interval $T - \tau$ the intensity of the vehicle leaving the decoupling zone is equal to zero. The difference $Z = X - Y$ is the number of vehicles remaining on the section (arriving but not having time to leave it during one cycle of the traffic light object). The random variable Z, as the difference between two Poisson random variables, has Skellam's distribution law [34]:

$$f\left(k, \lambda T, \int_0^{\tau} \mu(t)dt\right) = P(V = k) = e^{-\left(\lambda T + \int_0^{\tau} \mu(t)dt\right)} \left(\frac{\lambda T}{\int_0^{\tau} \mu(t)dt}\right)^{\frac{k}{2}} I_{|k|}\left(2\sqrt{\lambda T \times \int_0^{\tau} \mu(t)dt}\right). \tag{27}$$

where $I_{|k|}(x)$ is a modified Bessel function of the first kind, with $k = 0, \pm 1, \pm 2, \dots, \pm\infty$.

In reality, the value of Z cannot take negative values. Therefore, the distribution function of a random variable Z can be written as

$$\begin{cases} P(Z = 0) = e^{-\left(\lambda T + \int_0^{\tau} \mu(t)dt\right)} \sum\limits_{k=0}^{-\infty} \left(\frac{\lambda T}{\int_0^{\tau} \mu(t)dt}\right)^{\frac{k}{2}} I_{|k|}\left(2\sqrt{\lambda T \int_0^{\tau} \mu(t)dt}\right), at\ k \le 0, k \in Z \\ P(V^* = k) = e^{-\left(\lambda T + \int_0^{\tau} \mu(t)dt\right)} \left(\frac{\lambda T}{\int_0^{\tau} \mu(t)dt}\right)^{\frac{k}{2}} I_k\left(2\sqrt{\lambda T \int_0^{\tau} \mu(t)dt}\right), at\ k > 0, k \in Z \end{cases} \tag{28}$$

During traffic simulation experiments in the junction area, the exit from the WHSD to Shuvalovsky Ave. with the parameters of the TP and the traffic light facility, it was found that traffic congestion on the section between the TP exit and the traffic light line is consistently formed at an input flow rate of 900 vehicles/h, which amounts to 30 vehicles per 2 minutes. The TP is able to pass up to 1700 vehicles/h without jamming. Thus, the congestion is caused as a consequence of the sub-optimal operation of the traffic light. The task is to determine an optimal traffic light solution that allows passage of up to 1700 vehicles/h without jamming on the section between the TP exit and the traffic light.

Parameters		Current	Best
Direction	**Traffic light phase**		
Straight	Green	49	120
Straight	Red	66	40
Right turn	Green	95	16
Right turn	Red	72	13

Table 10.
The best settings for traffic light phases at the exit from the TP.

The integrated OptQest optimizer in AnyLogic was used to find an appropriate solution. The optimization result is presented in **Table 10**.

After the optimization of the traffic light phases at the TP exit, a number of simulations were carried out to assess the traffic capacity of the interchange. The results of the simulations showed that when the traffic light phases specified in **Table 10** are set, the capacity of the whole interchange coincides with the throughput capacity of the TP—1700 vehicles/h.

At any traffic intensity exceeding the value of 1700 vehicles/h, there is no congestion in front of the controlled intersection, because the traffic light phases are set so that the outgoing flow in a sufficiently long time corresponds to the traffic intensity of 1700 vehicles/h—the maximum intensity of the flow coming from the TP area. In this case, congestion will only occur in front of the TP.

6. Conclusion

Using the example of the WHSD toll road, the SMs of the different types of intraurban toll roads are constructed:

1. Toll road exit TP;

2. TP on the main course of a toll road;

3. TP at the exit from the toll road before the controlled intersection.

The SMs have both the parameters of the general simulation methodology and additional parameters affecting the intensity. As a result of the experiments, the traffic intensities at the TPs at which congestion starts to form were determined.

For each of the cases considered, the parameters of the gamma-law distribution of service times were determined. The results of the study show that when there is insufficient capacity, the TP ceases to operate efficiently and the user service time increases, regardless of the fare type.

The methods and SMs described above can be used in estimating the existing and projected capacity of the TPs in operation. With capacity data and model parameters obtained under toll road operating conditions, the toll road operator can more accurately evaluate the parameters for optimizing or upgrading the tolling system as well as other measures to increase TP capacity.

The described methodology is applicable to intraurban toll roads located in a balanced residential area with a high volume of pendulum migration. Also note that when the TPs are located in pronounced industrial-logistic areas of the city, as well as in the border areas between the city and the region, the analysis of TPs capacity may require additional assessment of traffic intensity under different conditions, taking into account daily, weekly and seasonal and irregular flow patterns.

Acknowledgements

The research was carried out at the expense of the Russian Science Foundation grant No 20-18-00365, https://rscf.ru/project/23-18-45035/

Author details

Alexander Talavirya[1*], Michael Laskin[2] and Alissa Dubgorn[3]

1 St. Petersburg State University, St. Petersburg, Russia

2 St. Petersburg Federal Research Center Russian Academy of Sciences, St. Petersburg, Russia

3 Peter the Great St. Petersburg Polytechnic University, St. Petersburg, Russia

*Address all correspondence to: a.talavirya@yandex.ru

References

[1] Punitha S. Design and evaluation of traffic delays in toll plaza using combination of queueing and simulation. Journal of Physics Conference Series. 2018;**1139**(1):012080. DOI: 10.1088/1742-6596/1139/1/012080

[2] Aksoy G, Celikoglu HB, Gedizlioglu E. Analysis of toll queues by micro-simulation: Results from a case study in Istanbul. Procedia-Social and Behavioral Sciences. 2014;**111**:614-623. DOI: 10.1016/j.sbspro.2014.01.095

[3] Aycin M. Simple methodology for evaluating toll plaza operations. Transportation Research Record Journal of the Transportation Research Board. 2006;**1988**(1):92-101. DOI: 10.1177/0361198106198800112

[4] Izuhara K, Matsui H, Fujita M, Sewaki Y, Kojima N. Theoretical study on optimal combination of gates for electronic toll collection system. In: 8th World Congress on Intelligent Transport Systems. 2001. pp. 1-11. Available from: https://trid.trb.org/view/688833

[5] Levinson D, Chang E. A model for optimizing electronic toll collection systems. Transportation Research Part A: Policy and Practice. 2003;**37**(4):293-314. DOI: 10.1016/S0965-8564(02)00017-4

[6] Yosritzal, Nurlaela S, Rizki M, Taki HM. Modelling toll traffic pattern-the Jagorawi toll case study. IOP Conference Series Earth and Environmental Science. 2018;**202**: 012022. DOI: 10.1088/1755-1315/202/1/012022

[7] Peng W, Jiandong Z, Yuan G, Sotelo MA, Zhixiong L. Lane work-schedule of Toll Station based on queuing theory and PSO-LSTM model. IEEE Access. 2020;**PP**(99): 1-1. DOI: 10.1109/ACCESS.2020.2992070

[8] Munawar A, Andriyanto A. Predicting queues and delays at toll plazas by computer simulation (case study-Cililitan toll plaza, Jakarta). Proceedings of the World Congress on Engineering. 2013;**1**:1-4

[9] Bari CS, Chandra S, Dhamaniya A. Dhamaniya, Estimation of system delay based toll equivalency factors at toll plazas using simulation. International Journal of Transportation Science and Technology. 2022:1-14. DOI: 10.1016/j.ijtst.2022.08.002/ [Article in press]

[10] Bari CS, Navandar YV, Dhamaniya A. Delay modelling at manually operated toll plazas under mixed traffic conditions. International Journal of Transportation Science and Technology. 2022;**11**(1):17-31. DOI: 10.1016/j.ijtst.2020.10.001

[11] Yogeshwar VN, Mahaveer S, Ashish D, Patel DA. Empirical analysis of level of service at toll plaza by using ordered probit model. Transportation Letters. 2019;**12**(10):1-9. DOI: 10.1080/19427867.2019.1694201

[12] Hewage D, Lalith R, Edirisinghe L. A study on electronic toll collection systems in Expressway in Sri lanka. In: Conference: 8th International Research Conference of KDUAt. Sri Lanka: Kothalawela Defence University; 2015

[13] Qian C, Yang V, Li P, Li S. Application of customer segmentation for electronic toll collection: A case study. Journal of Advanced Transportation. 2018;**2018**:1-9. DOI: 10.1155/2018/3635107

[14] Huang CV, Teaño MC, Wee HM. A comparative study of toll collection systems between Philippines and Taiwan. IOP Conference Series: Application of Simulation Modeling to Assess the Operation of Urban Toll Plazas Materials Science and Engineering. 2020;**730**:1-6. DOI: 10.1088/1757-899X/730/1/03 012034

[15] Tanino T, Sonoda H, Fujikawa K, Nonaka Y. Optimization of traffic control corresponding to the increase of ETC. In: 14th World Congress on Intelligent Transport Systems, ITS America. 2005;(3):2364-2371

[16] Zhu LH, Chen SD, Kong LJ, Liu MR. The influence of toll booths on highway traffic. Wuli Xue bao-Acta Physica Sinica. 2007;**56**(10):5674-5678

[17] Ito T, Hiramoto T. A general simulator approach to ETC toll traffic congestion. Journal of Intelligent Manufacturing. 2006;**17**(5):597-607. DOI: 10.1007/s10845-006-0023-3

[18] Roth G. Roads in a Market Economy. Gower Technical. 2008. p. 272. ISBN-13: 978-0291398147

[19] Lai CH, Hsiao PK, Yang YT, Lin SM, Lung SCC. Effects of the manual and electronic toll collection systems on the particulate pollutant levels on highways in Taiwan. Atmospheric Pollution Research. 2021;**12**(3):25-32

[20] Lee WH, Tseng SS, Wang CH. Design and implementation of electronic toll collection system based on vehicle positioning system techniques. Computer Communications. 2008; **31**(12):2925-2933. DOI: 10.1016/j.comcom.2008.05.014

[21] Vats S, Vats G, Vaish R, Kumar V. Selection of optimal electronic toll collection system for India: A subjective-fuzzy decision making approach. Applied Soft Computing. 2014;**21**: 444-452. DOI: 10.1155/2018/4949565

[22] Abuzwidah M, Aty MA. Safety assessment of the conversion of toll plazas to all-electronic toll collection system. Accident Analysis & Prevention. 2015;**80**:153-161. DOI: 10.1016/j.aap.2015.03.039

[23] Holguín-Veras J'e, Wang Q. Behavioral investigation on the factors that determine adoption of an electronic toll collection system: Freight carriers. Transportation Research Part C: Emerging Technologies. 2011;**19**(4): 593-605. DOI: 10.1016/j.trc.2010.09.010

[24] Fan SKS, Su CJ, Nien HT, Tsai PF, Cheng CY. Using machine learning and big data approaches to predict travel time based on historical and real-time data from Taiwan electronic toll collection. Soft Computing. 2018;**22**(17): 5707-5718. DOI: 10.1007/s00500-017-2610-y

[25] Weng J, Du B, Liu LL,Wang M, Rong J. ETC data based freeway traffic information mining techniques. In: 17th World Congress on Intelligent Transport Systems, 116982. Journal of Transportation Systems Engineering and Information Technology. 2010. DOI: 10.1016/S1570-6672(09)60033-7

[26] Komada K, Masukura S, Nagatani T. Traffic flow on a toll highway with electronic and traditional tollgates. Physica A: Statistical Mechanics and its Applications. 2009;**388**(24):4979-4990. DOI: 10.1016/j.physa.2009.08.019

[27] Hirai S, Xing J, Horiguchi R, Shiraishi T, Kobayashi M. Development of a network traffic simulator for the entire inter-urban expressway network in Japan. Transportation Research

Procedia. 2015;**6**:285-296. DOI: 10.1016/j.trpro.2015.03.022

[28] Karsaman RH, Mahendra Y, Rahman H, Sulaksono S. Measuring the capacity and transaction time of cash and electronic toll collection systems. Journal of Engineering and Technological Sciences. 2014;**46**(2):180-194. DOI: 10.5614/j.eng.technol.sci.2014.46.2.5

[29] Jehad AE, Ismail A, Borhan MN, Ishak SZ. Modelling and optimizing of electronic toll collection (ETC) at Malaysian toll plazas using microsimulation models. International Journal of Engineering & Technology. 2018;7(4):2304-2308. DOI: 10.14419/ijet.v7i4.14338

[30] Zhang C, He J, Wu J, Zhang H, Liu Z, Xing L. The analysis of ETC lane allocation in a toll collection station based on micro traffic simulation. In: APCIM & ICTTE 2018: Proceedings of the Asia-Pacific Conference on Intelligent Medical 2018 & International Conference on Transportation and Traffic Engineering 2018. ACM (Association for Computing Machinery); 2018. pp. 161-166. DOI: 10.1145/3321619.3321645

[31] Neuhold R, Garolla F, Sidla O, Fellendorf V. Predicting and optimizing traffic flow at toll plazas. Transportation Research Procedia. 2019;**37**:330-337. DOI: 10.1016/j.trpro.2018.12.200

[32] Northern Capital Highway LCC [Internet]. Online Cameras. North Section of WHSD. Bogatyrsky pr./Planernaya str., Exit. Available from: https://nch-spb.com/travel/cameras [Accessed: September 23, 2019]

[33] Talavirya A, Laskin M. Discrete-event modelling of the capacity of the toll collection exit point and the formation of congestion. International Journal of Simulation and Process Modelling. 2021;**17**(2–3):150-165. DOI: 10.1504/IJSPM.2021.121710

[34] Skellam JG. The frequency distribution of the difference between two poisson variates belonging to different populations. Journal of the Royal Statistical Society, Series A. 1946;**109**(3):296

[35] Laskin MB, Svistunova AS, Talavirya AY. Evaluation of the daily intensity of traffic at the road exit toll plaza of the interuban toll road. Systems Analysis in Design and Management: XXIV International Scientific and Practical Conference. 2020;**3**(2):257-274

[36] Gasnikov AV, Klenov SL, Nurminsky EA, Kholodov Ya A, Shamray NB. Introduction to Mathematical Modeling of Traffic Flows: tutorial. 2nd ed. Moscow. 2013. p. 428

[37] Ryzhikov YI. The theory of queues and inventory management. SPb.: Piter. 2001. p. 384

[38] Laskin M, Talavirya A. Estimation of traffic flow density based on simulation modeling of a toll plaza. Simulation modeling. Theory and practice. 2021:272-278. ISBN 978-5-905526-05-3

[39] Talavirya AU, Laskin MB. Simulation modeling during operation of the toll collection station on the intercity toll road. In: XIV International Scientific Conference "INTERAGROMASH 2021". Springer International Publishing; 2022:619-627. DOI: 10.1007/978-3-030-81619-3

Chapter 9

Immersive Innovation: Exploring Interactive Virtual Reality through Distributed Simulations

Jalal Possik, Adriano O. Solis and Gregory Zacharewicz

Abstract

This chapter explores the significant role of modeling and simulation techniques in various sectors, focusing particularly on distributed simulation (DS). The increasing importance of DS has been emphasized in response to evolving industrial, healthcare, and services settings. By leveraging DS, the integration of heterogeneous simulations enhances the effectiveness and efficiency of individual and classical simulations. In particular, this chapter introduces a DS that seamlessly combines two distinct simulation methods within a virtual reality (VR) environment. This integration enables users to fully immerse themselves in a 3D digital twin environment. Two case studies were conducted to evaluate the effectiveness of the developed DS system. The first case study focused on the implementation of DS in a hemodialysis unit, while the second case study examined its application in an intensive care unit. AnyLogic has been utilized for developing both discrete event and agent-based simulations, while the Unity platform has been employed for VR environment creation. In order to ensure smooth integration and synchronization, as well as address the demanding computational requirements, a network-based DS system has been implemented based on the high-level architecture—an IEEE standard for DS.

Keywords: distributed simulation, discrete event simulation, agent-based modeling and simulation, virtual reality, immersive technology

1. Introduction

Modeling and simulation (M&S) involves techniques widely used in various fields to enhance analysis and decision-making, improve system performance, as well as optimize operations. These techniques provide valuable insights into complex real-world systems by creating representations and conducting virtual experiments. While M&S offers numerous advantages, it does come with limitations, especially when dealing with complex models that require high computational resources and extensive interconnections between different components. To address the complexities associated with intricate models, the concept of distributed simulation (DS) has emerged as a promising solution in recent years. DS involves the integration of heterogeneous simulations, which serves to enhance the usability and efficiency of individual simulation components. This collaborative approach enables the synchronization and coordination of multiple simulation elements across diverse systems and platforms. By embracing DS, researchers and practitioners can effectively tackle challenges posed by complex models while leveraging the benefits of distributed computing and communication.

In order to harness the advantages of DS, the adoption of the high-level architecture (HLA) has become commonplace as a recognized international standard for facilitating interoperability and integration within DS environments. The HLA for DS emerged as a pivotal development in the field of computer simulation during the late 1990s. It was created as an open standard by the U.S. Department of Defense to address the increasing complexity and interoperability challenges associated with large-scale DSs. The origins of HLA can be traced back to the need for efficient simulation of complex systems such as military operations, where multiple simulations running on different computers need to seamlessly interact and exchange information in real-time. The HLA framework provides a standardized set of rules and protocols for achieving this interoperability, enabling simulations developed by different organizations, using different software and hardware, to work together seamlessly. By defining a common language for simulation components to communicate, HLA facilitates the integration and synchronization of diverse simulations, enabling the creation of comprehensive and realistic virtual environments. The HLA has been widely adopted in various domains, including defense, aerospace, transportation, and healthcare, among others, to support collaborative and DS environments (e.g., see [1–6]). Its open and extensible nature has contributed to its longevity and widespread use, making it a foundational framework in the field of DS. HLA enables the seamless integration of simulations developed using diverse tools and technologies. This integration allows for smooth communication and coordination among the DS components, promoting effective collaboration and ensuring efficient interoperability within the DS ecosystem.

In this chapter, we introduce a pioneering methodology that amplifies the interactive capabilities of virtual reality (VR) environments through the integration of M&S techniques using the DS paradigm. Our approach capitalizes on the utilization of the AnyLogic simulation software [7] in conjunction with the Unity platform [8] for the development of immersive digital twin VR environments. Through the seamless integration of these tools using DS principles, and leveraging the powerful capabilities of HLA, we showcase the seamless fusion of discrete event and agent-based simulations (ABSs) with VR.

The unique aspect of our approach lies in its application to two distinct case studies: a hemodialysis unit and an intensive care unit (ICU) both situated in a hospital located in Toronto, Canada. Staff members and trainees can immerse themselves in the VR environment using VR headsets, allowing them to explore and interact with a digital twin of the respective unit's environment. This integration of M&S and VR technology using DS opens up new possibilities for training, experimentation, and decision support in healthcare settings.

In the following sections, we will provide an overview of the advantages and limitations of M&S, discuss the concept of DS and its use of HLA, detail the integration of AnyLogic M&S with the Unity VR environment, and present the results and insights obtained from the case studies conducted in the hemodialysis unit and the ICU. Overall, this chapter serves to underscore the immense potential of integrating DS with VR technology in order to create captivating and interactive environments tailored for training and decision-making within complex systems, such as healthcare units. By merging the collaborative nature of DS with the immersive capabilities of VR, new avenues are opened for enhancing user engagement and facilitating realistic experiences in critical domains like healthcare. The seamless integration of DS and VR holds promise for revolutionizing training methodologies and augmenting decision-making processes within complex systems.

2. State of the art

Classical M&S techniques have been widely employed in various fields to enhance performance, ensure safety, refine processes, validate theories, train personnel, and explore new scenarios. By scientifically modeling systems, users can gain valuable insights into the outcomes of different conditions and actions. Simulation provides a powerful tool for fine-tuning performance metrics, identifying optimization opportunities, and predicting the consequences of potential interventions. Moreover, simulations can help uncover hidden patterns, assess risk factors, and support decision-making processes by providing a virtual environment for experimentation and analysis. This enables researchers, engineers, and practitioners to make informed choices, evaluate alternative strategies, and improve overall understanding of complex systems. From engineering and manufacturing to healthcare and finance, simulation techniques play a vital role in driving innovation, advancing knowledge, and achieving desired outcomes. By simulating real-world scenarios and exploring a multitude of possibilities, classical M&S techniques empower professionals to uncover new insights, optimize processes, and foster continuous improvement in a wide range of domains. However, the escalating complexity of these models has exposed the shortcomings of traditional simulation approaches in capturing the intricate details and nuances inherent in these intricate systems. Numerous studies in the literature have acknowledged the limitations of classical simulation techniques when dealing with highly intricate models and have put forth alternative solutions to tackle this challenge.

One prevalent solution that has gained traction in addressing the limitations of classical simulation techniques for complex models is the adoption of DS methodologies. DS involves integrating multiple simulations to enhance the effectiveness and efficiency of individual simulations by facilitating collaboration and synchronization of heterogeneous simulation components across different systems or platforms. Researchers have highlighted the advantages of DS in the literature. For instance, the application of DS in simulating large-scale manufacturing systems has been investigated, emphasizing its ability to distribute computational load across multiple machines and improve simulation performance while reducing the need for expensive server scaling [9, 10]. Similarly, the use of DS in simulating complex transportation networks, leveraging existing computational resources effectively and enhancing the efficiency of simulating large-scale systems, has been demonstrated [11]. The application of DS in healthcare systems has been studied in [12], showcasing how DS enables the integration of disparate simulation models to create a comprehensive representation of the healthcare environment. These studies [9–12] exemplify the potential benefits of DS in simulating complex models in manufacturing, transportation, and healthcare. DS offers a cost-effective approach by leveraging distributed resources, enabling scalability, and effectively handling large-scale systems across various domains. There are other domains where DS has also proven to be a valuable technology, such as in the defense [13, 14] and disaster management [15] domains.

The aforementioned studies provide valuable insights into the utilization of DS in various domains, underscoring its significance in addressing complex real-world challenges while facilitating informed decision-making.

The integration of M&S with VR technology through DS has garnered attention in numerous studies for simulating complex models. Researchers have explored combining discrete event simulation (DES) with VR using DS in various research endeavors, highlighting the synergistic benefits of this integration. These studies demonstrate the seamless fusion of DES with VR, enabling users to immerse themselves in a virtual environment and interact with simulated models.

For instance, in [16], the integration of DES with VR for simulating manufacturing processes has been investigated. A DS framework that combined the capabilities of DES with the immersive experience of VR was developed. This integration enabled decision-makers and trainees to navigate and interact with the virtual environment, gaining valuable insights into complex manufacturing systems. Similarly, a cross-disciplinary team of researchers focused on integrating DES with VR for simulating healthcare scenarios [17]. They developed a DS system that combined patient flow simulations with a VR environment, particularly in a hospital hemodialyisis unit. This integration provided an immersive and interactive experience, allowing healthcare professionals to explore different scenarios, evaluate system performance, and make informed decisions in a virtual setting. Furthermore, a design method using DES and VR for hot cell facility design has been developed [18]. The integrated DES-VR platform facilitated collaborative process simulation and design evaluation, improving material flow and obstacle checking compared to traditional approaches. The methodology proved effective for designing teleoperation-based nuclear facilities with varied equipment functions, structures, and degrees of automation. In summary, various studies have explored the integration of DS with VR, showcasing the seamless fusion of DES with VR to simulate complex models. These works [16–18] exemplify the immersive and interactive capabilities of the integration. By enabling users to explore and experience complex systems in a realistic and intuitive manner, the integration of DS with VR empowers decision-makers and trainees to gain valuable insights and make informed decisions.

The existing literature acknowledges the limitations of classical simulation techniques in dealing with complex models. As a result, researchers and practitioners have sought alternative solutions, with DS emerging as a promising approach. DS offers scalability and improved efficiency by leveraging the cost-effective distribution of resources, surpassing traditional simulation methods. Additionally, the integration of DS with VR presents significant potential, seamlessly combining DES with immersive VR environments. This integration enhances interactivity and realism in complex model simulations, opening new avenues for training and decision-making across various domains.

While previous studies have explored the benefits of integrating DES with VR for various applications, there is a notable gap in research regarding the integration of agent-ABS with DES and VR. ABS, which emerged in the 1990s, focuses on modeling individual agent behavior within a simulation, enabling a comprehensive representation of complex systems [19]. By incorporating ABS into the DES-VR framework, the dynamic interactions and decision-making processes of agents can be accurately captured, resulting in a more realistic simulation. This integration has the potential to significantly enhance the fidelity and accuracy of simulations by considering agent behavior and interactions within the virtual environment. The combination of these three technologies to create an immersive VR representation of a digital twin environment has thus far remained largely unexplored. This integration holds great promise in allowing users to not only interact with simulated models but also immerse themselves in a VR environment that faithfully reflects real-world dynamics. By leveraging the strengths of ABS, DES, and VR, researchers can usher in a new era of simulation, enabling users to gain deeper insights into complex systems and make informed decisions based on the digital twin representation. In this chapter, we present the methodology for unlocking the full potential of this integrated approach and demonstrate its applications within healthcare settings.

3. Materials and methods

The integration of DES, ABS, and VR represents a significant advancement in the field of simulation. This innovative approach aims to deliver a comprehensive and immersive simulation experience by leveraging the strengths of each paradigm. With DES, the complex and dynamic nature of events and system interactions can be accurately captured and analyzed, providing valuable insights into the behavior of the simulated system. Agent-based modeling assumes a crucial role in modeling the behaviours and decision-making processes of individual agents within the simulated environment, adding a layer of realism and complexity to the simulation. Furthermore, VR technology takes the experience to new heights by creating an interactive and immersive environment, enabling users to engage with the simulation in a more intuitive and realistic manner.

By carefully integrating these paradigms, this groundbreaking work seeks to unlock unparalleled potential in the realm of simulation. It provides a dynamic platform for exploring and understanding complex systems while enhancing decision-making processes. However, achieving seamless integration and interoperability among these heterogeneous components poses a significant challenge. This is where the HLA comes into play as the orchestrator. HLA acts as the mediator, coordinating the operations of the three components and facilitating data synchronization. It addresses interoperability issues at various levels, including data, middleware, operating system, and application heterogeneities. Through the efficient orchestration of these components, HLA ensures a cohesive and synchronized simulation experience, enabling researchers and practitioners to harness the full capabilities of this integrated simulation framework.

This integration has undergone extensive testing in two distinct healthcare environments within a hospital located in Toronto. Specifically, the first case involves the integration within a hemodialysis unit, while the second case focuses on the integration within an ICU. In the subsequent section, we delve into the technical aspects of integrating these three aforementioned technologies in these two unique healthcare settings.

3.1 High level architecture

The initial iteration of the HLA standard was introduced as HLA US DoD 1.3. Subsequently, it gained recognition from IEEE in 2000 and was adopted and renamed as HLA IEEE 1516 [1, 2, 4]. In 2010, the standard underwent revisions and updates, incorporating significant improvements, leading to the version known as HLA Evolved. Recently, the much-anticipated release of HLA4 made its debut, marking a significant milestone in the field of M&S. This latest version builds upon the foundation of its predecessors, incorporating new and advanced features to further enhance the capabilities of the HLA standard.

The HLA standard plays a pivotal role in advancing the field of DS by enabling the creation of simulations comprising diverse simulation components, referred to as "Federates." A federation consists of multiple federates, a runtime infrastructure (RTI), and a federated object model (FOM). The RTI provides a standardized set of services for data exchange, synchronization, and collaborative management. On the other hand, the FOM defines the objects/classes and interactions/parameters utilized for communication.

The primary objective of HLA is to ensure M&S compatibility and facilitate the reuse of models. These models are executed on separate computers equipped with different operating systems, implemented in various programming languages, and connected through a local area network (LAN) or wide area network (WAN). Despite these differences, all these components seamlessly integrate within a federation. Leveraging a publish/subscribe (p/s) mechanism based on FOM and HLA object management, these heterogeneous federates can efficiently exchange and synchronize simulation data.

Figure 1 illustrates the comprehensive global system architecture of the DS system. Within this architecture, two key components, namely DES and ABS, are utilized in the AnyLogic platform. These components seamlessly communicate with the VR environment, which is built using Unity.

In this integrated system, AnyLogic plays a crucial role by publishing the positions of each agent in the unit present within the simulation. This real-time tracking of agent positions allows for the precise monitoring of their movements and interactions within the simulated healthcare unit. Simultaneously, Unity, as part of the system, publishes the particle system results of these agents. This information is critical for AnyLogic, as it enables the real-time calculation of statistical studies related to COVID-19 transmission within the healthcare unit. Based on the p/s mechanism of the HLA standard, AnyLogic publishes the necessary information to the RTI. The RTI ensures that all subscribed components receive the specific data at a synchronized time. Similarly, Unity follows the same process, publishing relevant information to the

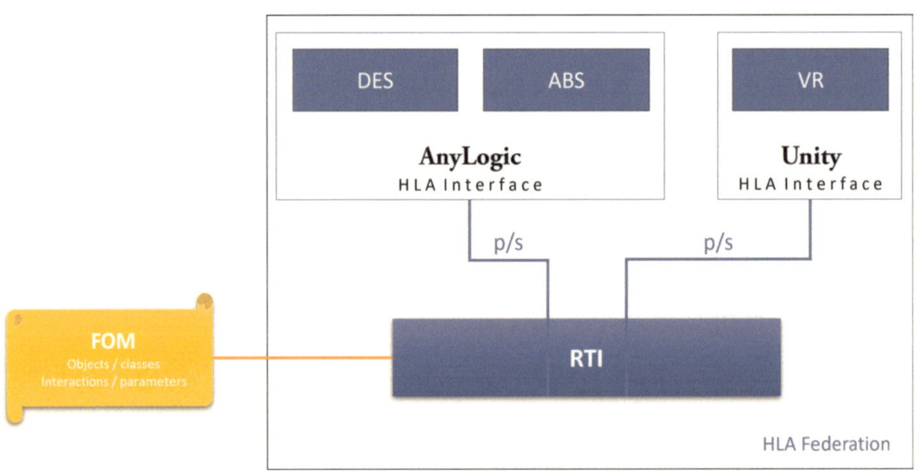

Figure 1.
DS system architecture.

RTI for dissemination. To enable communication between two distinct components, or federates, with different characteristics, an HLA interface has been specifically configured. This interface plays a vital role as an intermediary between the RTI and the HLA federate. By serving as a link, it facilitates seamless interaction and data exchange between these components, despite their heterogeneity.

In this scenario, time regulation and constraints are disabled, indicating that the time management mechanism of HLA is not currently being utilized. This means that the system does not enforce strict synchronization of time between the federates. However, this does not hinder the seamless communication between AnyLogic and Unity. Whenever AnyLogic or Unity publishes an object/class or interaction/parameter, the RTI triggers a callback function, transmitting the corresponding data to the connected federate. The Java pRTI library has been used to develop the HLA integration. The Java pRTI library served as an effective facilitator, offering a range of methods tailored to support the HLA-related functionalities. By employing this library, we were able to expedite the development process and enhance the efficiency of the HLA implementation. Algorithm 1 outlines the steps required to exchange data between AnyLogic and Unity using the PRTI library. AnyLogic sends agent positions to Unity, while Unity sends virus particle data to AnyLogic, enabling synchronized visualization and interaction between the two systems.

Algorithm 1: HLA configuration and Data Exchange Between AnyLogic and Unity

---

Input: AnyLogic model and Unity project with PRTI library integration
Output: Data exchange between AnyLogic and Unity

I. **Initialize AnyLogic-Unity Connectivity**

1. Import PRTI library into AnyLogic model and Unity project.
2. Initialize variables: **federationExecutionName**, **federateName**, **simulationTime**.

II. **Create and Connect to HLA Federation**

1. Create an instance of the RTI ambassador using the PRTI library in AnyLogic and Unity.
2. Connect AnyLogic and Unity to the HLA federation:
   a. Call the **connect()** function of the RTI ambassador in AnyLogic, passing **federationExecutionName**, **federateName**, and relevant parameters.
   b. Call the **connect()** function of the RTI ambassador in Unity, passing **federationExecutionName**, **federateName**, and relevant parameters.

III. **Join the Federation as Federates**

1. Join the federation as federates:

a. Call the **joinFederationExecution()** function of the RTI ambassador in AnyLogic, specifying **federationExecutionName**, **federateName**, and additional parameters.

b. Call the **joinFederationExecution()** function of the RTI ambassador in Unity, specifying **federationExecutionName**, **federateName**, and additional parameters.

IV. **Subscribe and Publish Object Classes/Attributes**

1. AnyLogic:

a. Subscribe to Unity's virus particle attributes:

- Call the **subscribeObjectClassAttributes()** function of the RTI ambassador in AnyLogic, providing the appropriate object class and attribute handles for virus particles.

b. Publish AnyLogic's agent positions:

- Call the **publishObjectClassAttributes()** function of the RTI ambassador in AnyLogic, specifying the object class and attribute handles for agent positions.

2. Unity:

a. Subscribe to AnyLogic's agent position attributes:

- Call the **subscribeObjectClassAttributes()** function of the RTI ambassador in Unity, providing the appropriate object class and attribute handles for agent positions.

b. Publish Unity's virus particle attributes:

- Call the **publishObjectClassAttributes()** function of the RTI ambassador in Unity, specifying the object class and attribute handles for virus particles.

V. **Enter Main Simulation Loop**

1. while simulation is not complete:

a. **Update Simulation Time**

- Update simulation time based on AnyLogic model time or Unity game time.

b. **Send Agent Position Updates from AnyLogic to Unity**

- Get agent positions from AnyLogic model.

- Call the **updateAttributeValues()** function of the RTI ambassador in AnyLogic, passing the object and attribute handles for agent positions and the updated values.

c. **Send Virus Particle Updates from Unity to AnyLogic**

- Get virus particle data from Unity.
- Call the **updateAttributeValues()** function of the RTI ambassador in Unity, passing the object and attribute handles for virus particles and the updated values.

d. **Receive Agent Position Updates in Unity**

- Call the **tick()** function of the RTI ambassador in Unity to receive the latest data for subscribed agent positions.
- Process the received updates and update the corresponding objects in Unity.

e. **Receive Virus Particle Updates in AnyLogic**

- Call the **tick()** function of the RTI ambassador in AnyLogic to receive the latest data for subscribed virus particles.
- Process the received updates and update the corresponding objects in AnyLogic.

VI. **Disconnect from HLA Federation**

1. Disconnect from the HLA federation:

a. Call the **resignFederationExecution()** function of the RTI ambassador in AnyLogic.

b. Call the **resignFederationExecution()** function of the RTI ambassador in Unity.

VII. **Cleanup Resources**

1. Release the resources used by the RTI ambassador and the PRTI library in AnyLogic and Unity.

This bidirectional communication ensures that relevant information is shared efficiently between AnyLogic and Unity, enabling both components to work in harmony within the DS system. By analyzing the data collected from both AnyLogic and Unity, the DS system facilitates the determination of important metrics such as the contact matrix in the healthcare unit. The contact matrix provides valuable insights into the rate of contacts between agents, aiding in the identification of potential hotspots for disease transmission. This information proves vital for hospital managers as it helps pinpoint the sources of disease transmission within the healthcare unit.

Equipped with this knowledge, hospital managers can take proactive measures to control and mitigate the spread of diseases. By identifying high-risk areas and focusing efforts on implementing targeted interventions, such as improved hygiene protocols or physical layout modifications, hospitals can effectively minimize the impact of disease transmission. Ultimately, this integrated DS system empowers healthcare facilities to make informed decisions based on accurate data, leading to improved patient safety and enhanced overall healthcare outcomes.

3.2 Discrete event simulation

DES is a widely utilized computational modeling approach employed to replicate the dynamic behavior and operational performance of systems across a span of time. It entails representing the system as a sequence of distinct events that take place at precise points in time, effectively capturing the interplay and interdependencies between various entities and resources. Each event signifies a noteworthy alteration in the system's state, such as an arrival, departure, or completion of a task. Through simulating the system's behavior under diverse scenarios and parameter configurations, DES enables the analysis, optimization, and evaluation of intricate systems without the need for resource-intensive and time-consuming real-world experimentation.

3.2.1 Hemodialysis case study

The DES technique was applied within the hemodialysis unit to model the dynamic processes and interactions involving hemodialysis patients, nephrologists and other healthcare professionals, and available resources. The simulation encompassed various discrete events, such as patient arrivals, assignment to hemodialysis machines, duration of dialysis sessions, and other relevant activities. Through accurate representation of event timings and dependencies, the simulation provided valuable insights into the unit's performance metrics, including patient waiting times, utilization rates of hemodialysis machines, and overall operational efficiency. By exploring diverse scenarios and resource allocation strategies, the DES facilitated the evaluation of different approaches and identified potential enhancements to optimize the effectiveness of the hemodialysis unit and enhance patient care. Detailed implementation of the hemodialysis model within the DES system, including entities, resources, parameters, assumptions, and model dynamics, is elaborated in **Table 1**.

3.2.2 ICU case study

In the context of the ICU management system, DES was leveraged to simulate the dynamic processes and interactions involving patients, healthcare professionals, and the available resources. The discrete events encompassed patient arrivals, consultations with medical professionals, administration of treatments, diagnostic procedures, and other relevant activities. By accurately modeling the timing and interdependencies of these events, the simulation provided valuable insights into the system's performance metrics, including patient waiting times, utilization rates of healthcare resources, and overall operational efficiency. Through the exploration of different scenarios and resource allocation strategies, the DES enabled the evaluation of various approaches and the identification of potential improvements to enhance the

Entities:
a. Patients: Individuals requiring hemodialysis treatment. b. Nurses: Healthcare professionals responsible for administering treatment and monitoring patients. c. Dialysis machines: Medical devices used for hemodialysis treatment. d. Supplies: Consumables and equipment necessary for the process. e. MRP (medical renal physicians): Specialists responsible for overseeing the treatment plan and making medical decisions. f. UT nephrologists: Nephrologists affiliated with the University Hospital responsible for consultation and treatment oversi/ght. g. Residents: Medical residents involved in the patient care process. h. Housekeepers: Staff responsible for maintaining cleanliness and order in the unit.
Resources:
a. Nurse capacity (C_nurse): The number of available nurses. b. Dialysis machine capacity (C_machine): The number of available dialysis machines. c. Supplies availability: The availability of necessary supplies and equipment. d. MRP availability: The availability of MRP for treatment plan review. e. UT nephrologist availability: The availability of UT Nephrologists for consultation and treatment oversight. f. Resident availability: The availability of residents for assisting in patient care. g. Housekeeper capacity (C_housekeeper): The number of available housekeepers.
Parameters and variables:
a. Arrival rate (λ): The average rate at which patients arrive at the unit. b. Service rate (μ): The average time it takes to complete a hemodialysis treatment for a patient. c. Consultation time (τ): The average time taken by a UT Nephrologist to consult and provide oversight. d. Treatment plan review time (θ): The average time taken by an MRP to review and update the treatment plan. e. Nurse availability (A_nurse): The proportion of time nurses are busy. f. Dialysis machine availability (A_machine): The proportion of time dialysis machines are busy. g. MRP availability (A_mrp): The proportion of time MRP are busy. h. UT nephrologist availability (A_nephrologist): The proportion of time UT Nephrologists are busy. i. Resident availability (A_resident): The proportion of time residents are busy. j. Housekeeper availability (A_housekeeper): The proportion of time housekeepers are busy.
Model assumptions:
a. Patients arrive according to a Poisson process with an average arrival rate of λ. b. The service time for each patient follows an exponential distribution with a mean of $1/\mu$. c. The consultation time for each patient with a UT Nephrologist follows an exponential distribution with a mean of $1/\tau$. d. The treatment plan review time for each patient by an MRP follows an exponential distribution with a mean of $1/\theta$. e. Patients join the corresponding queues when resources are not immediately available. f. The number of nurses, dialysis machines, supplies, MRP, UT nephrologists, residents, and housekeepers remain constant throughout the simulation.
Model dynamics:
Initialization: • Set simulation time to $t = 0$. • Generate the first patient arrival time based on the exponential distribution.
Patient arrival process:

• At time t, a patient arrives.
• If there is an available UT nephrologist for consultation:
  o The patient consults the UT nephrologist and receives treatment oversight.
  o Calculate the treatment completion time for the patient based on the service time.

- Update the availability of the UT nephrologist.
- If there are available nurses, dialysis machines, and supplies:
  - Assign a nurse, a dialysis machine, and necessary supplies to the patient.
  - Update the availability of the assigned resources.
- If any of the resources are not available:
  - The patient joins the corresponding queues.

Treatment process:

- When a nurse, dialysis machine, or supplies become available:
  - Assign the available resource to the patient at the front of the corresponding queue (if any).
  - Start the treatment for the patient, setting the treatment completion time based on the service time.
  - Update the availability of the assigned resource.

Consultation process:

- When a UT nephrologist becomes available:
  - Assign the UT nephrologist to the patient at the front of the consultation queue (if any).
  - Conduct the consultation and provide treatment oversight.
  - Update the availability of the UT nephrologist.
  - If there are available nurses, dialysis machines, and supplies:
    - Assign the necessary resources to the patient.
    - Update the availability of the assigned resources.
  - If any of the resources are not available:
    - The patient joins the corresponding queues.

Treatment plan review process:

- When an MRP becomes available:
  - Assign the MRP to the patient at the front of the treatment plan review queue (if any).
  - Conduct the treatment plan review and update the treatment plan if necessary.
  - Update the availability of the MRP.
  - If there are available UT nephrologists, nurses, dialysis machines, and supplies:
    - Assign the necessary resources to the patient.
    - Update the availability of the assigned resources.
  - If any of the resources are not available:
    - The patient joins the corresponding queues.

Resident support process:

- When a resident becomes available:
  - Assign the resident to assist in patient care for the patient at the front of the corresponding queue (if any).
  - Assist in patient care as directed by the nurses, UT nephrologists, or MRP.
  - Update the availability of the resident.

Treatment completion:

- When the treatment completion time for a patient is reached:
  - Release the nurse, dialysis machine, and supplies assigned to the patient.
  - Update the availability of the released resources.
  - If there are patients in the consultation, treatment plan review, or resident support queues:
    - Assign the next available resources to the respective patients.
    - Start the corresponding process for the assigned patients.

Housekeeping process:

- Housekeepers continuously monitor and maintain cleanliness and order in the unit.
- Update the availability of the housekeepers.

Performance metrics:

- Track and record the following metrics during the simulation:
  - Average patient waiting time in the queues.
  - Average patient treatment time.

∘ Resource utilization rates (nurses, dialysis machines, MRP, UT nephrologists, residents, housekeepers).
Simulation termination:
• Continue the simulation until a specified simulation time is reached or a predetermined number of patients have been treated.
Data analysis:
• Collect simulation outputs and calculate statistics for performance metrics. • Conduct sensitivity analysis and optimization techniques to improve the efficiency of the hemodialysis unit process.

Table 1.
Hemodialysis model.

effectiveness of the ICU management system and improve patient care. A comprehensive description of the ICU model implementation, including entities, resources, parameters, assumptions, and model dynamics, is provided in **Table 2**.

The DES played a crucial role as an initial step in the analysis, setting the foundation for further exploration using ABS. By simulating discrete events and their interdependencies, the DES provided valuable insights into system behavior and performance. However, to capture the intricate dynamics of individual agents, their behaviors, and decision-making processes, the subsequent subsection will delve into ABS. ABS offers a more nuanced approach by representing each entity as an autonomous agent, capable of interacting with other agents and responding to their environment. This advanced M&S technique enables a deeper understanding of emergent system behaviors, complex interactions, and the impact of individual agent actions. The following subsection will present the details of the agent-based modeling framework, shedding light on its advantages and its potential applications in the context of the system under study.

3.3 Agent-based simulation

In both the hemodialysis unit and ICU case studies, a unified agent-based model was employed to investigate disease transmission within the healthcare facility, with a particular focus on the challenging period characterized by the COVID-19 pandemic. This critical phase necessitated hospitals' proactive pursuit of effective measures to mitigate the spread of the virus and ensure the safety of patients and healthcare professionals. Leveraging the agent-based modeling approach, our objective was to gain valuable insights into the complex dynamics of disease transmission within these units while considering the unique characteristics of the pandemic context.

To effectively capture the behavior and interactions of human agents, including healthcare workers and patients, we implemented a hierarchical structure within the model. This hierarchical structure allowed for the extension and inheritance of human properties across agent types. Specifically, a specialized entity, referred to as the "human agent framework," served as the overarching representation of human characteristics and served as the foundation for modeling disease transmission.

Within the human agent framework, a dedicated module focused on disease transmission was developed. Drawing inspiration from established epidemiological models, we devised an expanded version of the SEIR (susceptible, exposed, infected,

Entities:

a. Patients: Individuals requiring intensive care treatment, including cardiac patients.
b. Nurses: Healthcare professionals responsible for patient care and monitoring.
c. Doctors: Medical professionals responsible for diagnosing and prescribing treatment.
d. Respiratory therapists: Healthcare professionals specializing in respiratory care.
e. Pharmacists: Professionals responsible for medication management and dispensing.
f. Lab technicians: Personnel responsible for conducting laboratory tests.
g. Housekeepers: Staff responsible for maintaining cleanliness and order in the unit.
h. Medical equipment technicians: Personnel responsible for maintaining and repairing medical equipment.
i. Family members: Relatives or guardians of patients providing support and information.

Resources:

a. Nurse capacity (C_nurse): The number of available nurses.
b. Doctor capacity (C_doctor): The number of available doctors.
c. Respiratory therapist capacity (C_therapist): The number of available respiratory therapists.
d. Pharmacist capacity (C_pharmacist): The number of available pharmacists.
e. Lab technician capacity (C_technician): The number of available lab technicians.
f. Equipment availability: The availability of medical equipment required for intensive care, including cardiac monitoring equipment.
g. Medication availability: The availability of necessary medications, including cardiac medications.
h. Housekeeper capacity (C_housekeeper): The number of available housekeepers.
i. Technician capacity (C_technician): The number of available medical equipment technicians.

Parameters and variables:

a. Arrival rate (λ): The average rate at which patients arrive at the ICU, including cardiac patients.
b. Service rate (μ): The average time it takes to complete treatment for a patient.
c. Consultation time (τ): The average time taken by a doctor to consult and prescribe treatment.
d. Respiratory therapy time (θ): The average time taken by a respiratory therapist to administer respiratory care.
e. Medication dispensing time (ψ): The average time taken by a pharmacist to dispense medication.
f. Laboratory test time (ω): The average time taken by a lab technician to conduct laboratory tests.
g. Nurse availability (A_nurse): The proportion of time nurses are busy.
h. Doctor availability (A_doctor): The proportion of time doctors are busy.
i. Respiratory therapist availability (A_therapist): The proportion of time respiratory therapists are busy.
j. Pharmacist availability (A_pharmacist): The proportion of time pharmacists are busy.
k. Lab technician availability (A_technician): The proportion of time lab technicians are busy.
l. Housekeeper availability (A_housekeeper): The proportion of time housekeepers are busy.
m. Technician availability (A_technician): The proportion of time medical equipment technicians are busy.

Model assumptions:

a. Patients arrive according to a Poisson process with an average arrival rate of λ.
b. The service time for each patient follows an exponential distribution with a mean of $1/\mu$.
c. The consultation time for each patient with a doctor follows an exponential distribution with a mean of $1/\tau$.
d. The respiratory therapy time for each patient follows an exponential distribution with a mean of $1/\theta$.
e. The medication dispensing time for each patient follows an exponential distribution with a mean of $1/\psi$.
f. The laboratory test time for each patient follows an exponential distribution with a mean of $1/\omega$.
g. Patients join the corresponding queues when resources are not immediately available.
h. The number of nurses, doctors, respiratory therapists, pharmacists, lab technicians, housekeepers, and medical equipment technicians remains constant throughout the simulation.

Model dynamics:

DOI: http://dx.doi.org/10.5772/intechopen.1002004

Initialization:

- Set simulation time to $t = 0$.
- Generate the first patient arrival time based on the exponential distribution.

Patient arrival process:

- At time t, a patient arrives, including cardiac patients.
- If there is an available doctor for consultation:
    - The patient consults the doctor and receives treatment prescription.
    - Calculate the treatment completion time for the patient based on the service time.
    - Update the availability of the doctor.
    - If there are available nurses, respiratory therapists, pharmacists, lab technicians, equipment, and medications:
        - Assign a nurse, a respiratory therapist, a pharmacist, a lab technician, necessary equipment, and medications to the patient.
        - Update the availability of the assigned resources.
    - If any of the resources are not available:
        - The patient joins the corresponding queues.

Treatment process:

- When a nurse, respiratory therapist, pharmacist, lab technician, equipment, or medication becomes available:
    - Assign the available resource to the patient at the front of the corresponding queue (if any).
    - Start the treatment for the patient, setting the treatment completion time based on the service time.
    - Update the availability of the assigned resource.

Consultation process:

- When a doctor becomes available:
    - Assign the doctor to the patient at the front of the consultation queue (if any).
    - Conduct the consultation and prescribe the treatment plan.
    - Update the availability of the doctor.

Respiratory therapy process:

- When a respiratory therapist becomes available:
    - Assign the respiratory therapist to the patient at the front of the respiratory therapy queue (if any).
    - Administer respiratory therapy to the patient.
    - Update the availability of the respiratory therapist.

Medication dispensing process:

- When a pharmacist becomes available:
    - Assign the pharmacist to the patient at the front of the medication dispensing queue (if any).
    - Dispense the required medications to the patient.
    - Update the availability of the pharmacist.

Laboratory testing process:

- When a lab technician becomes available:
    - Assign the lab technician to the patient at the front of the laboratory testing queue (if any).
    - Conduct the necessary laboratory tests for the patient.
    - Update the availability of the lab technician.

Treatment completion:

- When the treatment completion time for a patient is reached:
    - Release the nurse, respiratory therapist, pharmacist, lab technician, equipment, and medications assigned to the patient.
    - Update the availability of the released resources.
    - If there are patients in the consultation, respiratory therapy, medication dispensing, or laboratory testing queues:
        - Assign the next available resources to the respective patients.
        - Start the corresponding process for the assigned patients.

Housekeeping process:
• Housekeepers continuously monitor and maintain cleanliness and order in the unit. • Update the availability of the housekeepers.
Medical equipment maintenance process:

• When a medical equipment technician becomes available:
  o Assign the technician to inspect, maintain, or repair medical equipment as required.
  o Update the availability of the technician.

Family support process:
• Family members of patients provide support and information as needed.
Performance metrics:

• Track and record the following metrics during the simulation:
  o Average patient waiting time in the queues.
  o Average patient treatment time.
  o Resource utilization rates (nurses, doctors, respiratory therapists, pharmacists, lab technicians, housekeepers, medical equipment technicians).

Simulation termination:
• Continue the simulation until a specified simulation time is reached or a predetermined number of patients have been treated.
Data analysis:
• Collect simulation outputs and calculate statistics for performance metrics. • Conduct sensitivity analysis and optimization techniques to improve the efficiency of the ICU process.

Table 2.
ICU model.

and recovered) model, as depicted in **Figure 2** [12]. This enhanced model incorporated additional features to account for asymptomatic and presymptomatic infected healthcare workers, who served as potential sources of viral contamination and transmission. Their interactions with susceptible individuals within the defined physical space of the model played a crucial role in simulating the spread of the disease. Similar modeling approaches have proven effective in other simulations, validating their applicability and relevance.

Tracking the different stages of infection within the human agent framework, including periods of active viral transmission, was accomplished through a comprehensive state chart. This state chart facilitated the precise monitoring of disease progression, enabling the identification of crucial time points when human agents became potential sources of contamination and transmission. Symptomatic healthcare workers underwent testing, and upon confirmation of a positive diagnosis, they were isolated to prevent further virus transmission. During the quarantine or recovery period, these symptomatic agents were designated as inactive viral vectors and temporarily replaced within the model. Conversely, for ICU patients, the virus transmission component remained active throughout the simulation, reflecting the assumption that the ICU ward exclusively admitted COVID-19 patients. Healthcare workers entering the state of active viral transmission were governed by the same state chart within the human agent framework, ensuring consistent modeling of their behavior. Furthermore, the model recorded the sources of contamination for exposed/infected medical staff agents, facilitating the identification of significant contributors to viral spread.

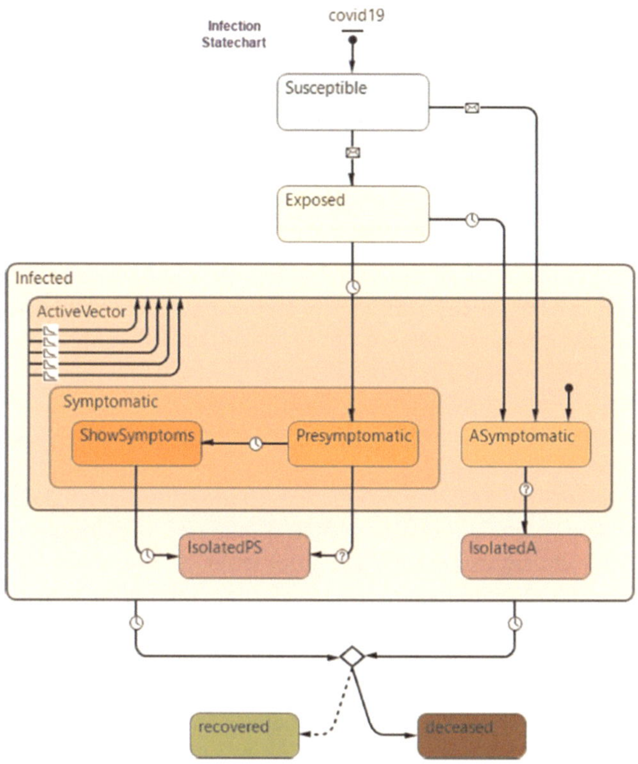

Figure 2.
Agent-based model for COVID 19 transmission inside the units.

By utilizing this unified agent-based model, we aimed to provide a comprehensive understanding of disease transmission dynamics within both the hemodialysis unit and the ICU, particularly during the challenging period of the COVID-19 pandemic. The integration of advanced modeling techniques and the meticulous representation of human agent behaviors enabled us to evaluate various scenarios and devise strategies for improving infection control measures, optimizing patient care, and safeguarding the well-being of healthcare workers.

4. Results and discussion

The aforementioned case studies involved building digital twin models specific to the units under investigation, enabling hypothesis testing and contextual exploration to understand their impact on system performance. To enhance the capabilities of the digital twin frameworks, we introduced a VR layer. This immersive layer comprises VR environments that faithfully replicate the physical units in a three-dimensional space. The VR environments showcase empty 3D models without agents, providing a platform to gain insights into unit behavior and dynamics across different scenarios.

Integration between the DES system, ABS system, and the VR layer was achieved through the utilization of HLA. The DES system facilitates real-time agent position transmission, enabling us to receive crucial information such as agent type, unique identifier, and position within the system. This data plays a pivotal role in accurately defining and locating agents within the VR model.

To achieve synchronization between the digital twin and the VR environment, we utilized Unity, a powerful game development engine. Leveraging the agent information obtained from the DES and ABS system, we imported the corresponding agent avatars into the VR environment. By aligning the agents' positions from the DES and ABS system with their virtual counterparts, we established a realistic and interactive experience, facilitating comprehensive monitoring and analysis of agent behavior within the simulated environment.

The immersion inside the units using VR headsets offers a unique and engaging experience, allowing researchers and stakeholders to gain a profound understanding of system dynamics. The VR environment provides a sense of presence and realism that traditional desktop simulations cannot replicate, enabling intuitive exploration and comprehension of complex system behaviors (**Figure 3**).

By immersing oneself in the virtual environment, users can visually inspect the units from various perspectives, observing intricate details of the spatial layout, environmental conditions, and architectural features. This level of immersion enhances contextual understanding, enabling researchers to identify potential design flaws, safety hazards, or operational inefficiencies that may not be evident in traditional simulations or physical environments. The interactive nature of the VR environment empowers researchers to manipulate objects, simulate interactions, and trigger events, replicating real-world scenarios within the digital twin. This interactivity facilitates hypothesis testing, validation of assumptions, and exploration of the consequences of different actions, leading to informed decision-making and system optimization. The immersive experience in the VR environment also enhances collaboration and communication among stakeholders. By donning VR headsets, researchers, domain experts, and decision-makers can convene within the same virtual space, inspect the units together, and discuss findings and insights. This collaborative environment promotes interdisciplinary discussions, knowledge sharing, and improved collective understanding of the system, ultimately facilitating more effective decision-making processes.

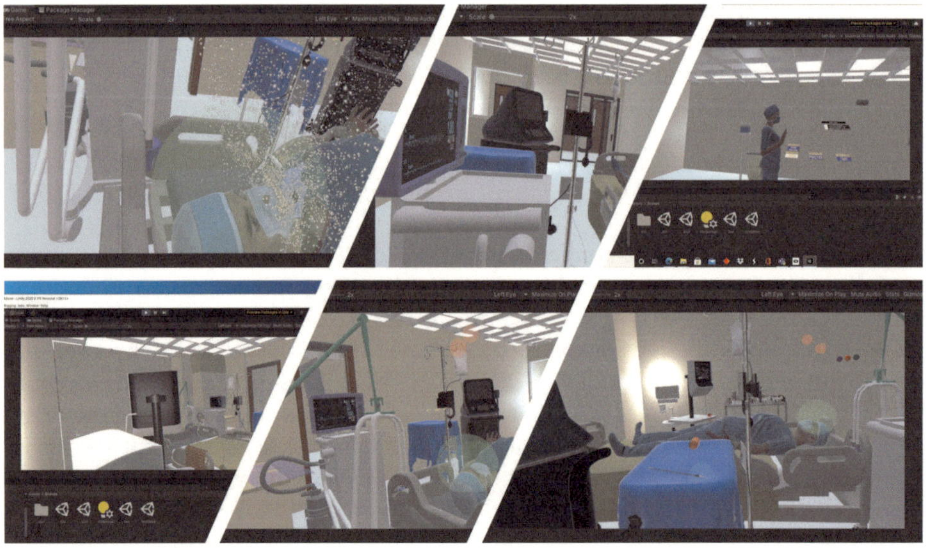

Figure 3.
An immersive experience inside the ICU.

While the immersion within the units using VR headsets offers significant advantages, there are considerations to address. Factors such as user comfort, mitigation of motion sickness, and the availability of high-quality VR hardware can impact the adoption and accessibility of this approach. Furthermore, ongoing advancements in VR technology should be considered to enhance the realism and fidelity of the virtual environment, further improving the accuracy and usefulness of the digital twin framework.

5. Conclusion

In conclusion, this research has made significant contributions to the advancement of digital twin frameworks by seamlessly integrating DES, ABS, VR based on HLA standard. The utilization of these technologies has allowed for the simulation and analysis of a wide range of systems, services, and processes in a scientifically rigorous manner. The construction of accurate digital twin models using DES and ABS has laid the groundwork for hypothesis testing, contextual exploration, and performance evaluation. DES enables the modeling of system dynamics based on discrete events, while ABS facilitates the representation of individual agent behaviors and interactions. This integration provides a comprehensive understanding of system-level behavior and the emergence of complex phenomena.

The inclusion of the VR layer within the digital twin framework has greatly enhanced the simulation experience by providing an immersive virtual environment. By faithfully replicating the physical units in a three-dimensional space, the VR environments enable researchers to visually and interactively explore the simulated systems, leading to a deeper understanding of their intricacies. The ability to experiment with different scenarios within the VR environment increases the realism and fidelity of the simulations, enabling more accurate analysis and decision-making. The seamless integration facilitated by HLA between the DES/ABS systems and the VR environment has enabled real-time data exchange, ensuring synchronization between the digital twin and the immersive virtual environment. By transmitting agent positions from the DES system to the VR layer, precise agent placement and visualization within the virtual environment are achieved, resulting in a coherent and lifelike representation of the system. This integration is crucial for tracking and analyzing agent behavior, providing valuable insights into system performance and facilitating evidence-based decision-making.

The scientific rigor and advancements presented in this research open up new avenues for further exploration and development of digital twin frameworks. Future research efforts can focus on refining the modeling capabilities of DES and ABS, incorporating more intricate agent behaviors and interactions. Furthermore, enhancing VR environments to include realistic avatars, environmental factors, and sensory feedback can further improve the fidelity and accuracy of the simulations. The practical implications of this research are substantial, particularly in domains such as healthcare, where digital twin frameworks can provide valuable insights into system performance, resource allocation, and optimization. By leveraging the combined power of DES, ABS, VR, and HLA, decision-makers can make informed choices, enhance operational efficiency, and mitigate potential risks.

In conclusion, this research highlights the potential of integrating DES, ABS, VR, and HLA within a digital twin framework for scientifically rigorous system simulation and analysis. The advancements presented in this study lay a solid foundation for further research and application in various domains, with the potential to revolutionize decision-making and optimization processes in complex systems.

Author details

Jalal Possik[1*], Adriano O. Solis[2] and Gregory Zacharewicz[3]

1 Lille Catholic University, Lille, France

2 York University, Toronto, Canada

3 IMT Mines Alès, Alès, France

*Address all correspondence to: jalal.possik@univ-catholille.fr

References

[1] Zeigler BP, Hall SB, Sarjoughian HS. Exploiting HLA and DEVS to promote interoperability and reuse in Lockheed's corporate environment. Simulation. 1999;**73**:288-295. DOI: 10.1177/003754979907300505

[2] Zacharewicz G, Frydman C, Giambiasi N. G-DEVS/HLA environment for distributed simulations of workflows. Simulation. 2008;**84**: 197-213. DOI: 10.1177/0037549708092833

[3] Possik J, Amrani A, Zacharewicz G. Development of a co-simulation system as a decision-aid in Lean tools implementation. In: Proceedings of the 50th Computer Simulation Multiconference (SummerSim '18); 9-12 July 2018; Bordeaux, France. San Diego, California: Society for Computer Simulation International; 2018. Article 21, pp. 1-12

[4] Possik J. Contribution to a methodology and a co-simulation framework assessing the impact of lean on manufacturing performance [PhD thesis]. Bordeaux, France: University of Bordeaux; 2019

[5] Possik J, D'Ambrogio A, Zacharewicz G, Amrani A, Vallespir B. A BPMN/HLA-based methodology for collaborative distributed DES. In: Proceedings of the 2019 IEEE 28th International Conference on Enabling Technologies: Infrastructure for Collaborative Enterprises (WETICE 2019); 12-14 June 2019; Capri, Italy. Piscataway, NJ: IEEE; 2019. pp. 118-123. DOI: 10.1109/WETICE.2019.00033

[6] Almaksour A, Gerges H, Gorecki S, Zacharewicz G, Possik, J. The use of the IEEE HLA standard to tackle interoperability issues between heterogeneous components. In: Proceedings of the 2022 IEEE/ACM 26th International Symposium on Distributed Simulation and Real Time Applications (DS-RT 2022); 26-28 September 2022; Alès, France. Piscataway, NJ: IEEE; 2022. pp. 175-178. DOI: 10.1109/DS-RT55542.2022.9932042

[7] AnyLogic Company. AnyLogic Simulation Software [Internet]. 2022. Available from: https://www.anylogic.com/ [Accessed: January 27, 2023]

[8] Unity Technologies. Unity Simulation Pro [Internet]. 2022. Available from: https://unity.com/ [Accessed: December 2, 2022]

[9] Gorecki S, Possik J, Zacharewicz G, Ducq Y, Perry N. A multicomponent distributed framework for smart production system modeling and simulation. Sustainability. 2020;**12**:6969. DOI: 10.3390/su12176969

[10] Possik J, Zouggar-Amrani A, Vallespir B, Zacharewicz G. Lean techniques impact evaluation methodology based on a co-simulation framework for manufacturing systems. International Journal of Computer Integrated Manufacturing. 2021;**35**:91-111. DOI: 10.1080/0951192X.2021.1972468

[11] Ahmed, MS, Hoque MA. Partitioning of urban transportation networks utilizing real-world traffic parameters for distributed simulation in SUMO. In: Proceedings of the 2016 IEEE Vehicular Networking Conference (VNC 2016); 08-10 December 2016; Columbus, OH, USA. Piscataway, NJ: IEEE; 2016. pp. 1-4. DOI: 10.1109/VNC.2016.7835942

[12] Possik J, Asgary A, Solis AO, Zacharewicz G, Shafiee MA,

Najafabadi MM, et al. An agent-based modeling and virtual reality application using distributed simulation: Case of a COVID-19 intensive care unit. IEEE Transactions on Engineering Management. 2022;**70**:2931-2943 DOI: 10.1109/TEM.2022.3195813

[13] Dahmann JS, Fujimoto RM, Weatherly RM. The Department of Defense high level architecture. In: Proceedings of the 29th Conference on Winter Simulation; 7-10 December 1997; Atlanta, GA, USA. New York, NY: Association for Computing Machinery; 1997. pp. 142-149. DOI: 10.1145/268437.268465

[14] Kim TG, Sung CH, Hong S-Y, Hong JH, Choi CB, Kim JH, et al. DEVSim++ toolset for defense modeling and simulation and interoperation. The Journal of Defense Modeling and Simulation: Applications, Methodology, Technology. 2011;**8**:129-142. DOI: 10.1177/1548512910389203

[15] Boukerche A. Smart disaster management and responses for smart cities: A new challenge for the next generation of distributed simulation systems. In: Proceedings of the 2019 IEEE/ACM 23rd International Symposium on Distributed Simulation and Real Time Applications (DS-RT 2019); 07-09 October 2019; Cosenza, Italy. Piscataway, NJ: IEEE; 2019. pp. 1-2. DOI: 10.1109/DS-RT47707.2019.8958693

[16] Hutabarat W, Oyekan J, Turner C, Tiwari A, Prajapat N, Gan X-P, Waller A. Combining virtual reality enabled simulation with 3D scanning technologies towards smart manufacturing. In: Proceedings of the 2016 Winter Simulation Conference (WSC 2016); 11-14 December 2016; Washington, DC, USA. Piscataway, NJ: IEEE; 2016. pp. 2774-2785. DOI: 10.1109/WSC.2016.7822314

[17] Possik J, Gorecki S, Asgary A, Solis AO, Zacharewicz G, Tofighi M, Shafiee MA, Merchant AA, Aarabi M, Guimaraes A, Nadri N. A distributed simulation approach to integrate anylogic and unity for virtual reality applications: Case of covid-19 modelling and training in a dialysis unit. In: Proceedings of the 2021 IEEE/ACM 25th International Symposium on Distributed Simulation and Real Time Applications (DS-RT 2021); 27-29 September 2021; Valencia, Spain. Piscataway, NJ: IEEE; 2021. pp. 1-7. DOI: 10.1109/DS-RT52167.2021.9576149

[18] Yu S, Han J. Virtual reality platform-based conceptual design and simulation of a hot cell facility. The International Journal of Advanced Manufacturing Technology. 2021;**116**:487-505. DOI: 10.1007/s00170-021-07139-7

[19] Siebers PO, Macal CM, Garnett J, Buxton D, Pidd M. Discrete-event simulation is dead, long live agent-based simulation! Journal of Simulation. 2010; **4**:204-210. DOI: 10.1057/jos.2010.14

Chapter 10

Integrated Multiscale Modeling-Simulation (MMS) and Machine Learning (ML)-Based Design and Development of Novel Technologies, Systems, and Processes

Seçkin Karagöz

Abstract

The development of novel technologies, systems, and processes is conventionally complemented by experimental testing. However, experimental tools for testing and examining the results are expensive, and their use is time-consuming. In this context, to accelerate the development, commercialization, utilization, and problem solutions of novel technologies, systems, and processes, the simultaneous use of computational and experimental tools such as hierarchical integrated machine learning (ML)-assisted multi-scale modeling-simulation (MMS) and experimental approaches is essential. These approaches greatly improve the entire technology development process by reducing cost and time and allow us to tackle problems that cannot be solved using theoretical or experimental methods alone. In this chapter, we describe ways in which integrated multiscale modeling-simulation and machine learning have been leveraged to facilitate the design and development of novel technologies, systems, and processes. We first provide a taxonomy of multiscale modeling-simulation and machine learning paradigms and techniques, along with a discussion of their strengths and limitations. We then provide an overview of opportunities and existing research using multiscale modeling-simulation and machine learning for the design and development of novel technologies, systems, and processes. Finally, we propose future research directions and discuss important considerations for deployment.

Keywords: machine learning, multiscale modeling, multiscale simulation, integrated MMS and ML, integrated MMS and ML-based design

1. Introduction

Traditionally, the development of novel technologies, systems, and processes is complemented by testing. Experimental tools for testing and examining the results are expensive, and their use is time-consuming. In recent years, thanks to the rapid growth in computational speed, it has been seen that the use of computer-aided methods in the systems' design phase contributes greatly to the reduction of cost and time for the entire technology development process. Moreover, the simultaneous use of computational and experimental tools allows us to tackle problems that cannot be solved using theoretical or experimental methods alone. Experimental methods can improve the computational model by incorporating new data, while computational tools can use this advanced model to analyze structure, properties, and optimization by examining a wide set of possible configurations.

To accelerate the development, commercialization, utilization, and problem solutions of novel technologies, systems, and processes, hierarchical integrated theoretical approaches such as machine learning (ML)-assisted multi-scale modeling-simulation (MMS), optimization, techno-economic analysis, and experimental approaches (**Figure 1**) are essential.

Multiscale modeling-simulation is a rapidly growing subject of study that encompasses a variety of disciplines, including physics, biology, chemistry, mathematics, statistics, chemical engineering, mechanical engineering, and materials science [1]. The most typical objective of multi-scale modeling and simulation is to anticipate the macroscopic behavior of an engineering system process from fundamental principles by computing information at a smaller scale and passing it to a model at a larger scale by leaving out degrees of freedom as one moves from smaller to larger scales (upscaling or bottom-up approach) [2, 3]. Machine learning (ML) is becoming widely acknowledged as a promising technology in a wide range of scientific applications. ML is a potent method for integrating multimodality and multi-fidelity data and revealing correlations between interconnected phenomena presents a unique opportunity in this regard [4].

Integrated multiscale modeling-simulation (MMS) and machine learning (ML) approach allows many systems and processes to be accurately simulated and many

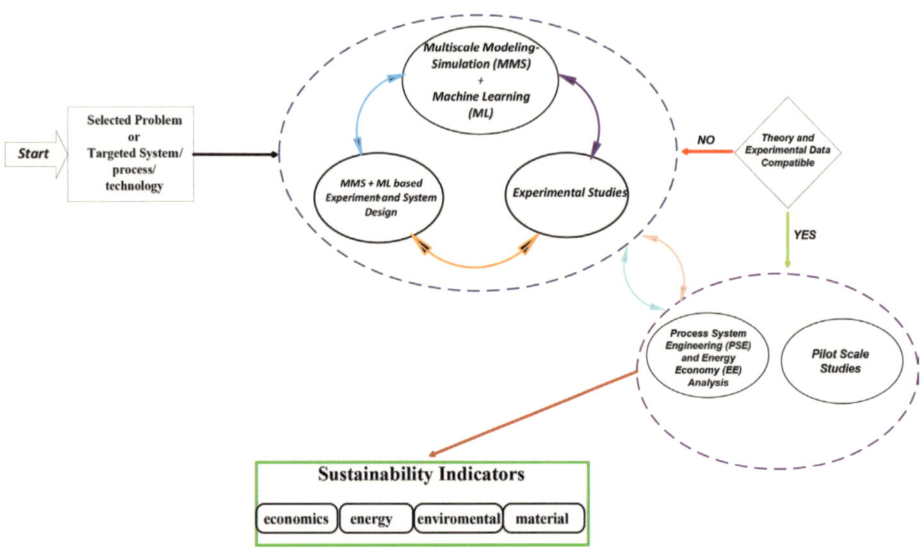

Figure 1.
Hierarchical integrated theoretical and experimental approaches to the technology-system-process development and problem solutions.

system and process parameters to be calculated without the need for experiments before performing expensive and time-consuming experiments and tests. This strategy significantly improves the entire technological development process by saving costs and time and also assists us tackle problems that cannot be solved using traditional theoretical or experimental approaches alone. Thus, in this chapter, we describe ways in which integrated multiscale modeling-simulation and machine learning have been leveraged to facilitate the design and development of novel technologies, systems, and processes. We begin with a taxonomy of multiscale modeling-simulation and machine learning paradigms and methodologies, followed by a discussion of their strengths and limitations. Following that, we present an overview of prospects and existing research employing multiscale modeling-simulation and machine learning for the design and development of novel technologies, systems, and processes. Finally, we offer future research areas and highlight critical deployment factors.

2. Modeling

The development of models is driven by the necessity to comprehend and anticipate the functioning and performance of particular component(s) of the system/process or the overall system/process behavior. The study of physical processes on models is known as simulation. In the most basic form, a model reproduces the studied system (SS=Original) while retaining its physical nature and geometric similarity to the original and differing only in parameters and variables. Depending on the objective of modeling, the modeling process takes into account a number of variables while rejecting others [5]. A model can be described as the following:

$$M \equiv \langle SS, SO, G, I, L \rangle \equiv \langle SS, SO, G, T, C, L \rangle$$

where SS is the studied system (Original) and the source of information; SO is the subject/observer who requires information on the SS for study or decision-making purposes; G is(are) goal(s); I is the infrastructure employed in the process of modeling and spans a set of technologies such as T (methods, means, algorithms *etc*.) and C (conditions of modeling: external and internal factors affect model development and functioning) and L is the language for describing gnoseological features of model-original relationships.

To create a model that is "fit for purpose," a wide range of tasks, knowledge, and skills need to be practiced. The steps in mathematical modeling are as follows [6, 7]: (1) Definitions: Problem(s) and models (model end-uses, model types, *etc*.) are identified. (2) Model conceptualization: The process of generating an initial notion through brainstorming with multiple stakeholders. It is crucial that the stakeholders comprise those who are intimately knowledgeable with the process or product, the modeling team, and an experienced modeler with the capacity to determine the veracity or relevance of stakeholder perspectives. (3) Model data requirements: In this stage of model development, several data concerns must be recognized for use in model development, model validation, and model deployment/use/reuse. (4) Model construction: Model construction is the process of expressing an abstracted description of a system so as to generate a simulation or model of its actual behavior. (5) Model simulation: Solution of the model equations. Modeling and simulation tools are used routinely to solve the set of model equations. Integrated conceptualization, modeling, and simulation systems are used by many modeling tools. (6) Model verification: Model verification jobs are distinct from model validation tasks in that they are fundamentally debug activities. Model verification is the process of comparing model implementation or equation code to the conceptual description or original model equations. It must be established whether the model accurately represents the conception. (7) Model validation: Model validation is a distinct activity from model verification, yet it is inextricably tied to it. A model that fails the verification test may have major validation issues. The following inquiry is referred to as validation: Is the model a reasonable representation of the real system? A variety of factors may play a role in answering this question. (8) Model deployment and maintenance: The process of acquiring the model for the application is only the beginning of the model's existence. The processes originally used to build the model continue to play an ongoing role in the model's long-term application. Several model classification approaches can be proposed for various purposes and applications.

2.1 Mathematical modeling approach classification

As depicted in **Figure 2**, mathematical modeling can be divided into three categories: mechanistic-white box, empirical-black box, and hybrid-gray box, and these, in turn, have subcategories [7–10].

Mechanistic-white box models provide some level of comprehension or explanation of the represented phenomena. A well-constructed mechanistic model is clear and extensible, essentially without limits. Mechanistic models are built on theories regarding how the system functions, the key components, and their interrelationships. These models make it possible to understand the input and output variables, as well as the factors involved in the modeling procedure. Mechanistic models tend to be more focused on research than on practical applications, though this is changing as our mechanistic models get more trustworthy. The evaluation of such models is crucial, albeit frequently and unavoidably subjective. Conventional mechanistic models are intricate and unfriendly.

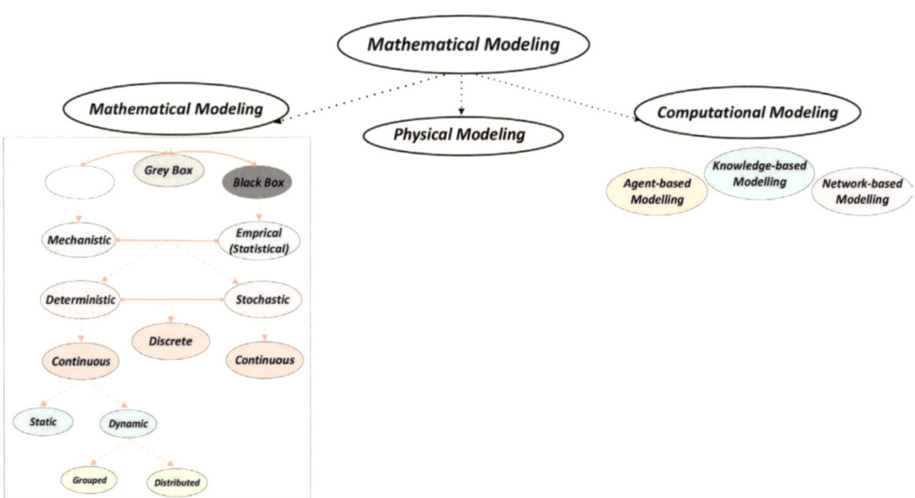

Figure 2.
Mathematical modeling approach classification.

Empirical-black box models describe a system's behaviors using mathematical or statistical equations without scientific content, constraints, or principles. This type may be the ideal form of model to develop depending on your goals. Its construction is based exclusively on experimental data and does not explain dynamic mechanisms, therefore its process is unknown. It is difficult to estimate an unknown function from observations of its values. The general recommendation in this regard is to estimate models of varying complexity and evaluate them using validation data. A regularized fit criterion is an effective approach to limit the flexibility of certain types of models. Finding a suitably flexible parameterization model is a critical challenge. Researchers typically use algorithms to predict physiological characteristics, such as support vector machines (SVM), back-propagation neural network (BPNN), artificial neural network (ANN), deep neural network (DNN), and the combination of wide and deep neural network (WDNN).

Hybrid-gray box models are intermediate models, defined as semi-empirical or semi-mechanistic. These models are also known as gray box or hybrid models since they combine empirical and mechanistic models. Both mechanical and empirical models can be deterministic or stochastic. Determinists make specific quantitative forecasts without regard for the accompanying probability distribution. This is appropriate in many circumstances; however, it may not be adequate for highly variable quantities or processes. Stochastic models, on the other hand, include a random element as part of the model so that the predictions have a distribution. Stochastic models are difficult to create, test, and falsify.

Both the deterministic and stochastic models, on the other hand, can take either a continuous or discrete form. It is referred to as a time-continuous model when a mathematical model is used to represent the relationship between continuous signals in time. Differential equations are a common tool for describing relationships of this kind. A discrete model is a type of model that establishes a direct relationship between the values of the signals at the times when they are sampled. Differential equations are the standard way to describe a model of this kind. Continuous models are categorized as dynamic and static. Dynamic models anticipate how quantities change over time; hence, a dynamic model is typically expressed as a set of ordinary differential equations with the independent variable time (t). On the other hand, static models do not produce predictions that are time-dependent. Finally, dynamic models can be categorized as grouped or distributed.

2.2 Discipline-domain approach classification of models

Two distinct academic disciplines study systems: process systems engineering (PSE) and energy economics (EE). Different factors such as theoretical foundations, technology aggregation, spatial and temporal scales, and model purposes all contribute to the categorization in the Discipline-Domain Approach of Models (seen in **Figure 3**). PSE typically simulates systems at the unit operation, processing plant, or supply chain scale. Each of these scales reflects a level of technology aggregation: At the plant size, different unit operations are aggregated to provide an overall conversion process, and the conversion process, together with the feedstock supply and product distribution network, are aggregated at the supply chain scale. The objective of modeling systems in PSE is to gain insight into their technological performance in order to facilitate optimal decision-making at the design, operations, and control levels. Consequently, the technological properties of the system's components are modeled endogenously (e.g., temperature, pressure, enthalpy,

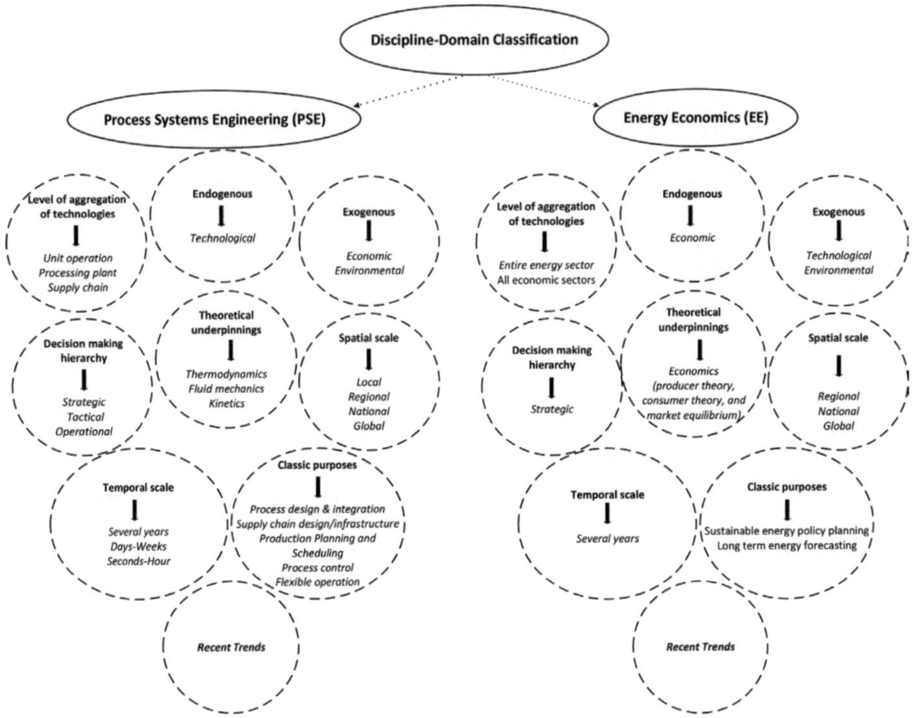

Figure 3.
Discipline-domain approach to the classification of models.

Gibbs free energy, process size). Depending on the optimization goal, economic, environmental, or social aspects (e.g., raw material prices, equipment prices, product demand, interest rates, global warming potential, resource depletion) can be modeled exogenously. At low levels of technology aggregation, modeling, and simulation have traditionally been utilized in PSE for strategic, tactical, and operational decision-making [8, 11, 12].

In contrast, energy economics approaches employ models with a high level of technology aggregation: All technologies including the entire energy sector on a regional, national, or worldwide scale may be analyzed (e.g., using bottom-up models), as well as other economic sectors including manufacturing, mining, and construction, *etc.* (e.g., in top-down models). Models of EE are grounded in economic principles including supply and demand, as well as market equilibrium. Thus, technological, environmental, or social elements may be treated exogenously, while the economic characteristics of the system's components are modeled endogenously. Modeling systems in EE have traditionally served to support policymaking at the regional, national, or global scales. Capital intensiveness, long gestation periods, and long payback periods all encourage long-term thinking, and addressing sustainability issues requires the planning of energy transformation pathways that may take decades to mature. These factors all contribute to the common use of long-time scales in EE models. However, the advent of intermittent-supply renewable technologies, as well as energy market deregulation, has resulted in the recent use of EE models for operational decision-making. The PSE models include supply chain component spatial distribution, while the EE model simply abstracts economic features. The PSE models just cover the supply chain, but the EE model often incorporates all national energy sources. Energy supply corporations and enterprises utilize PSE supply chain models for strategic, tactical, and operational decision-making, whereas national planners and public policy authorities use EE models [8, 13, 14].

2.3 Multi-scale approach classification of models

Multi-scale modeling and simulation is a rapidly growing subject of study that encompasses a variety of disciplines, including physics, biology, chemistry, mathematics, statistics, chemical engineering, mechanical engineering, and materials science. Multi-scale simulation allows for the connection of phenomena at multiple scales, including the quantum, molecular, mesoscopic, device, and plant scales. The primary challenge of multi-scale modeling is to integrate information from multiple simulation models in a consistent manner so that the behavior of the entire system can be predicted from its constituent models. Recent modeling advancements have made it possible to predict behavior extending from the quantum- and nano-scales to the macroscopic and continuum scales. Given the ability to predict the behavior of systems (such as materials) at multiple scales, the natural next step is to utilize these models to design systems at several scales, a process known as multiscale simulation-based design (MSBD). MSBD's primary challenge, in contrast to multi-scale modeling, is to effectively and efficiently utilize information generated by a diverse set of models that predict system behavior at multiple scales. System-level design is carried out concurrently with parts, sub-assemblies, and the corresponding material [15, 16].

Systems are designed using micro- to nano-scale analytical, experimental, and computational approaches to assess performance. Molecular-based approach to product and process engineering has been made possible by the significant rise in computer speed over the previous decades. Molecular and quantum modeling/simulations, such as molecular dynamics (MD), quantum Monte Carlo (QMC), and quantum mechanical (QM-ab initio and density functional theory (DFT) computations, have emerged as premier computing tools for science and engineering research. These calculations provide information to mesoscale continuum simulation system(s). In order to analyze the behavior of a system(s), bridges are constructed between models at multiple lengths and time scales in close conjunction with companion experiments. Properties at micro-scale surfaces depend on nanoscale interfaces. When building systems on different levels, this hierarchy must be taken into account. This increases the design's coupling, thereby increasing the problem's complexity. Although design complexity is a challenge in multiscale design, the benefit of designing systems at multiple scales is increased design freedom (i.e., greater flexibility in configuring the system to accomplish the desired behavior), which enables designers to achieve improved performance.

Figure 4 depicts the multiscale systems engineering concept, which focuses on the analysis and evaluation of systems across various scales of time, length, or both. Multi-scale systems analysis and evaluation necessitate the sequential or simultaneous use of key engineering components (modeling, design, synthesis, simulation, and optimization) across different time and length scales. This theme is made clear through (a) the application of coarse-grained models at higher levels, such as the plant or supply chain level, that capture phenomena occurring at lower molecular scales, (b) the incorporation of results obtained from smaller scales that are used as information in larger spatial and length scales, or (c) the construction of a holistic model that incorporates data from multiple temporal and spatial dimensions.

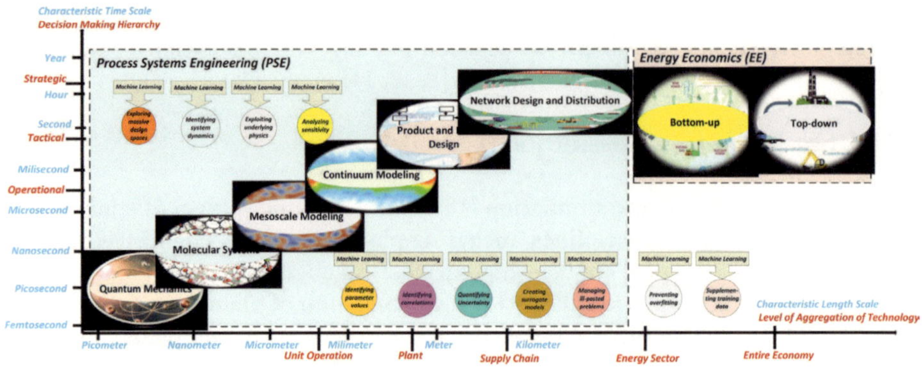

Figure 4.
Machine learning-assisted multiscale modeling across characteristic length vs time scale and level of technological aggregation vs decision making.

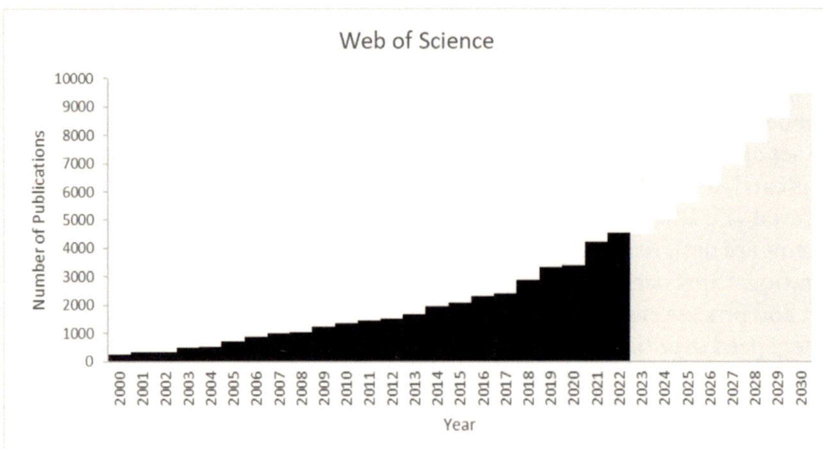

Figure 5.
The existing (black) and projected (gray) trend for research articles featuring the words multi-scale or multiscale in the title of the published literature.

There has been a significant rise in the number of articles published about multi-scale research during the past two decades. **Figure 5** shows the trend over time for the percentage of articles in Web of Science with the terms "multi-scale" or "multiscale" in the title. Based on the available data, a 100%" increase in total publication is projected by 2030.

2.3.1 Challenges

Successful modeling of multiscale systems must overcome the following challenges [17, 18]:

1. Accuracy-computational cost balance: One of the most difficult challenges in multiscale systems is balancing the need for accuracy with the computational expense. In general, employing a smaller-scale model to anticipate system perfor-

mance provides a more accurate depiction of the system. Running these models at smaller scales in a broad enough domain to capture greater impacts, on the other hand, is computationally expensive.

2. Appropriate system components to accurately model component interactions: Multiple components are frequently present in multiscale systems. All of these system components interact with one another. By reducing the number of modeled system components and interactions, the number of required calculations decreases, but the accuracy must also be improved. Therefore, modelers at each scale must take into account the appropriate system components in order to accurately simulate component interactions.

3. Simulating necessary physical phenomena pertinent to the system: Existing multiscale systems are predominantly multi-physical in nature. Different physical principles and mathematical equations govern these physical phenomena. These phenomena may be dependent on or interdependent with one another. The influence of diverse phenomena on the precision of the prediction of the overall system behavior varies. Consequently, in order to obtain a reasonable understanding of the system, it is necessary to model phenomena that are interrelated.

4. Modeling how scales interact with each other and connecting them so that they can be compatible physically: Multiscale modeling requires distinct physical principles, mathematical equations, and parameters to simulate phenomena at different length and time scales. Assumptions vary for each level. These models provide distinct insights into system behavior and must be merged to produce consistent system behavior. Model integration involves consistent mathematical and physical connectivity between scales. Because scales depend on each other, it's crucial to understand their relationships. Therefore, the success of a multiscale model depends on accurately simulating the relationships among the various scales, components, and physical processes involved.

5. Selection of appropriate models and model parameters at each scale: Modifying the model's scope and assumptions allows for the creation of simulation models with varying degrees of authenticity. A material system, for instance, could be modeled in one, two, or three dimensions. Different models may be appropriate for forecasting the behavior of the system under investigation. It is critical to choose the appropriate collection of models and assumptions. However, for multiscale systems, the condition is crucial since it must be examined at several scales, with information from one model feeding into another. The applicability of models is also determined by the compatibility of assumptions made in various models. Appropriate model selection has a significant impact on model correctness and execution time.

6. Managing large amounts of data at multiple abstraction levels: Each finer scale necessitates a more sophisticated theory and a deeper understanding of the system. Integrating information at various scales necessitates the use of software infrastructure as well as mathematical issues. Multiscale modeling entails issues such as synchronization of information generated by models at various scales, long run times, load balancing, capturing information at various levels of

abstraction in a consistent database, and integrating distributed computational models and hardware resources.

7. Uncertainty quantification and management, as well as its dissemination: Uncertainty quantification is essential for efficiently utilizing simulation model data. It is also critical to have information on the degree to which models may be trusted. Uncertainty is extremely prominent in multiscale models due to interactions between events at various scales, and quantifying this uncertainty in the models is challenging. Multiscale simulation models communicate uncertainty and information between models. The system-level simulation model may not be acceptable even though individual model uncertainty bounds are acceptable.

8. Targeted refinement of models: The accuracy of the overall multiscale simulation model is determined by the quality of the comprising models at different scales and how uncertainty is magnified owing to information flow from one model to another. As a result, in order to increase the entire model's accuracy, it is critical to identify the model with the greatest impact on total uncertainty and then refine that model in a focused manner.

9. Adaptive selection of details and resolution: Although uncertainty is an important part of multiscale modeling and should be managed, many multiscale models can be greatly reduced, lowering model execution time while maintaining accuracy. The goal of multiscale modeling is to take advantage of such scenarios and to choose appropriate levels of detail in the models.

2.3.2 Approaches

Parameterization and concurrent coupling are some of the approaches currently used for multiscale modeling to address some of these challenges.

Parameterization is a method of capturing information from lower-level models in the form of a set of parameters and their values. The parameters can be empirical or semi-empirical, and they can be used to estimate average physics behavior on a smaller scale. At some level, all models contain parameterization. No model can completely simulate a phenomenon using the first principles. The advantage of parameterization is that it is simple to account for occurrences at lesser scales; the downside is that it is inaccurate.

Coupling is the method of "on-the-fly" utilizing a model at one scale while performing calculations at another one. Because of this, the total simulation needs to make dynamic use of a hierarchy of increasingly detailed models at decreasing scales. As a direct result of this requirement, high-performance computer options are required. As a result, the availability of suitable computational tools is the primary determinant of the capability to match models across a wide range of sizes. Improved overall model accuracy comes at the tradeoff of increased complexity and computing cost when coupling multiscale models together.

There is a practical limit to the complexity of problems that can be addressed, even by the most powerful computers now in existence. Therefore, it is important to fulfill a balance between accuracy and computing cost when integrating multiscale models through the use of an optimal combination of parameterization and coupling.

3. Integration of multi-scale modeling with machine learning

3.1 Machine learning

Machine learning (ML) is becoming widely acknowledged as a promising technology in a wide range of scientific applications. The idea of machine learning (ML) refers to a type of data-driven programming that can automatically learn programs by observing existing ones. Machine learning (ML) is a subfield of artificial intelligence (AI) that focuses on developing methods to simulate human intelligence in machines by simulating perception, speech, movement, and logical reasoning. ML has a close relationship with statistics, while statistics are more concerned with finding the truth in the underlying data, ML is more concerned with task performance. ML is also closely related to optimization and control theory [4].

Although there are numerous types of ML techniques, the majority of ML algorithms are founded on only three elements: (1) A model or hypothesis class that defines the ML algorithm's functionalities. This can be thought of informally as the skeleton of the program that the algorithm generates. Frequently, these models have unrestricted parameters that can be tailored to the task at hand. (2) An objective function or loss function that outlines the behavior that the model should be aiming for and (3) A method for training or optimizing a model that defines how its parameters should be selected or modified to maximize performance on a given target.

Supervised-unsupervised-reinforcement learning, online learning, and transfer learning are the notable ML paradigms applicable to various contexts in which ML is utilized. Supervised learning describes methods that use a dataset of input/output pairs to learn a function. Supervised learning excels at image classification, automated speech recognition, and machine translation. In many circumstances, obtaining the labeled data is too expensive to employ supervised learning, or the system of interest has a decision-making process that cannot be well described by single input/output pairs. Unsupervised learning refers to techniques that seek structure within a set of inputs. Unsupervised methods are useful for analyzing and partitioning data, but the algorithm designer usually chooses key attributes, so the learned outputs may be artifacts of the algorithm rather than data attributes. The term "reinforcement learning" is used to describe a set of methods that, through the use of a sequential environment, seek to improve the performance of an agent. Instead of working over a predetermined dataset, as is the case in supervised and unsupervised learning paradigms, RL algorithms are deployed in a context where the method can influence the future state of some systems. Adaptive control and this setup share a lot of common features, but their structural assumptions about the system differ. Agent-based modeling (ABM) is similar to RL, except ABM requires manual specification of behavioral norms, while RL learns them automatically [19, 20].

In the online (or streaming) context, data points arrive one at a time, necessitating a prediction from the algorithm before the next data point is received. Because online learning models' parameters are updated as data is analyzed, they frequently necessitate different assessment criteria than offline algorithms (discussed above). The transfer learning paradigm, as well as the related disciplines of multi-task and meta-learning, focuses on the ability of machine learning models to adapt to new tasks. Modern ML algorithms are data-hungry, meaning they require a prohibitively huge amount of data for training to function well. Transfer learning techniques use a limited amount of data from the new task to fine-tune pre-trained ML models for low-data scenarios [4, 21].

On the other hand, deep learning, decision trees, support vector machines, ensemble learning, Bayesian models, and Gaussian processes are the major algorithmic approaches used within ML. One of the most popular ML methods, deep learning, uses layers of linear and nonlinear functions and gradient-based backp ropagation methods optimize these layers. Back propagation-trained composable layers can express complex functions and generalize effectively to fresh data. Generic feed forward networks, convolutional networks, recurrent networks, and graph networks are also specialized deep learning models. Deep learning is used in supervised, unsupervised, and reinforcement learning ML paradigms. Decision trees are a type of model that iteratively divides incoming data into subsets based on the relative relevance of certain features. Decision tree approaches evaluate patterns in input data to create trees with decision nodes, which can be followed to forecast a data point based on feature values. These optimization strategies use multiple different loss functions. One of the earliest ML techniques, decision trees have recently gained popularity because of its predictive accuracy and ease of interpretation when combined with ensemble methods. Another sort of traditional ML model, support vector machines (SVMs) use either a linear model class or its nonlinear extension, the kernel hypothesis class, to make predictions. These strategies often employ a (regularized) hinge loss as their loss function, making them a so-called max-margin classifier in terms of geometry. The resulting optimization problem can be efficiently handled on a global scale, with different approaches taken in practice for different instances of the problem. Although many of the aforementioned methods can fulfill the requirements, using a group, or ensemble, of ML algorithms is a frequent approach in situations where a superior level of performance is required for a given task. The outputs of these algorithms can be averaged together by using weighting. Bayesian learning, in the context of machine learning, is a set of techniques for modeling uncertainty in data and model parameters. Bayesian approaches are prevalent in many areas of ML, and Bayesian versions of the majority of the aforementioned methods are feasible; however, Gaussian processes (GPs) are the standard approach in this class. GPs (nonlinear regression models) use covariance or kernel functions to measure point similarity. GPs model predictions and estimate uncertainty based on the model's previous data and this makes GPs useful for Bayesian optimization [22].

3.2 Multiscale modeling meets machine learning

ML is a potent method for integrating multimodality and multifidelity data and revealing correlations between interconnected phenomena presents a unique opportunity in this regard. However, machine learning typically performs poorly with sparse data, disregards the fundamental laws of physics, and can lead to ill-posed problems or non-physical solutions. Classical physics-based simulation appears to be irreplaceable in this discipline. Multiscale modeling is an effective method for integrating multiscale, multiphysics data and identifying the mechanisms that explain the emergence of function. However, multiscale modeling alone frequently fails to effectively combine large datasets from various sources and resolutions. In this context, machine learning and multiscale modeling can naturally complement one another to produce robust predictive models that incorporate the underlying physics in order to manage ill-posed problems and investigate vast design spaces. Machine learning can incorporate physics-based knowledge in the form of governing equations, boundary

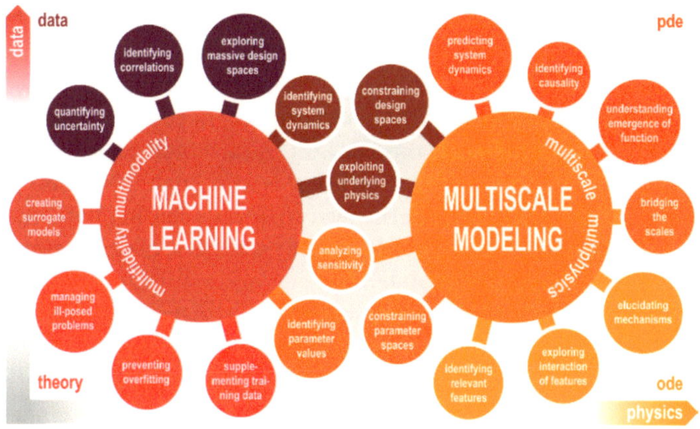

Figure 6.
The integration of machine learning and multiscale modeling on the parameter level [23].

conditions, or constraints to manage poorly defined problems and handle sparse and noisy data. On the other hand, to bridge the scales and comprehend the evolution of function, multiscale modeling can include machine learning to generate surrogate models, determine system dynamics and parameters, examine sensitivities, and quantify uncertainty [23].

Figure 6 depicts the integration of machine learning and multiscale modeling at the parameter level through constraining their spaces, identifying values, and analyzing their sensitivity, and at the system level through exploiting the underlying physics, constraining design spaces, and identifying system dynamics. Machine learning provides the necessary tools for supplementing training data, avoiding overfitting, dealing with ill-posed problems, developing surrogate models, and measuring uncertainty. Multiscale modeling incorporates fundamental physics to discover relevant features, investigate their interactions, elucidate mechanisms, bridge scales, and comprehend the genesis of function.

4. Applications of integrated MMS and ML for design and development of novel technologies, systems, and processes

In this section, we will list some application areas of integrated MMS and ML for the design and development of novel technologies, systems, and processes.

Biological, biomedical, and behavioral sciences: More information than ever before is being gathered in the fields of biological, biomedical, and behavioral sciences thanks to recent technological advances. In order to improve people's health, we urgently need more efficient methods of analyzing and interpreting this data. In the biological, biomedical, and behavioral sciences, it is a significant challenge to comprehend systems for which the underlying data are incomplete, and physics are not yet completely understood. In other words, with a complete set of high-resolution data, machine learning could be used to explore design spaces and identify correlations; with a validated and calibrated set of physics equations and material parameters, multiscale modeling could be used to predict system dynamics and identify causality [24].

The ultimate objective of incorporating machine learning and multiscale modeling is to provide quantitatively predictive insight into biological systems. Integrating machine learning and multiscale modeling has several applications in biological, biomedical, and behavioral systems. Metabolic networks, microbiology, immunology, cancer, neuroscience, biomechanics, and public health are some examples of these application areas. Drug side effects, for instance, have been studied using machine learning and genome-scale models [25]. Additionally, a recent study used machine learning in conjunction with multi-omics data, including proteomics and metabolomics, to accurately predict pathway dynamics, yielding both qualitative and quantitative predictions that can be used to direct synthetic biology projects [26]. To discover the metabolic dynamics modeled by coupled nonlinear ordinary differential equations that best suit the available time series data, supervised learning approaches are frequently employed. Using neural networks to map genotype to phenotype to bridge scales in cancer progression in specific microenvironments is another application [27]. Principal component analysis and neural networks have been used more frequently to study aging, Alzheimer's disease, chaotic dynamics of epileptic seizures, and memory formation [28], among other topics. Predicting response functions, such as stress–strain relations or cell-scale rules in continuum theories of development and remodeling [29], is where machine learning has the most potential to impact the field of biomechanics. Example: a recent study characterized the dynamic growth and remodeling during heart failure at the molecular, cellular, and cardiac levels by integrating machine learning and multi-scale modeling. The ordinary differential equations of disease spreading have recently been applied to describe the prion-like spread of neurodegenerative diseases, where the parameters could be found via machine learning from magnetic resonance imaging [30].

Material sciences: Many of the twenty-first century's difficulties, such as low-carbon energy and sustainability, are mostly material-related. Materials having precise chemical and physical qualities for effective energy storage and conversion are desperately needed to achieve human society's long-term development. For a long time, the development of innovative materials has been based on trial-and-error, implying a lengthy schedule and high cost that cannot match the needs for more advanced materials. Because of advances in theoretical and computational chemistry, quantum mechanics (QM) and molecular mechanics have matured as methodologies for predicting quantitative structure–property connections prior to investigations. With rapid advances in high-performance computers, high-throughput computational screening has greatly sped materials science research, allowing the properties of thousands of molecules to be computed. Density functional theory (DFT) is widely used for the computation of material structures and properties, and it has accelerated the development of materials databases with calculated properties for a large number of systems, including the Materials Project (MP) database, the AFLOWLIB consortium, the Open Quantum Materials Database (OQMD), and MaterialGo (MG) [31–34]. Researchers can use cutting-edge supercomputers and algorithms to calculate molecules with thousands of interacting ions and electrons using QM approaches. The enormous computing cost of QM-based approaches, on the other hand, limits their use to large-scale complex systems. Furthermore, using QM approaches to exhaust all potential systems is unfeasible.

The Materials Genome Initiative (MGI) is bringing huge materials data, and greater attempts have been made to collect materials attributes and develop materials databases. Accelerating materials design requires massive data management and use. Current materials research requires fast and effective analysis of vast data to identify hidden rules. With the rapid increase of materials databases, the progressive adoption of ML toolkits like TensorFlow, Pytorch, and scikit-learn, the creation of workflow

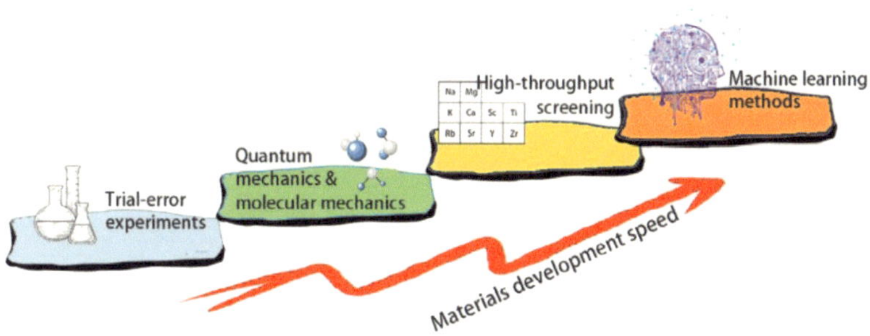

Figure 7.
Development in methods to accelerate new materials discovery.

toolkits like Atomate, and the progress of algorithms, ML is being used more in materials research [35, 36]. Using big data, ML approaches have generated several discoveries in energy storage and conversion materials like catalysts and batteries (**Figure 7**) [37–39].

Mathematical sciences: Models that are based on ordinary or partial differential equations might be as simple as a single equation or as complex as a vast system of equations or stochastic differential equations. This implies that the number of parameters is typically large and can easily reach hundreds or more. There are many applications of ordinary and partial differential equations that integrate machine learning and multiscale modeling for systems.

Energy systems: Integrated MMS and ML have huge application potential for the development and operation of sustainable energy systems. The aforementioned global challenges can be divided into three broad categories: (1) energy security and increasing energy demand, (2) energy affordability, and (3) cleaner energy generation (reducing carbon emissions) [40, 41]. In the current state, to meet the increasing energy demand in sustainable, stable, and safe manners, synergic energy supply/demand phenomena in economic development have become increasingly prominent. Producing clean and cheap power by eliminating the economic and energy penalties requires either the development of new technologies or the improvement of existing energy systems based on sustainability standards and alternative fuels [42, 43]. However, the development of sustainable energy systems, do not negatively influence the economic and societal benefits, is the challenging global target of the energy sector. The incorporation of sustainability into the design of energy systems helps to attain this goal by avoiding or limiting negative impacts. However, monitoring sustainability performance and decision-making are critical and challenging processes for defining sustainability levels. Thus, the development of a methodological approach with proper sustainability indicators ensures the capability of gathering and abstracting complex process operations and provides straightforward analysis and communication [44].

As can be seen in **Figure 8**, each of the aforementioned global issues propels advancement in a variety of emerging and mature technology domains and research fields. Each of these technical domains and research fields will necessitate the discovery and development of ideas on smaller scales, followed by an assessment of their commercialization potential, integration of sustainability, and lastly supply chain optimization. This sequential flow displays the multi-scale nature of the described challenges, technology domains, and research topics, as well as the opportunity for multi-scale systems engineering to have a beneficial impact. To sum up, to eliminate

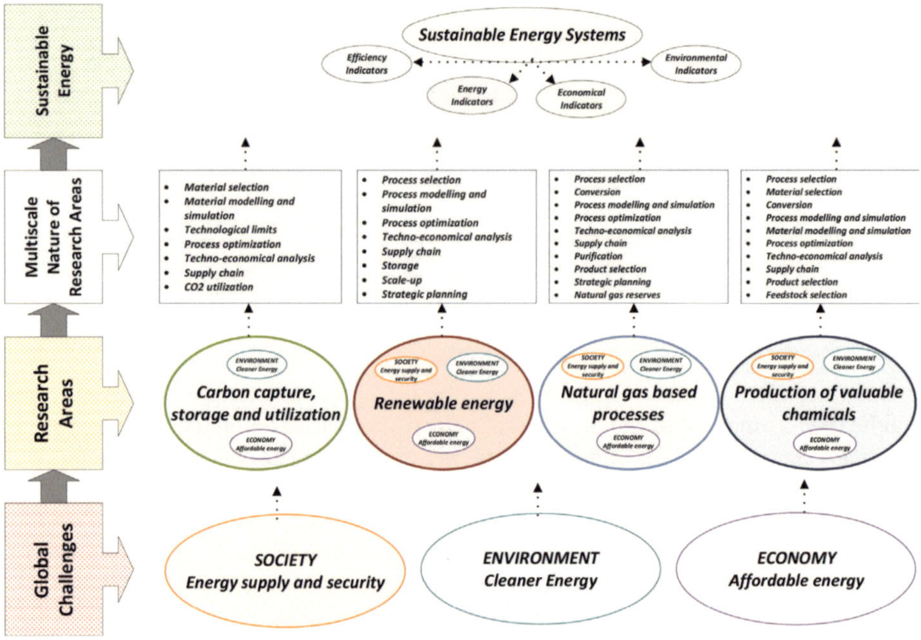

Figure 8.
The summary of the sequential flow from global challenges to sustainable energy systems.

the existing problems and reach the global goals, the research activities need a robust, efficient, reliable methodological approach that spans multi-scale modeling and simulation (MMS)-based analysis-design-synthesis-optimization, and integration of machine learning (ML). To bridge the gap between atomistic, mesoscopic, and macroscopic systems of interest, powerful computational techniques have been developed. As a result of this development, multi-scale systems engineering studies have produced significant outcomes for the oil, natural gas, and coal industries, as well as the renewable energy and environmental fields.

Electricity supply and demand, optimizing energy systems, maximizing renewable power generation, reducing fossil fuel impacts, predictive maintenance and fault detection, planning sustainable energy infrastructure, managing energy systems data, developing next-generation sustainable energy technologies, and informing policy are some of the application areas of integrated MMS and ML in energy systems.

Digital twin: The high-tech industry is growing rapidly, and new technologies like Digital Twins (DT) are emerging. DT technology is an innovative interactive system that can control interactions between the actual and virtual worlds. It's a center of learning on a worldwide scale. It is integrated with other technologies and used in various industries such as smart factories in industrial production, digital model of life in the medical field, smart city building, aerospace field, immersive shopping in commercial field, *etc*. Digital Twins are largely explained conceptually, with few practical applications. DT is a technology that incorporates several disciplines, and it presently lacks a unified definition, which is still being developed and evolved. However, according to the NASA definition, "DT is a comprehensive multi-physical, multi-scale, probabilistic simulation system for vehicles or systems. It uses the best physical model to describe the historical use of equipment to reflect the life of its corresponding physical equipment."

DT has the following characteristics: (1) Concentration: Digital mainframes retain all data collected throughout a physical system's lifetime under centralized and unified control, streamlining data exchange in both directions. (2) Integrity: For complex systems, DT integrates all subsystems, which is the foundation for high-precision modeling, while real-time data monitoring can further enrich and expand the model, allowing the model to incorporate all system knowledge and (3) Models can be dynamically updated based on sensor data characterizing the physical system's surroundings or status, updated models can dynamically direct actual operation, and real-time interaction between physical systems and digital models allows models to grow and evolve. Integrated MMS and ML is a highly effective and promising tools that utilized a variety of techniques to create a Digital Twin.

Of course, the application areas are not limited as discussed above, the areas can be extended and detailed.

5. Conclusions

At the intersection of machine learning and multiscale modeling, a plethora of fascinating new applications are being created. Learning by machine and modeling at multiple scales naturally complement one another and are mutually beneficial to one another. Both machine learning and multiscale modeling have the ability to forecast system dynamics, which allows for the identification of causation. Machine learning can search enormous design spaces to find correlations. The combination of machine learning with multiscale modeling has the potential to have a significant impact in a wide variety of fields, including the biological, biomedical, and behavioral sciences, the material sciences, the mathematical sciences, energy systems, digital twins, and other related fields. Applications unquestionably become more complex as time passes, necessitating an increased awareness of the inherent constraints of overfitting and data bias on the part of developers. The need for increased transparency, rigor, and repeatability will be a significant obstacle to overcome in order to make headway in this subject.

Acknowledgements

The author acknowledges the support of the Qatar University grants QUST-1-CENG-2023-914 and QUST-1-CENG-2023-915.

Author details

Seçkin Karagöz
Department of Chemical Engineering, College of Engineering, Qatar University, Doha, Qatar

*Address all correspondence to: ekleczyk@targetrwe.com

References

[1] Salciccioli M, Stamatakis M, Caratzoulas S, Vlachos DG. A review of multiscale modeling of metal-catalyzed reactions: Mechanism development for complexity and emergent behavior. Chemical Engineering Science. 2011;**66**(19):4319-4355. DOI: 10.1016/j.ces.2011.05.050

[2] Floudas CA, Niziolek AM, Onel O, Matthews LR. Multi-scale systems engineering for energy and the environment: Challenges and opportunities. AIChE Journal. 2016;**62**(3):602-623. DOI: 10.1002/aic.15151

[3] Vlachos DG. A review of multiscale analysis: Examples from systems biology, materials engineering, and other fluid–surface interacting systems. Advances in Chemical Engineering. 2005;**30**:1-61. DOI: 10.1016/S0065-2377(05)30001-9

[4] Donti PL, Kolter JZ. Machine learning for sustainable energy systems. Annual Review of Environment and Resources. 2021;**46**:719-747. DOI: 10.1146/annurev-environ-020220-061831

[5] Volkova VN, Kozlov VN, Mager VE, Chernenkaya LV. Classification of methods and models in system analysis. In: 2017 XX IEEE International Conference on Soft Computing and Measurements (SCM). St. Petersburg, Russia: IEEE; 2017. pp. 183-186

[6] Heitzig M, Sin G, Sales-Cruz M, Glarborg P, Gani R. Computer-aided modeling framework for efficient model development, analysis, and identification: Combustion and reactor modeling. Industrial and Engineering Chemistry Research. 2011;**50**:5253-5265. DOI: 10.1021/ie101393q

[7] García-Rodríguez del LC, Prado-Olivarez J, Guzmán-Cruz R, Rodríguez-Licea MA, Barranco-Gutiérrez AI, Perez-Pinal FJ, et al. Mathematical modeling to estimate photosynthesis: A state of the art. Applied Sciences. 2022;**12**:5537. DOI: 10.3390/app12115537

[8] Subramanian ASR, Gundersen T, Adams TA. Modeling and simulation of energy systems: A review. Process. 2018;**6**:238. DOI: 10.3390/pr6120238

[9] Yoro KO, Daramola MO, Sekoai PT, Wilson UN, Eterigho-Ikelegbe O. Update on current approaches, challenges, and prospects of modeling and simulation in renewable and sustainable energy systems. Renewable and Sustainable Energy Reviews. 2021;**150**:111506. DOI: 10.1016/j.rser.2021.111506

[10] Guidotti R, Monreale A, Ruggieri S, Turini F, Giannotti F, Pedreschi D. A survey of methods for explaining black box models. ACM Computing Surveys. 2019;**51**:1-42. DOI: 10.1145/3236009

[11] Foley AM, Gallachóir BÓ, Hur J, Baldick R, McKeogh EJ. A strategic review of electricity systems models. Energy. 2010;**35**:4522-4530. DOI: 10.1016/j.energy.2010.03.057

[12] Ventosa M, Baíllo Á, Ramos A, Rivier M. Electricity market modeling trends. Energy Policy. 2005;**33**:897-913. DOI: 10.1016/j.enpol.2003.10.013

[13] Elia JA, Baliban RC, Floudas CA. Nationwide energy supply chain analysis for hybrid feedstock processes with significant CO2 emissions reduction. AICHE Journal. 2012;**58**:2142-2154. DOI: 10.1002/aic.13842

[14] Lopion P, Markewitz P, Robinius M, Stolten D. A review of current challenges and trends in energy systems modeling. Renewable and Sustainable Energy Reviews. 2018;**96**:156-166. DOI: 10.1016/j.rser.2018.07.045

[15] Olson GB. Computational design of hierarchically structured materials. Science. 1997;**277**(5330):1237-1242. DOI: 10.1126/science.277.5330.1237

[16] Olsen GB. Pathways of discovery designing a new material world. Science. 2000;**228**(12):933-998

[17] Miller RE. Direct coupling of atomistic and continuum mechanics in computational materials science. International Journal of Computational Materials Science. 2003;**1**(1):57-72. DOI: 10.1615/IntJMultCompEng.v1.i1.60

[18] Rudd RE, Broughton JQ. Concurrent coupling of length scales in solid state systems. Physica Status Solidi (b). 2000;**217**:251-291. DOI: 10.1002/3527603107.ch11

[19] Busoniu L, de Bruin T, Toli'c D, Kober J, Palunko I. Reinforcement learning for control: Performance, stability, and deep approximators. Annual Reviews in Control. 2018;**46**:8-28. DOI: 10.1016/j.arcontrol.2018.09.005

[20] Silver D, Huang A, Maddison CJ, Guez A, Sifre L, et al. Mastering the game of go with deep neural networks and tree search. Nature. 2016;**529**(7587):484-489. DOI: 10.1038/nature16961

[21] Zhuang F, Qi Z, Duan K, Xi D, Zhu Y, et al. A comprehensive survey on transfer learning. Proceedings of the IEEE. 2020;**109**:43-76. DOI: 10.1109/JPROC.2020.3004555

[22] Goodfellow I, Bengio Y, Courville A. Deep Learning. Cambridge, MA: MIT Press; 2016

[23] Alber M, Buganza Tepole A, Cannon WR, De S, Dura-Bernal S, Garikipati K, et al. Integrating machine learning and multiscale modeling—Perspectives, challenges, and opportunities in the biological, biomedical, and behavioral sciences. NPJ Digital Medicine. 2019;**2**(1):115. DOI: 10.1038/s41746-019-0193-y

[24] Lytton WW et al. Multiscale modeling in the clinic: Diseases of the brain and nervous system. Brain Information. 2017;**4**:219-230. DOI: 10.1007/s40708-017-0067-5

[25] Shaked I, Oberhardt MA, Atias N, Sharan R, Ruppin E. Metabolic network prediction of drug side effects. Cell Systems. 2018;**2**:209213. DOI: 10.1016/j.cels.2016.03.001

[26] Costello Z, Martin HG. A machine learning approach to predict metabolic pathway dynamics from time-series multiomics data. NPJ Systems Biology Applications. 2018;**4**:19. DOI: 10.1038/s41540-018-0054-3

[27] Gerlee P, Kim E, Anderson ARA. Bridging scales in cancer progression: Mapping genotype to phenotype using neural networks. Seminars in Cancer Biology. 2015;**30**:3041. DOI: 10.1016/j.semcancer.2014.04.013

[28] Ognjanovski N, Broussard C, Zochowski M, Aton SJ. Hippocampal network oscillations drive memory consolidation in the absence of sleep. Cerebral Cortex. 2018;**28**(10):1-13. DOI: 10.1093/cercor/bhy174

[29] Ambrosi D, Ateshian GA, Arruda EM, Cowin SC, Dumais J, Goriely A, et al. Perspectives on biological growth and remodeling. Journal of the Mechanics and Physics of Solids. 2011;**59**:863-883. DOI: 10.1016/j.jmps.2010.12.011

[30] Weickenmeier J, Jucker M, Goriely A, Kuhl E. A physics-based model explains the prion-like features of neurodegeneration in Alzheimers disease, Parkinsons disease, and amyotrophic lateral sclerosis. Journal of the Mechanics and Physics of Solids. 2019;**124**:264-281. DOI: 10.1016/j.jmps.2018.10.013

[31] Jain A, Ong SP, Hautier G, et al. Commentary: The materials project: A materials genome approach to accelerating materials innovation. APL Materials. 2013;**1**:011002. DOI: 10.1063/1.4812323

[32] Calderon CE, Plata JJ, Toher C, et al. The AFLOW standard for high-throughput materials science calculations. Computational Materials Science. 2015;**108**:233-238. DOI: 10.1016/j.commatsci.2015.07.019

[33] Kirklin S, Saal JE, Meredig B, et al. The open quantum materials database (OQMD): Assessing the accuracy of DFT formation energies. NPJ Computational Materials. 2015;**1**:15010. DOI: 10.1038/npjcompumats.2015.10

[34] Jie J, Weng M, Li S, et al. A new MaterialGo database and its comparison with other high-throughput electronic structure databases for their predicted energy band gaps. Science China Technological Sciences. 2019;**62**:1423-1430. DOI: 10.1007/s11431-11019-19514-11435

[35] Pedregosa F, Varoquaux G, Gramfort A, et al. Scikit-learn: machine learning in python. Journal of Machine Learning Research. 2011;**12**:2825-2830

[36] Mathew K, Montoya JH, Faghaninia A, et al. Atomate: a highlevel interface to generate, execute, and analyze computational materials science workflows. Computational Materials Science. 2017;**139**:140-152. DOI: 10.1016/j.commatsci.2017.07.030

[37] Beck DAC, Carothers JM, Subramanian VR, Pfaendtner J. Data science: accelerating innovation and discovery in chemical engineering. AIChE Journal. 2016;**62**:1402-1416. DOI: 10.1002/aic.15192

[38] Fujimura K, Seko A, Koyama Y, et al. Accelerated materials design of lithium superionic conductors based on first principles calculations and machine learning algorithms. Advanced Energy Materials. 2013;**3**:980-985. DOI: 10.1002/aenm.201300060

[39] SendekAD, Yang Q, Cubuk ED, et al. Holistic computational structure screening of more than 12000 candidates for solid lithium-ion conductor materials. Energy & Environmental Science. 2017;**10**:306-320. DOI: 10.1039/C6EE02697D

[40] Pichardo PA, Karagöz S, Manousiouthakis VI, Tsotsis T, Ciora R. Techno-economic analysis of an intensified integrated gasification combined cycle (IGCC) power plant featuring a combined membrane reactor-adsorptive reactor (MR-AR) system. Industrial & Engineering Chemistry Research. 2019;**59**:2430-2440. DOI: 10.1021/acs.iecr.9b02027

[41] Karagöz S, da Cruz FE, Tsotsis TT, Manousiouthakis VI. Multi-scale membrane reactor (MR) modeling and simulation for the water gas shift reaction. Chemical Engineering and Processing-Process Intensification. 2018;**133**:245-262. DOI: 10.1016/j.cep.2018.09.012

[42] Karagöz S, Chen H, Cao M, Tsotsis TT, Manousiouthakis VI. Multiscale model based design of an energy-intensified novel adsorptive

reactor process for the water gas shift reaction. AICHE Journal. 2019;**65**(7):e16608. DOI: 10.1002/aic.16608

[43] Karagöz S, Tsotsis TT, Manousiouthakis VI. Multi-scale model based design of membrane reactor/separator processes for intensified hydrogen production through the water gas shift reaction. International Journal of Hydrogen Energy. 2020;**45**(12):7339-7353. DOI: 10.1016/j.ijhydene.2019.05.118

[44] Karagöz S. A methodological sustainability assessment to process intensification (MSAtoPI) by reactive-separation systems. Fuel. 2023;**348**:128562. DOI: 10.1016/j.fuel.2023.128562

Chapter 11

Table Tennis and Physics

Ralf Schneider, Lars Lewerentz, Stefan Kemnitz and Christian Schultz

Abstract

Table tennis is a fascinating sport with a lot of physics elements. This chapter will concentrate on the forces acting on a table tennis ball. Insights from molecular dynamics simulations clarify the basic properties of table tennis balls interacting with surfaces and their high coefficient of restitution. The table tennis ball trajectory is discussed considering the different force balance contributions. The sum of the gravitational force and the forces from aerodynamics, namely drag and lift, determine the flight path. Extensive numerical modeling is used to study the impact of changes in ball size and net height to the game characteristics. Half a billion different initial conditions like hitting location, initial spin and velocities were analyzed to reach sufficient statistical significance for the different cases. An advanced statistical analysis of the database generated by the simulation is presented.

Keywords: physics modeling, trajectory analysis, statistics, rule changes, table tennis

1. Introduction

In this chapter, some aspects of physics in table tennis will be discussed. Insights from molecular dynamics and finite element simulations clarify the basic properties of table tennis balls interacting with surfaces and their high coefficient of restitution. The table tennis ball trajectory is calculated considering the different force balance contributions. Statistical analysis allow a better prediction of the consequences on the game for possible rule changes.

Historically, there exists a strong link between physics and table tennis. Werner Heisenberg was an enthusiastic table tennis player. He is one of the founders of quantum physics and established around 1930 in his working group at the University of Leipzig regular meetings where scientific discussions, which could not be resolved, were decided by table tennis [1]. Other members of the group were less motivated, for example Werner Hückel, a famous chemist, did not like table tennis, "because one had to pick up the ball in the corners of the room" [2]. Werner Heisenberg was once beaten by an Asian PhD student. This motivated him so much that on the boat trip to China in 1929 for some lectures he spent most of the time to improve his table tennis skills. He intended to be able to compete against the Chinese players and wanted no further defeat by his PhD students anymore [3].

2. Interaction of table tennis balls with surfaces

In this section, physics effects of the interaction of table tennis balls with surfaces will be discussed. A table tennis ball can be described as a hollow sphere with uniform material properties. Vibrational mode frequencies and coefficient of restitution (CoR) were studied for a hollow elastic sphere with molecular dynamics simulations of spherical clusters of up to 13,500 atoms [4].

In the molecular dynamics simulations, a hollow sphere was studied. This thin but rigid shell represents a table tennis ball and experiences a gentle, long duration of the collision, which does not allow vibrational modes to be excited. A thicker shell shows some excitation of vibrational modes, which corresponds to an energy loss. Collision time is speed-dependent for a solid sphere, but it appears to become independent of speed for a thin spherical shell. Acceleration and consequently force are linearly proportional to compression c for a thin shell sphere. This is much different from the situation for a solid sphere where one gets in agreement with Hertz a force proportional to compression $c^{3/2}$ [5].

Scaling MD simulations to a table tennis ball, the duration of its bounce with a hard table should be 0.7 ms with dominant Fourier components up to about 1400 Hz. The gap between this frequency and the lowest vibrational frequency of the ball (5300 Hz) means that excitation of vibrational modes is not an important mechanism for energy loss, which explains its high CoR with values above 0.8. The CoR values reduce as the drop height is increased, i.e. as the impact speed increases [6].

A spinning ball struck head-on does not rebound along its incident path [7, 8]. The rebound angle and spin depend in a nontrivial manner on the coefficient of friction between the ball and the surface and on the elastic properties of the ball and the surface.

Studies of elastoplastic collisions are of interest also in engineering [9, 10]. Finite element simulations calculated the CoR for a multi-body system involving a revolute joint to simulate the influence of different material yield strength ratios on the CoR [11]. This established a new CoR model with a maximum error of 5% compared with experiment.

This model calculates the CoR as

$$C_\mathrm{r} = 1 - 0.1\ln\left(\frac{V_1}{V_\mathrm{y}}\right)\left(\frac{\frac{V_1}{V_\mathrm{y}} - 1}{59}\right)^{0.156} \tag{1}$$

with the initial drop velocity V_1 and the yield velocity V_y as

$$V_\mathrm{y} = 3.194\left(\frac{\left(1.61K\sigma_\mathrm{y}\right)^5 R^{*3}}{E^{*4} m^*}\right)^{\frac{1}{2}} \tag{2}$$

with the coefficient K

$$K = \begin{cases} \dfrac{k-1}{2+k^{1.65}} & \sigma_\mathrm{plate} < \sigma_\mathrm{ball} < 3\sigma_\mathrm{plate} \\ 1.246 & 3\sigma_\mathrm{plate} < \sigma_\mathrm{ball} \\ 1 & \sigma_\mathrm{ball} < \sigma_\mathrm{plate} \end{cases} \tag{3}$$

in which $k = \frac{\sigma_{\text{ball}}}{\sigma_{\text{plate}}}$. σ is the yield strength of the materials, $m^* = \frac{m_1 m_2}{m_1+m_2}$ the equivalent mass, and $R^* = \left(\frac{1}{R_1} + \frac{1}{R_2}\right)^{-1}$ the equivalent radius. The elasticity moduli E of both colliding bodies are combined with their Poisson's numbers μ to $E^* = \left(\frac{1-\mu_1^2}{E_1}\frac{1-\mu_2^2}{E_2}\right)^{-1}$. Applying this model to the situation of plastic and celluloid table tennis ball impinging on the table gives the fit functions shown in **Figure 1** together with the experimental data of [12]. The material properties used for the calculations are $E_1 = 1400$ MPa, $\mu_1 = 0.3$, and $\sigma = 50$ MPa for the celluloid table tennis ball [4], $E_1 = 1197$ MPa, $\mu_1 = 0.24$, and $\sigma = 51$ MPa for a table tennis ball made from ABS plastic [13] and $E_2 = 5992$ MPa and $\mu_2 = 0.3$ for the table [11].

The model describes quite well the experimental results for plastic balls. For celluloid balls, the model does not match so well because one observes a kind of transition in the data. The celluloid balls behave at smaller initial velocities similarly to plastic balls, but their CoR drops more quickly for initial velocities larger than 20 km/h.

Measurements of the impact force acting on table tennis balls [14] show that the small spherical cap surrounding the initial contact point plays a major role in the interaction. Buckling of the cap in a table tennis ball occurs for impact speeds above about 5 m/s, but that is a relatively small speed compared to typical speeds of about 20–30 m/s in the game. Due to the impact force on the cap large amplitude oscillations are excited. The vibrations of the cap when it buckles to reduce the CoR from 0.9 to 0.8 or even below. This is expected from the MD simulations because energy losses appear. The vibration frequency is relatively high, typically around 10 kHz, and impact duration increases with ball speed. For most other ball types, the impact duration decreases as the ball speed increases. One possible explanation for the different results for the celluloid balls is the onset of stronger, non-linear buckling, which then leads to the larger drop of the CoR at higher velocities compared to plastic balls.

Also, the friction coefficient was found higher for plastic balls when the initial horizontal contact point velocities are lower [12]. Hence, a service with back-spin,

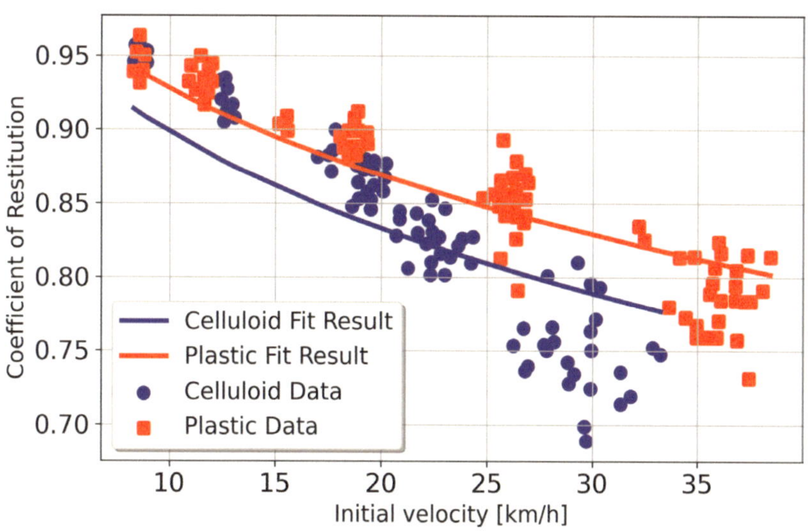

Figure 1.
Coefficient of restitution as a function of initial velocity for plastic and celluloid balls (from [12]). The fit functions calculated from the model of [11] are also shown.

which has lower speed, is experiencing more deceleration upon collision with the table than celluloid balls. On the contrary, fast plastic balls with a lot of topspin accelerate more upon collision.

Even more complex than the interaction of the table tennis ball with the table is the interaction with different rubbers. These are optimized for different purposes, for example maximum speed and spin, minimum speed and spin, spin variations, and others. This optimization was done originally, mostly experimentally, to quantify the rubber effect (e.g. [15]). Later, more and more 3D finite element analysis was done to improve the understanding of the interaction. A 3D finite element model was developed where the polymeric time-dependent dissipating compliant behavior was measured with dynamic mechanical analysis and compression tests. The calculations confirm that the friction between the ball and the polymeric layer plays a key role in the non-linear energy dissipation process, alongside the rate-dependent behavior and architecture of the polymeric constituents [16, 17].

Obviously, the interaction of table tennis balls with the table and the racket is a rather complex mechanism. Different materials and deformations play important roles and create a non-linear, multi-scale process, where microscopic interactions determine the macroscopic behavior.

3. Table tennis ball trajectories

In this section, the trajectory of table tennis balls defined by the forces of gravitation and aerodynamics will be discussed. After introducing the basic physics, a statistical approach will be used to study the influence of potential rule changes on the characteristics of table tennis.

3.1 Basic physics

The trajectory of a table tennis ball is determined by its equation of motion. The equation of motion needs a mathematical description of the acting forces.

The gravitational force of the earth and aerodynamic forces determine the flight trajectory of a table tennis ball. The gravitational force

$$\vec{F}_G = m \cdot \vec{g} \tag{4}$$

accelerates a ball of mass m towards the center of the earth with the gravitational constant $g = 9.81\ \text{m/s}^2$ and results in a parabolic trajectory. Subsequently, air drag and lift create more complex trajectories. Air drag is the friction force created by the interaction with the background medium and is directed against the direction of the movement of the ball. If a hand is held out of a driving car, this force will push it back by the interaction of the air molecules with the hand surface. The drag force increases with larger velocity and with the cross-sectional area:

$$\vec{F}_D = -0.5 \cdot C_D \cdot \rho \cdot A \cdot v \cdot \vec{v} \tag{5}$$

with the density of air ρ, the cross-sectional area $A = r^2 \cdot \pi$ for a ball with radius r, the ball velocity v, and an air drag coefficient C_D. This coefficient can be experimentally determined, e.g. in wind tunnel experiments.

The second important aerodynamic force, the airlift, was discovered by Heinrich Gustav Magnus (1802–1870). The "Magnus effect" describes the observation that a rotating ball deviates from its flight path. This is caused by the interaction of the flow with the surface of the spinning ball, which creates a co-rotating air layer at the surface of the ball. The ball rotates on one side with the airflow created by the spinning ball, on the other side opposite to it. This creates a pressure imbalance. On the counter-rotation side, the total velocity of the airflow is reduced, because both velocities compensate partly. On the co-rotation side, a larger flow velocity is created, because both velocities add up. Larger velocities in a flow translate into lower pressure and the pressure differences created by this produce the Magnus force, which deviate the ball from its original path. The air lift force with an air lift coefficient C_L is

$$\vec{F}_L = 0.5 \cdot C_L \cdot \rho \cdot A \cdot v \cdot \vec{e}_\omega \times \vec{v}. \tag{6}$$

The airlift force acts perpendicular to the velocity $\vec{v}$ and to the axis of rotation $\vec{e}_\omega$ (see **Figure 2**).

The coefficients of air drag and lift for a rotating ball (see **Figure 3**) as a function of the ratio of spinning velocity to translational velocity are implemented into the computer code as a fit of experimental data [18–22].

During a topspin shot with forward rotation, the lift force acts downwards, during a backspin with backward rotation it acts upwards.

Figure 4 shows the parabolic solution for the free flight of a 40-mm table tennis ball with only gravitational force acting. Activating the drag force results in an asymmetric, shorter trajectory. Accounting only for the lift force in the case of topspin flattens the trajectory even more and this is even amplified if both aerodynamic forces act together. In contrast, the lift force in the case of backspin creates a higher, but again asymmetric trajectory, because both drag and lift force act non-uniform.

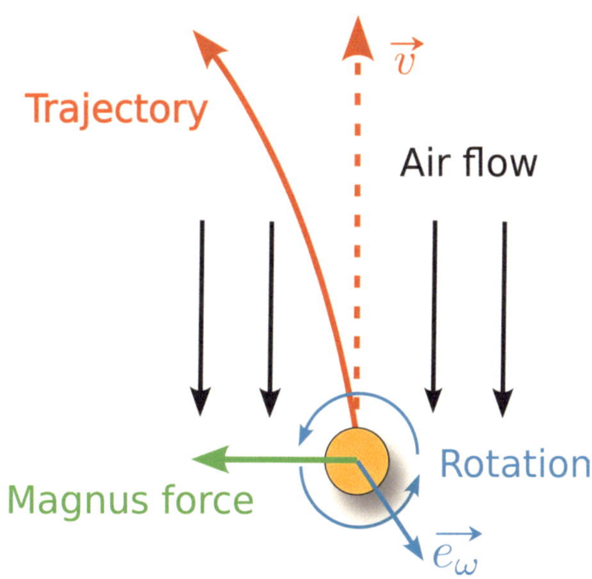

Figure 2.
Forces on a table tennis ball.

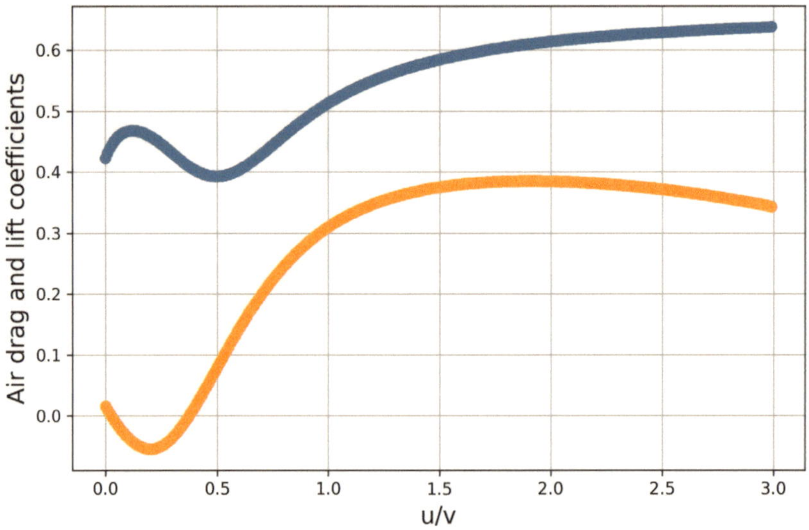

Figure 3.
Air drag coefficient C_D (upper blue curve) and air lift coefficient C_L (lower orange curve) as a function of the ratio of spinning velocity u to the translational velocity v.

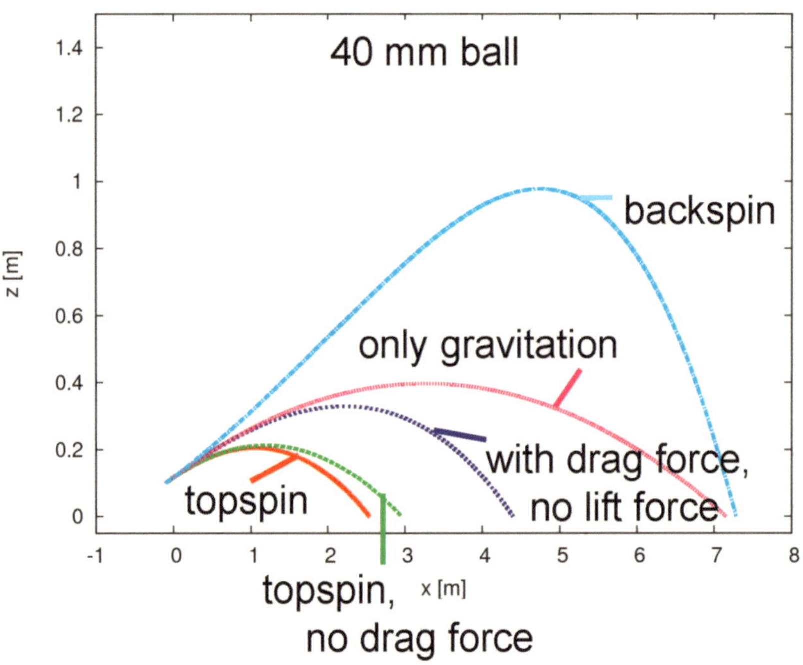

Figure 4.
Influence of the different forces on a table tennis ball trajectory.

Figure 3 shows for low spinning velocities a negative value of the air lift coefficient, which is called the inverse Magnus effect. For table tennis balls, the inverse Magnus effect has been experimentally confirmed [23]. During a trajectory of a flying ball, the air lift coefficients can change from positive to negative coefficients and even back again. This can produce swirling balls, not only discussed in table tennis, but also in soccer and volleyball. For table tennis balls the effect exists, but results only in deviations of some millimeters because the airlift coefficient is only negative where the value of the coefficient itself is already quite small. Table tennis players complain about swirling balls with long pimples, but this is an effect created by the pre-programming of the players in the training. Most players are used to playing against a strongly rotating ball from a normal rubber sponge. However, the long pimples reduce the rotation of the ball and modify its trajectory. The ball has less lift and falls down earlier and by this, the player misses the ball.

4. Statistical analysis of table tennis trajectories

One big problem table tennis faces today is the fact that the speed of the game is nowadays so high that it is very hard for spectators to follow the balls [24–26]. As a consequence, the medial appeal of table tennis seems to go down in terms of TV hours, at least outside Asia. Rule changes to counteract this and to allow a better visual tracking during the rallies, were a larger ball, different counting system, stricter limits for rubbers, and new service rules [24, 25] in order to slow down the game and to reduce the impact of spin on the game. Additional rule changes, like higher nets, are discussed. Such rules or technical changes have strong impacts on the techniques and strategies of the players, because they have to adapt their individual training programs. This creates hesitation of the players towards new rules.

One example for the effect of rule changes is the change of ball size. Today's 40-mm ball is 2 mm larger and 0.2 grams heavier than the 38-mm ball used before. The larger cross sectional area produces a larger air drag and should reduce the maximum velocities [27].

The mass distribution of the larger ball is shifted further away from the center compared with the 38-mm ball. This creates a larger inertial moment and reduces the spin. Measurements show a velocity and spin reduction of about 5–10% [28, 29] for the 40-mm ball.

Interestingly, this is not observed in table tennis match analysis, because the players compensated the effects of the size increase by larger exertions [28, 30].

This put much more emphasis on the fitness of the individual player, because the stroke forces are nowadays not only created by the arms but by the whole body. Especially, the more intense use of the legs compensate the ball size increase in modern table tennis. A similar effect is the more effective wrist usage to support spin production. Using only the forearm gives much less spin for the 40-mm ball compared with the 38-mm ball. The greater forces needed for the compensation of the larger ball effects require much better athletics of the players. The more complex and longer movements for this compensation create also the danger of technical mistakes [31].

Possible changes to further slow-down the game are a larger 44 mm ball or an increased net height (up to 3 cm) for the standard ball of 40 mm. The larger 44-mm ball with small weight could suppress high velocities, but the players could again compensate for this by improving their physical fitness to perform stronger shots.

The second alternative actually discussed is to increase the net height. This is expected to change the general characteristics of table tennis, because it limits fast spins and shots, but also modifies the service. To be able to find a scientifically based decision, a database is missing. Empirical work of short and limited tests with some players exists to study the impact of such changes on the players and the game [32].

One has to be aware of the fact, that the players in these tests have trained for the existing situation (ball size, ball weight, net height) and modifications needed for the new situation take too long for the players to be automatized in the training to be considered. An alternative approach is to solve numerically the equation of motion of table tennis balls for a large number of given initial positions, velocities, and spins in three spatial dimensions (x,y,z) and to establish this a statistical database. Such a trajectory analysis was done for the 38-mm-ball with a weight of 2.5 g, used in tournaments until end of 2000, the actual 40-mm-ball with 2.7 g, and a 44-mm-ball with a weight of 2.3 g, which was tested already in Japan. For the 40-mm-ball an increase of the net height of 1 and 3 cm was analyzed, too.

A first example is given to study differences of the balls for a typical top-spin shot. Initial conditions were chosen as starting point (−0.7, 0.8, 0.1) m, velocities (10.0, 0.1, 2.0) m/s and spins (10, 120, 10) 1/s. The difference between a 38-mm and a 40-mm ball is rather small: the 38-mm ball flies only 3 mm shorter. In contrast, the difference between a 40-mm and a 44-mm ball is more pronounced (see **Figure 5**), because the 44-mm ball is lighter and experiences stronger friction forces due to its larger size.

Trying to get to the same trajectory with the 44-mm ball requires—as expected—a larger initial velocity of (10.7, 0.1, 2.1) m/s. The rotation velocities are unchanged. Therefore, in terms of necessary force for the player to execute the same shot, the smallest force is needed for the 38-mm ball, followed by the 40-mm ball (because it is heavier). Due to the large change in initial velocities, despite its smaller mass, the 44-mm ball requires even more force.

A database is created to quantify the influence of such changes. Modifications in technique, tactics, strength and fitness are not considered in this analysis. For a huge number of initial conditions, the effect on successful strokes is studied by solving the equation of motion. This delivers the maximum amount of possible strokes for different conditions in terms of statistical distributions. One obtains the best possible adaptation to the changes, independent of what this would mean for the players in terms of changes in their training.

For our system studied here, the drag and the Magnus force both influence the trajectory depending on the actual velocity. If a particle on a closed path comes back to

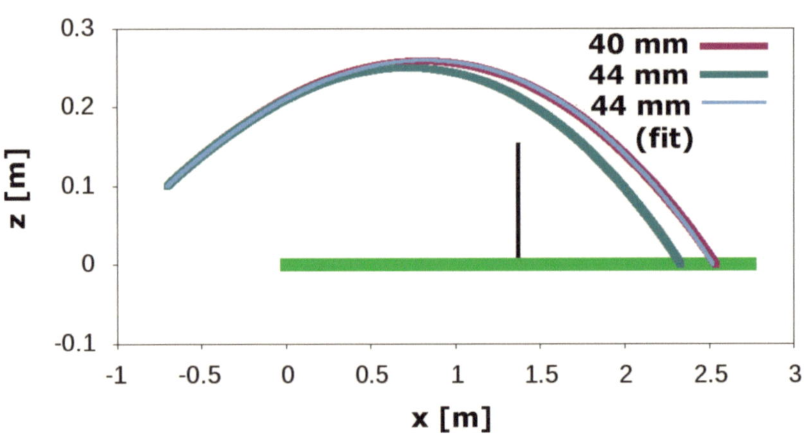

Figure 5.
Trajectory of a 40-mm and a 44-mm ball for the same initial conditions. Using a fit algorithm the starting velocities and rotation velocities of the 44-mm ball are modified to get as close as possible to the 40-mm trajectory.

its origin, the drag will slow it down. Energy is not conserved. Therefore, the Euler method was chosen as a fast method for the calculation of half a billion trajectories for each case within the precision limit of 1 mm for the time step of 10^{-4} s.

The Euler solver has also the advantage of algorithmic simplicity, which allows an easy transfer onto a Graphics Processing Unit (GPU) to speed up the code.

Our implementation is described in detail in [33].

For statistical analysis of the effects of ball sizes and net heights on trajectories of table tennis balls, a Monte Carlo procedure was used. Many different initial conditions were solved; hitting locations were varied from 30 cm above the table to 3 m behind the table. The position relative to the table corner was kept constant at 0.381 m, which is one quarter of the width of the table tennis table, because the exact location of the hitting point is not important. The ball was started with an initial height between 0.4 m and −0.4 m. The initial velocity was sampled in the following way: the horizontal angle was varied between the limiting angles of the starting point to the net posts, the elevation angle was chosen randomly. The normalized spin vector was created randomly representing topspin, backspin and sidespin. Experimentally, it was proven that the spinning of the ball is constant during the flight [34].

For the decision on possible rule changes, fast shots are in particular of interest. This was introduced into the algorithm by limiting the height where a ball passes the net to 30 cm. Translational velocities were chosen from 20 to 200 km/h, and spinning velocities from 0 to 150 turns/s (which is equal to 9000 turns/min). These are experimental values for 38-mm balls [35].

In a successful trajectory, the ball stays within the height limit when it passes the net and hits the other side of the table tennis table.

Trajectories for $5 \cdot 10^8$ initial conditions are calculated for the different cases. The total run-time on a Linux Cluster with 32 cores was 640 hours. Using GPU computing with CUDA on a Dell Precision T7500 Desktop with a NVIDIA Quadro FX3800 results in only 3 hours for the same calculation [36].

Calculation of trajectories in table tennis mostly concentrated on individual cases without statistical analysis [12].

New interest in the fast calculation of table tennis trajectories is also motivated by the research on robots [37, 38] and the programming of computer games [39].

5. Analysis of concepts for slowing-down the game

Data sets of successful hits for the different cases (38 mm ball, 40 mm ball, 44 mm ball, 40 mm ball with 1 and 3 cm higher net) are created. **Figure 6** shows as a function of initial kinetic energy the number of successful trajectories.

There is a clear ranking of the different scenarios visible: the largest number of successful hits for the same initial kinetic energy is produced by the 44-mm ball, because it has lighter weight and higher air drag. The 40-mm ball is very similar to the 38-mm ball. Changes of the balls are compensated by other parameter changes. A higher net affects strongly the ball hits above the table limiting there the number of successful trajectories. In **Figure** 7 the number of successful trajectories as a function of final velocities are shown. Again, the results for the 38 and 40 mm ball differ only marginally. For the 44-mm ball one gets more successful trajectories compared to the 38- and 40-mm ball for higher initial velocity, the distributions for the final velocities are nevertheless very close again. Very high velocities above 35 m/s are suppressed earlier for the 44-mm ball, because the larger size increases the drag forces.

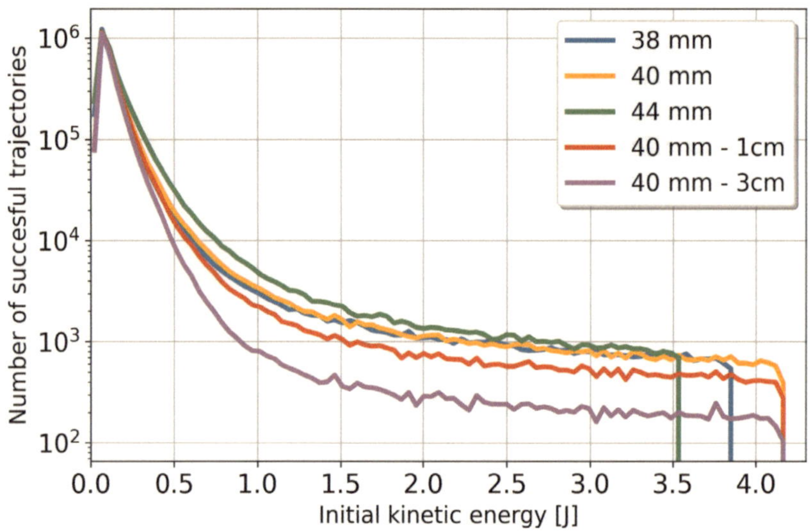

Figure 6.
Number of successful trajectories as a function of initial kinetic energy for the different cases studied.

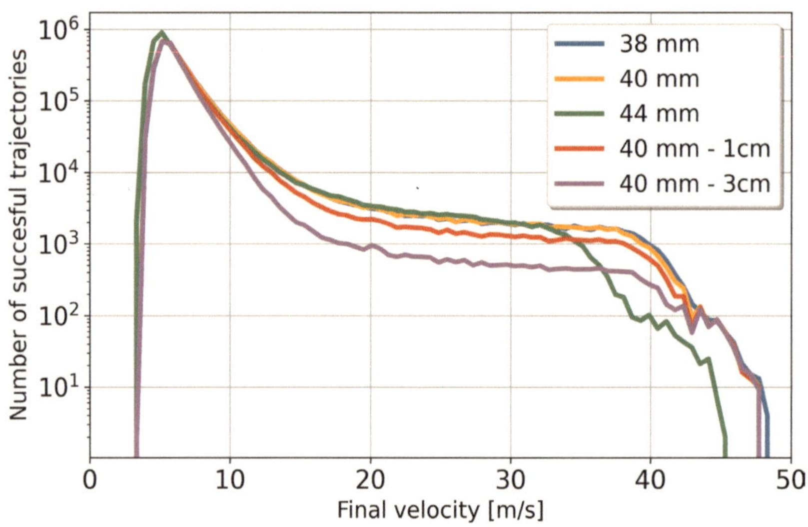

Figure 7.
Number of successful trajectories as a function of final velocity for the different cases studied.

To influence the trajectories one needs larger spins for the 44-mm ball compared to the 40-mm ball. However, the players could compensate for this again by improving their physical fitness to perform stronger shots. For very low velocities the impact of the air drag is not yet important resulting in larger number of successful trajectories. The strong reduction of successful strokes for higher nets is linked with rather small top spin components, but rather larger sidespin components. This would change the characteristics of long rallies in table tennis. For these cases, not only reduced velocities are important, but the tactical possibilities will be modified by reducing the number of spinless short trajectories. This will not only slow down the game but will increase the importance of diagonal play with longer reaction times for the opponent than fast parallel balls. This should produce longer and more attractive rallies. Unfortunately, the game characteristics would change strongly, because the players have much fewer technical and tactical alternatives for successful shots. Specifically, the influence of the service will be strongly reduced. The International Table Tennis Federation will have to decide if such strong modifications to the existing sport are worth a possible gain in attractiveness of table tennis to the TV audience.

These results are supported by advanced statistical analysis using the Python packages pandas [40] and scikit-learn [41] of the 21-dimensional data base [42]. A principal component, multi-dimensional histograms, and a cluster analysis confirmed the previous findings and identified the pronounced sidespin component for the cases with higher nets.

6. Conclusions

Table tennis is a fascinating sport, where many elements of physics are important. The interaction of the table tennis ball with the table or the rubber is a highly complex, non-linear, and multi-scale problem.

The trajectories of table tennis balls are determined by the gravitational and aerodynamic forces. Physics based analysis of trajectories allow reliable predictions of possible rule changes for the distribution of successful shots. This is an important input for the discussion, but here also the impact on the changed characteristics of the game on players and the audience has to be considered.

Author details

Ralf Schneider*†, Lars Lewerentz†, Stefan Kemnitz† and Christian Schultz†
University of Greifswald, Institute of Physics, Greifswald, Germany

*Address all correspondence to: schneider@uni-greifswald.de

† These authors contributed equally.

References

[1] Konrad Lindner. Jugendliches Genie - Carl Friedrich von Weizsäcker als Student in Leipzig, 2016. Available from: https://www.leipzig-lese.de/persoenlichkeiten/w/weizsaecker-carl-friedrich-von/jugendliches-genie-carl-friedrich-von-weizsaecker-als-student-in-leipzig/. [Accessed: July 9, 2022]

[2] Hückel E. Ein Gelehrtenleben. Ernst und Satire. Weinheim: Verlag Chemie; 1975

[3] Konrad Lindner. Gehen Sie zu Heisenberg! Edward Teller als Doktorand in Leipzig, 2015. Available from: https://www.leipzig-lese.de/persoenlichkeiten/t/teller-edward/gehen-sie-zu-heisenberg-edward-teller-als-doktorand-in-leipzig/. [Accessed: July 09, 2022]

[4] Daniel Murray. Rebound of Hollow Elastic Spheres, 2002. Available from: https://www.geocities.ws/cofrest/md3.htm. [Accessed: July 09, 2022]

[5] Hertz H. Ueber die Berührung fester elastischer Körper. 1882;**1882**(92): 156-171

[6] Haron A, Ismail KA. Coefficient of restitution of sports balls: A normal drop test. IOP Conference Series: Materials Science and Engineering. 2012;**36**: 012038

[7] Cross R. Bounce of a spinning ball near normal incidence. American Journal of Physics. 2005;**73**(10):914-920

[8] Sarafian H, Lobe N. Angular impulse and spinning bouncing ball. World Journal of Mechanics. 2017;**7**(7):177-183

[9] Zhang X, Vu-Quoc L. Modeling the dependence of the coefficient of restitution on the impact velocity in elasto-plastic collisions. International Journal of Impact Engineering. 2002; **27**(3):317-341

[10] Ahmad M, Ismail KA, Mat F. Impact models and coefficient of restitution: A review. In ARPN Journal of Engineering and Applied Sciences. 2016;**11**:6549-6555

[11] Liu X, Chen W, Shi H. Improvement of contact force calculation model considering influence of yield strength on coefficient of restitution. Energies. 2022;**15**(3):1041

[12] Inaba Y, Tamaki S, Ikebukuro H, Yamada K, Ozaki H, Yoshida K. Effect of changing table tennis ball material from celluloid to plastic on the post-collision ball trajectory. Journal of Human Kinetics. 2017;**55**(1):29-38

[13] Dielectric Manufacturing. ABS - Food Safe Plastic, 2022. Available from: https://dielectricmfg.com/knowledge-base/abs/. [Accessed: July 09, 2022]

[14] Cross R. Impact behavior of hollow balls. American Journal of Physics. 2014; **82**(3):189-195

[15] Araki S, Sato S, Yamazaki H. Collisional properties of ball-racket interaction in terms of normal and tangential coefficients of restitution. In International Journal of Table Tennis Sciences. 1996;**3**:17-49

[16] Rinaldi RG, Manin L, Bonnard C, Drillon A, Lourenco H, Havard N. Non linearity of the ball/rubber impact in table tennis: Experiments and modeling. Procedia Engineering. 2016;**147**:348-353. The Engineering of SPORT 11

[17] Rinaldi RG, Manin L, Moineau S, Havard N. Table Tennis Ball Impacting Racket Polymeric Coatings: Experiments

and Modeling of Key Performance Metrics. Applied Sciences. 2019;**9**(1):158. DOI: 10.3390/app9010158

[18] Achenbach E. Experiments on the flow past spheres at very high reynolds numbers. Journal of Fluid Mechanics. 1972;**54**(3):565-575

[19] Bearman PW, Harvey JK. Golf ball aerodynamics. Aeronautical Quarterly. 1976;**27**(2):112-122

[20] Davies JM. The aerodynamics of golf balls. Journal of Applied Physics. 1949; **20**(9):821-828

[21] Maccoll JW. Aerodynamics of a spinning sphere. The Journal of the Royal Aeronautical Society. 1928; **32**(213):777-798

[22] Mehta RD. Aerodynamics of sports balls. Annual Review of Fluid Mechanics. 1985;**17**(1):151-189

[23] Konishi Y, Okuizumi H, Ohno T. Piv measurement of a flying table tennis ball. Procedia Engineering. 2016;**147**:104-109. The Engineering of SPORT 11

[24] Djokic Z. Ittf scored a goal (changes of rules in table tennis during 2000–2003). In: Proceedings of the 10th International Table Tennis Sports Science Congress. Zagreb, Croatia; 2007. pp. 336-341

[25] Kondric M. Did Changes of Rules in 21st Century Brought Further Development to Table Tennis Game? Academia Letters; 2021

[26] Nelson R. Es geht um die zukunft unseres sports. In: Deutscher Tischtennis-Sport. Vol. 10. Münster: Philippka-Verlag; 1997

[27] Bai KX, Hong X, Hu P, Yin H. Technical contrastive analysis after ping-pong diameter altering. In: Proceedings of the 9th ITTF Sports Science Congress. Shanghai, China; 2005. pp. 27-30

[28] Li J, Zhao X, Zhang C. Changes and development: Influence of new rules on table tennis techniques. Sports Science Research. 2005;**26**(3):55

[29] Iimoto Y, Yoshida K, Yuza N. Rebound characteristics of the new table tennis ball; differences between the 40 mm (2.7 g) and 38 mm (2.5 g) balls. International Journal of Table Tennis Sciences. 2002;**5**:233-243

[30] Liu YX. Comparative analysis and research of the impacts by 40 mm ball on the first-3-stroke skills of shake-hand looping style of world-class male table tennis players. In: Proceedings of the 9th ITTF Sports Science Congress. Shanghai, China; 2005. pp. 27-30

[31] Kondrič M, Medved V, Baca A, Kasović M, Furjan-Mandić G, Slatinšek U. Kinematic analysis of top spin stroke with balls of two different sizes. In: Scientific Approach in Table Tennis and Tennis in Slovenia. 2007. pp. 53-60

[32] Takeuchi T, Kobayashi Y, Hiruta S, Yuza N. The effect of the 40mm diameter ball on table tennis rallies by elite players. In International Journal of Table Tennis Sciences. 2002;**5**:278-284

[33] Schneider R, Kalentev O, Ivanovska T, Kemnitz S. Computer simulations of table tennis ball trajectories for studies of the influence of ball size and net height. International Journal of Computer Science in Sport. 2013;**12**(2)

[34] Zhang Y, Zhao Y, Xiong R, Wang Y, Wang J, Chu J, editors. Spin observation and trajectory prediction of a ping-pong ball. In: 2014 IEEE International

Conference on Robotics and Automation (ICRA). 2014. pp. 4108-4114

[35] Wu H. Analysis of the training for the chinese table tennis superiority from 1959 to 1989. Sport Science. 1993;**3**:48-58

[36] Nickolls J, Buck I, Garland M, Skadron K. Scalable parallel programming with cuda: Is cuda the parallel programming model that application developers have been waiting for? Queue. 2008;**6**(2): 40-53

[37] Koç O, Maeda G, Peters J. Online optimal trajectory generation for robot table tennis. Robotics and Autonomous Systems. 2018;**105**:121-137

[38] Serra D, Satici AC, Ruggiero F, Lippiello V, Siciliano B. An optimal trajectory planner for a robotic batting task: the table tennis example. In: International Conference on Informatics in Control, Automation and Robotics. Vol. 3. Scitepress; 2016. pp. 90-101

[39] Florian'Floyd' Mueller and Martin Gibbs. A table tennis game for three players. In: Proceedings of the 18th Australia conference on Computer-Human Interaction: Design: Activities, Artefacts and Environments. 2006. pp. 321-324

[40] Wes McKinney. Data Structures for Statistical Computing in Python. In Stéfan van der Walt and Jarrod Millman, editors, Proceedings of the 9th Python in Science Conference, pages 56–61, 2010.

[41] Pedregosa F, Varoquaux G, Gramfort A, Michel V, Thirion B, Grisel O, et al. Scikit-learn: Machine learning in Python. Journal of Machine Learning Research. 2011;**12**:2825-2830

[42] Schneider R, Lewerentz L, Lüskow K, Marschall M, Kemnitz S. Statistical analysis of table-tennis ball trajectories. Applied Sciences. 2018; **8**(12)